Lecture Notes in Computer Science 6090

Commenced Publication in 1973
Founding and Former Series Editors:
Gerhard Goos, Juris Hartmanis, and Jan van Leeuwen

Danny Weyns   Sam Malek
Rogério de Lemos   Jesper Andersson (Eds.)

Self-Organizing Architectures

First International Workshop, SOAR 2009
Cambridge, UK, September 14, 2009
Revised Selected and Invited Papers

 Springer

Volume Editors

Danny Weyns
Katholieke Universiteit Leuven
Leuven, Belgium
E-mail: danny.weyns@cs.kuleuven.be

Sam Malek
George Manson University
Fairfax, VA, USA
E-mail: smalek@gmu.edu

Rogério de Lemos
University of Kent
Canterbury, UK
E-mail: r.delemos@kent.ac.uk

Jesper Andersson
Linnaeus University
Växjö, Sweden
E-mail: jesper.andersson@lnu.se

Library of Congress Control Number: 2010930054

CR Subject Classification (1998): D.2, I.2, H.4, C.2, C.2.4, H.5

LNCS Sublibrary: SL 2 – Programming and Software Engineering

ISSN   0302-9743
ISBN-10   3-642-14411-X Springer Berlin Heidelberg New York
ISBN-13   978-3-642-14411-0 Springer Berlin Heidelberg New York

springer.com

© Springer-Verlag Berlin Heidelberg 2010
Printed in Germany

Typesetting: Camera-ready by author, data conversion by Scientific Publishing Services, Chennai, India
Printed on acid-free paper   06/3180

Preface

Self-adaptability has been proposed as an effective approach to automate the complexity associated with the management of modern-day software systems. Self-adaptability endows a software system with the capability to adapt itself at runtime to deal with changing operating conditions or user requirements.

Researchers in self-adaptive systems mostly take an architecture-centric focus on developing top–down solutions. In this approach, the system is monitored to maintain an explicit (architectural) representation of the system and based on a set of (possibly dynamic) goals, the system's structure or behavior is adapted. Researchers of self-organizing systems mostly take an algorithmic/organizational focus on developing bottom–up solutions. In this approach, the system components adapt their local behavior or patterns of interaction to changing conditions and cooperatively realize system adaptation. Self-organizing approaches are often inspired by biological or natural phenomena. With the term "self-organizing architectures" (SOAR) we refer to an engineering approach for self-adaptive systems that combines architectural approaches for self-adaptability with principles and techniques from self-organization.

Whereas both lines of research have been successful at alleviating some of the associated challenges of constructing self-adaptive systems, persistent challenges remain, in particular for building complex distributed self-adaptive systems. Among the hard challenges in the architectural-centric approach are handling uncertainty and providing decentralized scalable solutions. Some of the hard challenges in the self-organizing approach are connecting local interactions with global system behavior, and accommodating a disciplined engineering approach. The awareness grows that for building complex distributed self-adaptive systems, principles from both self-adaptive systems and self-organizing systems have to be combined. For instance, Web-scale information systems, intelligent transportation systems, and the power grid are all innately decentralized systems, but control in local sub-systems may be highly centralized. Engineering such complex systems puts forward questions such as: What kind of bottom–up mechanisms can be exploited in order to deal with uncertainty but at the same time provide the required assurances? How to derive and exploit tactics, architectural patterns, and reference architectures to realize robust, scalable, and long-lived solutions?

The general goal of SOAR is to provide a middle ground that combines the architectural perspective of self-adaptive systems with the algorithmic perspective of self-organizing systems.

The papers in this volume include both selected and thoroughly revised papers from the WICSA/ ECSA 2009 SOAR Workshop and invited papers.

The papers cover a broad range of topics related to self-organizing architectures, including self-adaptive architectures, decentralized architectures, nature-inspired approaches, and learning approaches. We hope that the papers in this volume stimulate further research in self-organizing architectures.

May 2008

D. Weyns
S. Malek
R. de Lemos
J. Andersson

Organization

SOAR 2009 was organized in conjunction with the Working IEEE/IFIP Conference on Software Architecture (WICSA) and the European Conference on Software Architecture (ECSA), Cambridge, UK, September 14, 2009.

Program Co-chairs

Danny Weyns	Katholieke Universiteit Leuven, Belgium
Sam Malek	George Mason University, USA
Rogério de Lemos	University of Kent, UK
Jesper Andersson	Linnaeus University, Sweden

Program Committee

Nelly Bencomo	Lancaster University, UK
Yuriy Brun	University of Southern California, USA
David Garlan	Carnegie Mellon University, USA
Kurt Geihs	University of Kassel, Germany
Holger Giese	Hasso Plattner Institute at the University of Postdam, Germany
Jorge J. Gómez Sanz	Universidad Complutense de Madrid, Spain
Tom Holvoet	Katholieke Universiteit Leuven, Belgium
Mark Klein	Software Engineering Institute, Carnegie Mellon, USA
Marco Mamei	Universitá di Modena e Reggio Emilia, Italy
Hausi A. Müller	University of Victoria, Canada
Flavio Oquendo	Université de Bretagne-Sud, France
Van Parunak	Vector Research Center, Ann Arbor, USA
Onn Shchory	IBM Haifa Research Lab, Israel
Mirko Viroli	Università di Bologna, Italy

Website

http://distrinet.cs.kuleuven.be/events/soar/2009/

Acknowledgements

We are grateful to the WICSA/ECSA 2009 organizers for hosting SOAR. We thank the PC members for their critical review work. Finally, we thank the Springer staff for supporting the publication of this volume.

Table of Contents

Elements of Self-adaptive Systems – A Decentralized Architectural Perspective*

Carlos E. Cuesta[1] and M. Pilar Romay[2]

[1] Dept. Computing Languages and Systems II,
Rey Juan Carlos University, Móstoles, 28933 Madrid, Spain
carlos.cuesta@urjc.es
[2] Dept. Information and Communication Systems Engineering,
Saint Paul-CEU University, Boadilla del Monte, 28668 Madrid, Spain
pilar.romay@gmail.com

Abstract. Software is evolving towards a greater complexity and variability, with a continuously changing environment. In this context, self-adaptive systems are acquiring a great relevance. Their architectures are inherently dynamic and by definition, also reflective. However, their self-referential nature might compromise their compositionality, and even the use of the architectural approach. This work intends to decide on its suitability, by considering its fundamentals in detail. After some initial definitions, the nature of both self-adaptation and self-organization is discussed, and the implicit modular structure is determined. Then a tentative taxonomy of elements in self-adaptive architectures is provided, which is also discussed in a decentralized setting. To support our initial hypothesis about the suitability of architectures, the algebraic properties of their composition is studied in detail. Then, the suitability of a reflective approach in this context is considered, and then a concrete example of an autonomic system is described, using a reflective architectural description language. The chapter concludes discussing the suitability of this approach, and how the architectural perspective of self-adaptation does not actually imply a centralized topology.

1 Introduction

The complexity of software systems has been increasing for the last decades, and it keeps reaching new levels. To tackle this complexity, a number of strategies have been tested. One of the first was to consider the global properties of systems, at a high level of description – that is the purpose of Software Architecture.

A recent approach states that systems should now realize a part of its internal functions themselves, without requiring external assistance; therefore relieving the user from this added complexity. This is the idea behind so-called *self-managed*, *autonomic* and *self-adaptive* systems, as well as, from a different perspective, of *self-organizing* structures, and their features, which collectively receive the name of *self-* systems* [3].

* This work has been partially funded by National Research Projects MULTIPLE (TIN2009-13838) and *Agreement Technologies* (CONSOLIDER CSD2007-0022), both from the Spanish Ministry of Science and Innovation; and by Research Project IDONEO (PAC08-0160-6141), from the autonomous Government of Castilla-La Mancha.

D. Weyns et al. (Eds.): SOAR 2009, LNCS 6090, pp. 1–20, 2010.

Their study affects many areas within Software Engineering, to the extent that some authors [17] have even stated that these issues define the future of the field.

A different concern has to do with the organization of self-* systems. It is often assumed that adaptive architectures imply a *centralized* setting, while self-organizing systems provide a *decentralized* approach. However, decentralized structures define flexible and efficient (even sometimes optimal) architectures; therefore, it would be often desirable to be able to implement the behaviour of a self-adaptive system on top of a self-organized (i.e. decentralized) infrastructure.

Though these systems have been mostly studied at the implementation level, we suggest along with many others [17] that it is both convenient and feasible to study them at the architectural level, therefore combining both strategies to tackle complexity.

Therefore, in this chapter we propose to follow this program to cover all our objectives in the study of *self-adaptive architectures*:

- First, we provide a general perspective of self-* systems and their properties, concentrating on the implications of these properties in their modular nature, which must be considered from an architectural perspective (sections 2 and 2.1).
- That said, the possibility of a decentralized setting is briefly introduced (section 2.2).
- Then, an architectural reasoning is used to identify the kinds of elements that we could find in an adaptive architecture - providing the basis for the remainder of the text. Again, the impact on decentralization is discussed (sections 3 and 3.1).
- Afterwards, to check the feasibility of the architectural approach, the compositionality of this taxonomy is examined. A semi-formal algebra of adaptive components is used for that purpose (section 4).
- Supported by these results, we hypothesize that a *reflective* approach would benefit from the inherently reflexive nature of self-adaptation (section 5).
- Finally, gathering all this, a reflective architectural perspective is used to describe a concrete example of a self-adaptive system, both proving the feasibility of the approach (section 6) and of using a decentralized setting (section 6.1).

At the end of the chapter, all our concerns –adaptation, architecture, decentralization and reflection– would have been considered as different aspects of the same problem, and the feasibility of the architectural approach would have been demonstrated.

2   On Self-* Systems: Some Basic Definitions

First, we will describe our basic element, the *system*, in the usual way. The term is used in the most generic meaning, i.e. that of Systems Theory, in which every element is a system. Formally, let's define *systems* $(c, s) \in \mathcal{S}$ as bounded active entities, which can be *composed* with others to create a more complex entity. This larger entity is also bounded, and therefore it is also a system, specifically a *composite system*. The components of this composite are termed as *subsystems* with respect to it.

Whether atomic or composite, all these elements are always systems; the use of the term is intended as generic as possible. In architectural terms, which will also be used throughout this chapter, the term system would be equivalent to that of *component* [1]. Sometimes, in the literature, the term *system* is reserved to refer to the resulting *architecture*, i.e. the largest composite component which composes all of the others. Here, this connotation is avoided: the term *system*, as the term *component*, refers to any element, either atomic or composite.

When talking about self-∗ systems, we are referring to systems which manifest one or several of certain specific properties or *attributes*. Those are *high-level*, in the sense that they are human-scale observable –i.e. measurable–, and system-wide, using our very generic definition of systems and their scope. Their distinguishing feature is that a self-∗ system claims to have full –or partial– automatization or control of those attributes. For the remainder of this chapter, we would refer to any such high-level property as a *self-attribute*, or ς-attribute for short.

The most important ς-attributes we could consider include *self-monitoring (ςm), self-tuning (ςt), self-configuration (ςC), self-optimization (ςO), self-protection (ςp),* and *self-healing (ςh).* The literature has used a number of others, some of them synonyms for these, but this roughly covers most of the self-∗ spectrum. Along these, we should also consider another set of related terms, which also refer to properties of self-∗ systems, but are not strictly ς-attributes. The most important among them are *self-organization (ςo), autonomy (ςa), self-management (ςM),* and *self-adaptation (ςA).* Again, there are a number of similar terms, and some of them are partial synonyms, such as *self-assembly,* or *self-awareness.* These terms are often used to designate either combinations or generalizations of the attributes in the first list [16,18]. However, we consider them from a different perspective: they describe high-level, general properties of several software systems, which sometimes take the concrete form of the small-scale attributes. Their characterization as attributes is not intrinsic to their nature: in fact, they are just *observations,* i.e. different ways to observe the consequences of the same large-scale property. In short, the semantics of small-scale ς-attributes is given by the observer. According to this vision, recent research prefers to concentrate in the global properties, whose features are analyzed and classified in a set of *dimensions* [2].

The most general of these properties is *self-adaptation* (ςA), which refers to the capability of a system to automatically react to changes in its environment. This feature, along with *autonomy* (ςa) and *self-management* (ςM), has a behavioural connotation; in fact, these two can be considered as specific facets of the first one. Indeed, when a system is adaptive, it must be able to react autonomously (ςa ⊃ ςA), and probably also to manage [17] its own internal details (ςM ⊃ ςA).

By contrast, *self-organization* (ςo) has a structural connotation, and it refers to the capability of the system to modify its internal structure and topology, possibly altering its own properties in the process. Its relationship to ςA is much more complex, as none of them implies nor precludes the other. Both of them can imply *emergence,* i.e. the non-designed appearance of some spontaneous (emergent) feature. It has been defined both for adaptation [15] and self-organization [22,23].

Finally, autonomy (ςa) has also known a wide popularity with the definition of autonomous agents [25] and, particularly, autonomic systems [16]. In the second context, it has been consistently described as the *combination* of several small-scale ς-attributes [16,18]. But for the remainder of this chapter, it would be just considered as particular case of self-adaptation (using both terms in a very wide sense).

Let's also define an heuristic pseudo-function of *measure* $\phi : [0..1]$ on systems for every ς-attribute, to measure to which extent this ς-attribute is met. It will be often limited to small-scale ς-attributes.

Of course, a practical definition of such a pseudo-function should also imply the definition of a method to provide that measurement. However, there are two reasons not to do that. First, we are trying to provide an abstract, *generic* definition, in order to be able to talk in general terms regardless of the considered ς-attribute. And second, a practical reason: we need to be able to explore the consequences of the definition of self-aware structures without the need to depend on a particular ς-attribute or a particular method to measure this ς-attribute.

Moreover, the measure function ϕ itself is also specifically provided in order to be able to abstract from the concrete ς-attribute at hand. In this study, we don't want to consider the implications of concrete attributes ($\varsigma m, \varsigma t, \varsigma C, \varsigma O, \varsigma p, \varsigma h$). Therefore, we would just consider an abstract attribute ($\varsigma \chi$), and use the measure function to provide a quantification for this abstract value.

In summary, for a certain system s, the measure function $\phi(s, \varsigma \chi)$ expresses the degree in which a certain ς-attribute, $\varsigma \chi$, is met by the system.

Let's also define, for this measure ϕ and every ς-attribute considered, two (again heuristic) values, namely the *upper threshold* (Ω) and *lower threshold* (η). Their intent is to provide a way to delimit the intuitive expression of these ς-attributes. Therefore, we would say that a system *has* the attribute when it measures greater than the lower threshold, but it would only have it *fully* when it is over the upper threshold. The purpose of thresholds is to provide the scale for any relevant attribute, thereby avoiding the negative effects caused by excessive abstraction.

Logically, in the most strict definition, for every ς-attribute, the upper limit must also be the maximum, i.e., $\Omega = 1$. Only the *complete* control of a ς-attribute would account as a full coverage. However, the opposite is not true. Usually, the lower limit, η, would be greater than zero; this just means that, under a certain threshold, a very small capability to control an attribute is equivalent to no control at all.

2.1   On Modularity of Self-* Systems

Once we have defined the metric on the influence of a ς-attribute and the corresponding thresholds, it can also be used to define another pseudo-function on scope, *width* (ω). This function delimits the "spatial" extension of a certain ς-attribute within a system. Therefore it determines which fragment or part of the system has the required degree of self-control for this ς-attribute.

$$\omega[s, \varsigma \chi] \subseteq s \in \mathcal{S} : \; \eta_{\varsigma \chi} \leq \phi(\omega[s, \varsigma \chi], \varsigma \chi) \leq \Omega_{\varsigma \chi}$$

The above expression implies that the *width* of a certain ς-attribute on a system can also be defined as a *subsystem* of that system. This cannot be derived from the above, but it is actually implied by the Boundary Condition, which is now introduced.

The Boundary Condition can be deduced as a corollary of the Principle of Recursive Systems Construction [13], and it provides the basis for modularity and for the definition of the constructive structure itself.

Definition 1 (Boundary Condition). *The evolutionary boundaries emerging in a system which stabilizes tend to converge with existing interfaces of components.*

Basically, the reasoning is the following. An *evolving* system is continuously changing until it reaches some degree of *stability*. The Principle states that such stable systems are recursively constructed by the recombination of stable subsystems.

This also implies that those stable subsystems are clearly separated of their environment, by the very notion of stability: they *maintain their shape* where the rest continues evolving. Therefore, the pace of change *does* define boundaries, identifying stable areas, assimilable to *loci* [21] in dynamic systems.

These "bounded areas of stability" can be considered as *separate components* with respect to the system's evolution, as they maintain a coherent dynamic *status* [21]. Considering that evolution must happen at the same place as communication, it is easy to deduce that their interfaces will set the guidelines for those boundaries. In summary, the frontiers defined by the system's evolution coincide, with the already existing interfaces of components – which is precisely the Boundary Condition.

Using these basic definitions, in the next sections we will explore some properties of self-* systems, and their architectures, in greater detail. For example, once the Boundary Coundition has related the inherently modular *nature* of the system with the modular *structure* of its design, we can state that the Principle of Recursive System Construction implies a recursive definition which can be equated to the notion of component, as indicated at the beginning of this section.

Therefore, any component is a system, but a composition of components (that is, a configuration) is also a system. But, this has always been the case in Software Architecture [1]: any configuration of components defines a composite component, which can also be instantiated as part of a larger configuration. Therefore, the term *architecture* will be used in this chapter mostly as a synonym of *system*, with the explicit connotation about referring always to composite systems. There can be still certain issues regarding composition; these will be discussed in section 4.

2.2 Self-organization for Self-adaptation

Self-organization is often defined [22] as the process in which the internal structure of a system increases in complexity without being managed by an external source. Here "complexity" is understood as an statitical or information-theoretic measurement, but it mostly coincides with its intuitive meaning.

Self-organizing systems are, by nature, not only *self-reconfiguring* –i.e. the structure of internal connections changes easily,– but also *self-replicating*: as the structure grows in complexity, it also grows in size; and the new elements are created from the existing ones, either by replication or mutation. Such growing structures must be supported by mechanisms able to accomodate the formation of a network of basic elements, while at the same time ensure that the network is constrained to follow certain patterns, so that it acquires a concrete shape.

Like in almost any other organization, those constraints are provided by means of two kinds of elements: controls and protocols. Adaptive structures, being typically *centralized* [2], are classic examples of the first kind: most of them manifest explicit control loops, inspired in regulators of classic control theory. And, on the other hand, Galloway [11] describes *protocols* as the way in which *decentralized* structures (such as self-organizing systems themselves) are also controlled. This defines a wide spectrum of regulation, in which both self-organizing and self-adaptive architectures are simultaneously harnessed by atomic, unary *controls* (norms, limits, locks, loops) and multiple, connective *protocols* (hubs, bridges, channels, spaces).

From an architectural point of view, self-organization has essentially an structural perspective –from the definition–, while self-adaptation is essentially behavioural –the system changes, either to do the same function under different circumstances, or to do something entirely different–. Both perspectives blend in the architecture: none of them *requires* or *excludes* the other; more than that, they are clearly complementary. Self-adaptation does not imply self-organization, nor vice versa; but they are often related, even at the theoretical level.

For instance, in the context of Systems Science, particularly in Cybernetics, the notion of *self-organization* was introduced by Ross Ashby. His definition depends on the Law of Requisite Variety, which can in turn be formulated [13] as the next logical step after the aforementioned Principle of Recursive Systems Construction. Moreover, to be stable, such a self-organized system must obey the Good Regulator Theorem [5], which states that the system *must contain a model of itself*; i.e, it must be *self-aware* [3]. Stafford Beer extended this result in his Viable System Model [4], where this condition is required to provide *autonomy* – which implies adaptivity.

In summary, we can conclude that, though ςo and ςA are not strictly required by each other, they are still deeply intertwined.

3   A Taxonomy of Elements in Self-adaptive Architectures

To study the conception of an adaptive architecture, it is essential to identify the features which, as a common rule, distinguish them from a conventional architecture. At the structural level, this necessarily implies the existence of certain specific elements, with special-purpose functions, directly related to the system self-managing capabilities.

This section presents a *tentative* taxonomy for elements of this kind. The five roles which a component can play in an adaptive architecture are identified and delimited. This taxonomy is inspired in the one provided by Cuesta *et al* [7] for dynamic architectures, which can be considered analogous to some extent. This is not strange: every adaptive system is necessarily dynamic, even structurally dynamic. However, though both taxonomies are related, they still have essentially different details and approaches.

As our intention is to apply this taxonomy to a great variety of systems and architectures, referring to different ς-attributes and their combinations, we have chosen to use *neutral* names, based upon the Greek alphabet, for the identified categories of elements. We would also provide intuitive names for these categories; but these could be less "intuitive" These are the following, namely:

Alpha (α). Every element which is in charge of itself, that is, which is self-contained with respect to the selected ς-attribute, without the need to resort to any kind of external control. For example, in a self-managed system, an α-element (an *alpha*, for short) would be each element capable of managing itself.

In terms of the measure function defined in section 2, an alpha-element α can be characterized as follows:

$$\phi(\alpha, \varsigma\chi) \geq \Omega_{\varsigma\chi}$$

In summary, an alpha is an **autonomous element**, in the classic sense. However, this term usually implies the connotation of dealing with a simple, even atomic element (the obvious example is an *autonomous agent*). But in the context of an adaptive architecture, many of the α-elements would be probably *composite*, the compound of at least a controller (a γ-element) and a controlled element (a ϵ-element). In fact, when the α-element is a fragment with sufficient size and structure, it can be better characterized as an **autonomic subsystem**[1]. The final goal, in this case, is that the architecture itself, considered as a composite component (i.e., the whole system), must also be an alpha, that is, an *autonomic system*.

Beta *(β)*. Every element which is (at least) **partially autonomous**, that is, which has self-controlling capabilities without requiring external assistance, but which is not (necessarily) self-contained with respect to the considerered ς-attribute. Self-control can be *partial* in two different ways; it can refer to *structural* features $(\beta\lceil_S)$ or *behavioural* aspects $(\beta_B\rceil)$. The former means that the element controls just *a part of* itself; while the latter means that element's behaviour does not fully cover the considered ς-attribute. Of course, a concrete β-element might be partial in both ways $(\beta\lceil_S|_B\rceil)$. For example, in a self-monitoring (ςm) system, a beta could be a composite element which monitorizes just a part of its components $(\beta\lceil_S)$, or which just observes a part of its own activities or their effects $(\beta_B\rceil)$. This distinction is somewhat similar to the difference between a self-organized (ςo) system, which is essentially a structural attribute, and a self-managed (ςM) system, which is essentially behavioural. Moreover, the second attribute can be considered more specific than the first one, and it is often supported by it (perhaps even $\beta\lceil_S \Leftarrow \beta_B\rceil$).

In terms of the measure function and threshold values described in section 2, a beta-element β is obviously defined as follows:

$$\phi(\beta, \varsigma\chi) \geq \eta_{\varsigma\chi}$$

Note that $\Omega > \eta$ for every ς-attribute $\varsigma\chi$, and therefore every α-element is also a β-element, by definition. It is equally obvious that every system which is *fully* self-managed must necessarily also be a member of the set of (partially) self-managed systems. In summary, alphas are just a special case of betas $(\alpha \subset \beta)$.

The obvious evolution of a beta would be to become an alpha, extending its control to become self-sufficient. But the limits between them are fuzzy, as it is difficult to decide when a certain scope is closed (just refer to section 2).

Gamma *(γ)*. Every element which is in charge of another, that is, which exerts some kind of *control* on another, with respect to the considered ς-attribute. In summary, a **controller** element. For instance, in a self-monitoring (ςm) system, a γ-element

---

[1] The term *autonomic* is used here in the generic sense, and not in the specific sense (ςa), as it refers to any ς-attribute or combination of them (including of course ςa itself).

would be a *monitor*, and in a self-configuring (ςC) system, a *configurator*. Both examples are also well-known cases at the architectural level.

Every gamma has this quality *in relation to* another element: a controller implies a controlled element (in principle, but not necessarily an ϵ-element). Therefore, its own existence guarantees the creation of a *composite*. This composition would be typically realized by interaction, though there are some other interesting alternatives (reification or superimposition, in particular).

Gammas define the essence of self-* systems. They describe the classic external *control element*, once it gets integrated within the architecture. In fact, we can assume, without loss of generality, that every beta –and therefore, also every alpha– is created by composition of a gamma with another element.

Delta *(δ)*. Every element in charge of control activities, which are necessary for the system's self-management; but which does not directly control itself any other element of the system. This means that it is a supporting or **auxiliary element** for the adaptive system. From a different point of view, it can also be seen as an element which is exerting some kind of *indirect* control. Its activity always complements that of one (or several) γ-elements, and therefore it indirectly affects, in turn, the elements controlled by those γ.

Every element of an adaptive system which is in charge of some specific task, relevant for the system as a whole, is a good example of a δ-element. For instance, a router, a broker, or even a calculation element can be deltas, if they are a part of the right subsystem. Though their function is essential to system-level self-management, their inclusion in this taxonomy does not imply any special complexity. They often need not to be aware of the role they play. In fact, their relevance for the architecture is just provided by its location within it.

Epsilon *(ϵ)*. Every element which lacks any kind of self-control (with respect to the considered ς-attribute). Theforore, it is a **conventional component**, which could equally be a part of a *non*-self-* system. Within an adaptive system, every epsilon is a candidate to be controlled, directly or indirectly, by some gamma.

In terms of the measure function defined in section 2, an epsilon-element ϵ can be characterized as follows:

$$\phi(\epsilon, \varsigma\chi) < \eta_{\varsigma\chi}$$

This means that the degree of exerted self-control is non-existent, or so low (under our perception) that it can be considered equal to zero.

Of course, this taxonomy does not intend to be definitive. It is very likely that every element of an adaptive architecture can be classified under one of these five categories; but it is also possible that some of them requires further refinement, perhaps even identifying some additional category.

This sort of taxonomy makes possible to reason about the complexity and structure of an adaptive architecture. However, to be actually useful, it must be possible to apply it to extract several conclusions, including even the consistency of the architectural approach itself. This will be the purpose of the discussion in section 4.

3.1   Discussion – A Decentralized Perspective

As already indicated, the taxonomy was specifically conceived for *self-adaptive* systems, in the wider sense. Therefore, it tries to identify the elements and structures which make possible for a such a system to react to changes in the environment, pointing out the need of specific control mechanisms which provide the required behaviour. These controls play usually a *central* role in the adaptation process, to the extent that the organization of self-adaptive architectures is often described as centralized.

But, as already noted, we are also interested in the structure of *self-organizing* systems, which manifest a decentralized setting [2], where there is no single point for managing the adaptations. Though the taxonomy was intended for self-adaptation, it is therefore relevant to consider to which extent it can be applied in a decentralized perspective, in order to decide if it needs to be extended with an additional set of abstractions, and the potential nature of those (replicators, attractors, mutators, etc).

First, we should note that our approach is still *architectural*, as originally intended. This means that it does not preclude either a centralized or decentralized setting: both of them should be able to be described with a similar effort. Therefore, the architectural elements identified by the taxonomy should appear in both kinds of systems.

In any case, it is clear that the most relevant feature is the existence of γ-elements, i.e. *controllers*. Obviously, the idea that any conventional component can be actually a gamma implies the potential for a decentralized setting: every gamma is an independent adaptation center. Therefore, the approach seems to be inherently distributed.

Moreover, an architectural approach must have the connections into account. Indeed, the central (or not) nature of control depends directly on the *topology* of the network of connections.

Connections can be provided by interaction or containment: in fact, compositional hierarchies affect clearly the scope of control. Both controllers and controlled elements can define their own hierarchies, and these are transitive. Any gamma (say γ_1) defines an area of control (say $\kappa[\gamma_1]$), which is centralized; and when one of the controlled elements is also a gamma (γ_2), its own area of influence gets included in the influence of the container ($\kappa[\gamma_2] \subset \kappa[\gamma_1]$). When this hierarchy defines a top-level gamma (γ_0) which directly or indirectly influences all of the system ($\kappa[\gamma_0] = s$), then we have a *global* controller, and the approach is **centralized**. By contrast, when the different areas of influence don't overlap ($\kappa[\gamma_i] \cap \kappa[\gamma_j] = \emptyset$), we can say that the approach is **decentralized**. Of course, this also depends on the kind of interaction.

To conclude, we should note that while probably this taxonomy does not cover every aspect of a self-organizing architecture, and would possibly require some extension in that context, it is still relevant from a decentralized perspective.

4   On the Compositionality of Self-∗ Systems

When an architectural approach has been chosen, it is both relevant and necessary to consider whether we are applying the right abstraction level. Not only because several aspects which are abstracted could still happen to be significant, but specially because before assuming the advantages of that approach, we should decide if an self-adaptive system is *compositional* indeed.

To question whether the system is compositional implies to question if the composition operation, within the system, has the right constructive features. To achieve that, the essential property is that the operation must behave the same way if the operators are simple or composite elements. In summary, to compose elements and include them as a part of a a larger system, composition needs not to know their internal structure, only their external interface.

We can summarize this property –i.e. that composition is an operation which can be iteratively applied to a set of elements, implicitly building a hierarchy– by stating that this operation has an *algebraic* structure. On the contrary, when this is not the case, it is necessary to take into account the internal structure of a complex element before composing it in a larger composite construction.

The algebraic nature of a traditional compositional system can almost trivially proved. Most of the existing approaches consider just one kind of elementary module, the classic component (c). In this case we just have to provide a "pure" composition operation, which we could express as an algebraic sum ($\oplus$). Therefore:

$$(c \oplus c) = c \tag{1}$$

This means that, as a general rule, "the composition of two components results always in another component". For the remainder of this chapter, we will refer to this rule under the name of the **General Composition Principle**. The Principle can be also applied to the resulting component, then effectively building a compositional hierarchy in a constructive process. Such an approach is obviously algebraic.

By using an architectural approach, we are implicitly assuming that self-∗ systems are compositional; otherwise this approach would not work. However, every self-∗ system is inherently *reflexive*, by definition. For instance, in a self-managed architecture, a part of it is necessarily used to manage the system *itself*. Of course this does not mean that the implementation is necessarily *reflective*, in terms of meta-levels and reification – though this will be our final approach in this chapter. But it clearly leads to an *apparently* non-algebraic nature, which would contradict our previous assumption.

To find an answer to this question, a good approach is to see if we are able to define, at least in an intuitive and non-strictly formal way, an *algebra of adaptive systems*.

4.1 An Algebra of Adaptive Elements

Our purpose in this section is, then, to check if it is possible to to describe every intuitive, archetypical property of a self-∗ system using an algebraic approach. If the answer is positive, then we could say that the system is *compositional*, thereby justifying the adequacy of our architectural approach.

This "algebra" would not strictly be an algebra of adaptive *elements*, but an algebra of element *kinds*. The idea is that we will use the categories defined in the taxonomy of the previous section, as they intend to summarize the specific properties of an adaptive architecture. If these features are consistently maintained by the algebraic combination of these categories, then we will be able to conclude that the nature of self-∗ systems is indeed compositional.

We will need to define two operations. The first one will be *pure composition*, defined as the *sum* ($\oplus$) operation – the same we already used above. It is defined as the juxtaposition of two components, gathering them together and connecting them to return a composite element. Ideally, we would like to generalize the General Composition Principle from Equation (1) to the five categories of our taxonomy. But, only one of the combinations happens to be self-evident:

$$(\epsilon \oplus \epsilon) = \epsilon \tag{2}$$

That is, the composition of two components without any self-adaptive features (two epsilons) results in a component which is *not* self-adaptive either (another epsilon). But this immediate result is not so easily translated to the rest of the categories. Let's consider the case of betas and alphas. In the first case the rule is true, but perhaps in a non-anticipated way. In the second case, we could find even the opposite situation. What we actually have is:

$$(\beta \oplus \beta) = \beta \tag{3}$$
$$(\alpha \oplus \alpha) \not\supset \alpha \tag{4}$$

The second axiom implies that the General Principle is also maintained in the case of betas. As the basic intuition says, composing two betas results in another beta – the union of elements with partial control yields a partially controlled composite. But the justification in this case is a bit more elaborate. In particular, it might happen that these two betas are *complementary* – their combination could turn the partial control into a full control scheme, resulting then in an alpha ($\beta \oplus \beta \rightarrow \alpha$). However, this does not contradict the Equation (3), as we already have that $\alpha \subset \beta$, by definition: full control is a special case of partial control.

Also, the General Principle is *not* always true in the case of alphas, as expressed the Inequation (4): the composition of two alphas does not imply a synergy between them, and therefore the result could not be an alpha. Indeed, this Inequation defines one of our basic rules, and we will refer to it as the **Autonomic Composition Principle**: *the union of autonomous components does not always yield an autonomic composite*. In fact this is quite evident –just consider the case in which two autonomous elements simply don't cooperate– but it still contradicts an extended (false) intuition. Of course, the result of the composition *is* a beta, as the Equation 3 still applies.

But, as already mentioned, *sum* ($\oplus$) is not enough. The specific features of adaptive architectures require a second operator, which defines an even deeper union of two components. This operator, the *product* ($\otimes$), describes the effect of a controller (γ) over a controlled element (in the simplest case, an epsilon ϵ). This union might have two different outcomes, as described in these rules:

$$(\gamma \otimes \epsilon) = \beta \tag{5}$$
$$(\gamma \otimes \epsilon) \rightarrow \alpha \tag{6}$$

We will refer to the Equation (5) as the **Weak Condition for Autonomic Control**, and to the Production (6) as the **Strong Condition for Autonomic Control**. These Conditions don't contradict each other; again, they are obviously compatible, as $\alpha \subset \beta$. And while the first equation (5) is always true, the second production (6) just happens when certain conditions (defined for the production considered as a rewriting rule, and potentially different for every considered $\varsigma\chi$) are met.

Using just these basic rules as axioms, and considering their mathematical properties (commutativity, associativity, etc.) we would be able to develop a complete algebra which summarizes the features of a self-* system. The idea is that, when any apparent conflict appears, it can be described and examined in terms of this algebra, to learn if some basic intuition gets compromised. If we are able to find a successful formulation for any challenging question which seems to contradict compositionality, we should be able to conclude that, indeed, *self-* systems are compositional.*

For instance, let's consider how our algebra solves one such challenge:

Example 1. *From the definition, we can assume that if we insert a non-managed element into a fully autonomic system, the result is always a partially controlled system, i.e. $(\alpha \oplus \epsilon) = \beta$. But sometimes it is also true that the new element gets included in the scope of control, and then the result is again a fully autonomic system, i.e. $(\alpha \oplus \epsilon \rightarrow \alpha)$. Can this behaviour be expressed in algebraic terms?*

The first part is true, by definition, and it can be easily deduced:

$$(\alpha \oplus \epsilon) = ((\gamma \otimes \epsilon) \oplus \epsilon) = (\gamma \otimes \epsilon \oplus \epsilon) \equiv (\gamma \otimes (\epsilon \oplus \epsilon) = (\gamma \otimes \epsilon) = \beta$$

But, the second part is much more complex. In fact, if it is true, it seems to be *non-algebraic*: when the scope of control extends to include the new ϵ-element, the modularity of the original α-element appears to be compromised.

However, what happens could be something like that:

$$(\alpha \oplus \epsilon_n) =$$
$$((\gamma \otimes \epsilon_i) \oplus \epsilon_n) \rightarrow$$
$$((\gamma \otimes (\epsilon_o \oplus \epsilon_x) \oplus \epsilon_n) \,\triangleright$$
$$(\gamma \otimes ((\epsilon_o \oplus \epsilon_n) \oplus \epsilon_x)) = \alpha$$

That is, the original α is expanded by using the Strong Condition (which is always true in this direction). And the elements of the composite are capable of reorganizing such that the controller (γ) extends the scope of control over the new construct. The controversial step is this reorganization – this is indicated by the $\triangleright$ symbol above. But, in fact, this operation *can* be described in an algebraic way: though not very conventional, it is quite similar to the *scope extrusion* operation for in the π-calculus, which is totally analogous while extending the scope [24].

In summary, from this (partial) analysis we can conclude that our hypothesis for self-adaptive systems can be described algebraically; and then their nature can be assumed to be compositional, therefore justifying the architectural approach.

5   Notes on the Relationship to Reflection

In previous sections we have already noted that the notion of a self-* system (and in fact, every ς-attribute) implies an obviously *reflexive* nature. While there is not any direct conceptual connection (against what the word could suggest), it is still natural to question if the use of *reflective* mechanisms could simplify their definition – indeed, reflection was specifically designed for systems to reason about themselves.

Moreover, many existing solutions for self-adaptation are supported by reflective frameworks, both at the technological [14] and the conceptual [12,20] levels.

The answer is quite simple. Adaptation can be easily implemented by using reflection [9], though the converse is not necessarily true. There are many non-reflective implementations of self-adaptive systems, and therefore adaptivity does not require reflection. However, when it is present, it greatly simplifies many of the most complex operations which are implicit in the approach.

In fact, the connection between both set of concepts can be easily made, by just comparing the taxonomy in section 3 to the kinds of elements in a reflective architecture, as described in [7,8]. Briefly, using the standard definition of reflection [19], a reflective architecture can be conceived an architecture "able to reason and act upon itself". This means that a part of it (*meta-level*) is devoted to watch over the other part (*base level*). These levels are connected by a *reification* relationship; those meta-level components directly affected by reification are named as *meta-components*. The discussion of Figure 3, in the next section, should make these definitions clearer.

Therefore, epsilons (ϵ) are conventional (base-level) components, by definition. Control as exerted by gammas (γ), on a reflective implementation, necessarily implies a reification; therefore, γ-elements are always *meta-components*. And, using the same deductive process, deltas (δ) are complementary *meta-level components*. The (two-level) *composite* created by a meta-component and every base-component which it is reflected upon creates a beta (β). And when the control exerted in this beta is complete enough, it is also an alpha (α). As this last element depends directly on the identification of ς-attributes, it needs additional information (specific to adaptive features) to be characterized within an architecture. Therefore, there is not a "pure" reflective correspondant of alphas, and they are just described as a subset of betas.

The example in the next section shows this (arguably complex) correspondence in a much simpler way, and it is more thoroughly discussed in section 6.1.

In the previous sections, we have discussed the convenience of an architectural approach to describe self-adaptive systems; and, as already noted, a reflective approach would also find many common features. However, the combination of both perspectives –that is, to use a *reflective architectural approach*, in which the architectural definition itself is reflective– is very unusual. For this reason, this is the approach that we will use in the remainder of this chapter, and specifically in the next section.

6   Case Study: A Self-adaptive Upgrading Architecture

In this section we are going to examine in detail a case study dealing with a *self-adaptive* system – in fact, as we are going to use the architectural approach, we will describe an adaptive architecture. Our purpose is to provide self-adaptivity using self-organized (at least to some extent) structure; in particular, we would provide an *apparently* centralized adaptive specification, which would be *actually* based on a decentralized architecture. This appearance will be caused by a shift in abstraction, which will be supported by reflective mechanisms.

Specifically, we are going to provide a classic and well-known case study of an autonomic architecture, as a typical example of a self-adaptive system, described here from a reflective perspective. That example will be Kephart's upgrading system from [16]. Then, our goals in this section, by examining this example, are:

1. To provide examples for the items in the taxonomy from section 3, in the context of a medium-sized system. At the same time, it will also be used to *map* the categories in the taxonomy to those of the reflective approach.
2. To describe in detail the reflective approach, and how it can be used at the architectural level – and this is not only limited to self-adaptive systems.
3. To show how a reflective architectural perspective is able to define a non-centralized specification for a self-adaptive system, even providing a different approach to decentralized systems.

To provide an architectural description of the example system, and to be able to provide it considering the reflective perspective, we will use a reflective ADL, specifically $\mathcal{P}i\mathcal{L}ar$. This language was originally defined in [7]; it achieved its mature form and final syntax in [6,8], and has been later extended to provide specific insights in works such as [9,10]. This ADL was originally conceived as a theoretical construct, with the purpose to integrate the concept of *reflection*, from an structural point of view, at the architectural level.

The case study in this section will show that $\mathcal{P}i\mathcal{L}ar$, without the need to include any new extension, is also capable of describing self-adaptive (at least, autonomic) architectures using its reflective syntax. At the same time, this example would serve as an evidence to support the hypothesis exposed in section 5, namely that a reflective approach to the description of adaptive architectures is indeed feasible, and that it is capable of providing a decentralized perspective of the self-adaptive system.

The sample architecture, whose initial configuration is presented in Figure 1, is directly inspired in the main example of the well-known article by Kephart and Chess [16] about autonomic computing, included here as a special case of a self-adaptive system. The case study describes an autonomic *upgrading* system. Specifically, the architecture must provide autonomous support and control for the upgrading of its components.

The desired behaviour of this autonomic system can be summarized as follows. First, the system must be able to support the possibility of dynamically upgrading components. This means that it must be capable of receiving a new version of a component (type) and substitute all its instances in the architecture. After that, the autonomic behaviour must happen. This means that the part of the system in charge of this behaviour begins an interactive process; for the sake of simplicity, this part will be referred to as the *"autonomic subsystem"*, but this subsystem does not actually need to be a conventional module, as we will see.

Then, the autonomic behaviour begins when a battery of regression tests is applied onto the newly instantiated components, in order to verify that the systems' global behaviour is still consistent. The results of these tests will be checked by an *analyzer* element, which must decide if their results are satisfactory (i.e. the behaviour is consistent) or, on the contrary, any kind of problem is identified – and located.

Figure 1 provides a graphic depiction of the initial system architecture. The specification in $\mathcal{P}i\mathcal{L}ar$ is quite straightforward from this topology, and therefore it is not

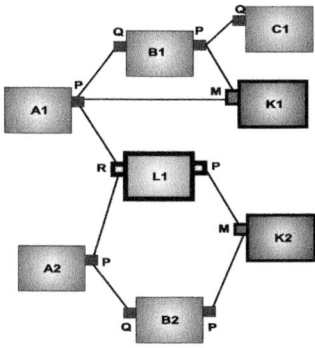

Fig. 1. Initial Configuration of the Upgrading Architecture

provided. It basically would define the five component types, and their connections, defining a global composite, the System.

The original version of this case study [16] assumes, following Kephart's MAPE-K model, that all components in the system are *autonomous* entities; i.e. all of them are α-elements. Our presentation uses a different approach. Basic components, as described in Figure 1 are just conventional modules (ϵ-elements); but this basic architecture is transformed into a self-adaptive system by adding a self-managing *layer*, which will be described in the following (see Figure 2). This layer is introduced by using reflective means, supporting the arguments of section 5.

Therefore, Figure 2 describes the aforementioned self-managing layer, which implements the desired autonomic behaviour, using $\mathcal{P}i\mathcal{L}ar$. The original architecture in the previous two Figures is not altered at all: we just introduce several new elements, but they will be located *at the meta level.* Their specification is similar to the conventional components we already had; but the reflective support in $\mathcal{P}i\mathcal{L}ar$ makes possible to situate them at the right level of abstraction.

Of course, this part of the description is much more complex, as it includes many more behavioural details, and also uses extensively the reflective support. We would not explain here in detail the theoretical foundations of architectural reflection in $\mathcal{P}i\mathcal{L}ar$; probably this would only make more difficult to understand the example. Let's just say that, like in any other reflective system, two different levels of abstraction are considered. The *base level*, where conventional behaviour happens, and the *meta level*, where reflective operations, i.e. those which operate on the system itself, are realized. In $\mathcal{P}i\mathcal{L}ar$, the relationship which binds these two levels in an architectural description is named *reification*; it is also implicitly defined between any component and its "materialized" (*reified*) type.

Therefore, for every component we are able to define base-behaviour, bound to instances and observed through the interface, and meta-behaviour, bound to the types. That said, it should be clear that the meta-level is the right place to define any kind of self-managing behaviour.

The structure defined in Figure 2 for System extends the architecture from Figure 1 – that is, it provides a meta-level description which extends the base description we

16   C.E. Cuesta and M.P. Romay

```
\component System (
  \metaface ( port U | port RT | port DA )
  \config  ( \bind (
    self.U = Updater.S | self.RT = Tester.S |
    self.DA = Analyzer.S |
    Tester.T = Analyzer.T ) )
  \constraint (
    SelfUp def= rep ( U?(nomT,defT);
      loop i ( avatarSet(nomT) )
        ( loop b ( bound(i) )
          ( new(j:defT); link (j,b);
            del(i) ) ) )

    SelfTest def= rep ( RT?(nomT,rtest);
      for ci ( avatarSet(nomT) )
        ( tau(ci,rtest,res); cons(res,VRes) );
      RT!(VRes);
      DA!(avatarSet(nomT)); SelfConn )

    SelfConn def= rep ( DA?(ci);
      DA!(bound(ci)) ) ) )
```

```
\component Updater (
  \metaface ( port S | import OT | import NT )
  \constraint (
    Updating def= rep ( OT?(nomT);
      NT?(defT); S!(nomT,defT) ) ) )

\component Tester (
  \metaface ( port S | export T | import RT )
  \constraint (
    Testing def= rep ( RT?(nomT,rtest);
      S!(nomT,rtest); S?(VRes); T!(VRes) ) ) )

\component Analyzer (
  \metaface ( port S | port T | export D )
  \constraint (
    Analyzing def= rep ( S?(vComp);
      loop ci ( vComp )( S!(ci); S?(vBnd);
      T?(VRes); tau(vBnd, VRes, ok);
      if ( not(ok) ) ( D!(ci) ) ) ) ) )
```

Fig. 2. Meta-Level: Autonomic Upgrading System

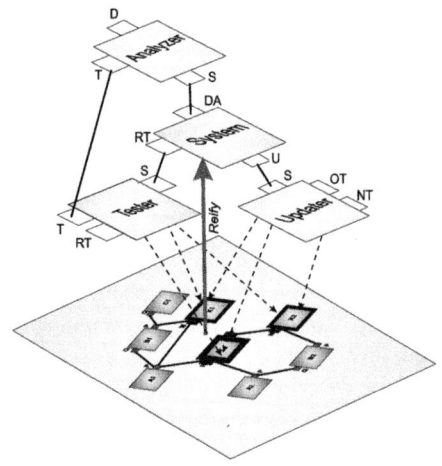

Fig. 3. Reflective Architecture with Autonomic Upgrading

already had. In short, this specification describes the meta-interface which would communicate with the new meta-level components (Updater, Tester and Analyzer), and the constraints imposed on these new ports, defined as concurrent processes according to $\mathcal{P}i\mathcal{L}ar$ semantics, and which define the desired behaviour.

We can briefly summarize that behaviour as follows. First, Updater receives a message with the name of a component, and the definition of a new version. Both data are sent to the System, which must now upgrade the component. New instances for the new version are created, and connected following the same pattern. Later, Tester receives battery of tests for the new component, which is also communicated to the System.

Table 1. Summary of Kinds of Components in the Case Study

Component Type	In the Architecture	In the Taxonomy
A	Base component	Epsilon (ϵ)
B	Base component	Epsilon (ϵ)
C	Base component	Epsilon (ϵ)
K	Base component	Epsilon (ϵ)
L	Base component	Epsilon (ϵ)
Updater	Meta-level Component	Delta (δ)
Tester	Meta-level Component	Delta (δ)
Analyzer	Meta-level Component	Delta (δ)
System	Base component	Epsilon (ϵ)
System	Meta-component	Gamma (γ)
System	Base- & Meta-component	Alpha (α)

Then, the process *SelfTest* executes these tests in every instance. Simultaneously, the system provides the Analyzer with the locations for every related instance.

Finally, the Analyzer receives both data sets, and request the set of connections for every instance (*SelfConn*), examining them individually. This way, every failure detected in the tests is checked in its origin, identifying its exact location in the architecture, and the final results are externally notified.

Figure 3 provides a depiction of the place occupied by the different components in the two-level architecture, which express clearly their roles in terms of the taxonomy of section 3. However, in order to clarify this information, and to provide a simpler explanation of the example, their classification is summarized in Table 1.

In summary, all the base components (A, B, C, K and L) are epsilons (ϵ). Meta-level components Updater and Tester are conceived and implemented as deltas (δ), as they perform an independent computational task, which just happen to use meta-information as data. The remaining meta-level component, Analyzer, is clearly a delta (δ): it just receives a set of (meta-)data and process them, regardless of the place where its computation happens. Finally, the global composite, System is an epsilon at the base level, and obviously its reification at the meta-level is a gamma; therefore their composition is a beta (β). However, as the self-adaptive behaviour of this β-element gathers all the autonomy we were trying to provide in the system, we must consider this system is *fully autonomous*, and then it is actually characterized as an alpha (α).

6.1   Discussion: On the Decentralized Approach

The example presented here is a typical example of self-adaptive behaviour (in the form of autonomy), and as such, it has a centralized presentation – in the sense that this behaviour depends on a set of "central" managing elements. This seems to be one of the defining patterns of self-adaptive systems, as opposed to the decentralized approach which defines self-organizing system.

However, this does not mean that the *topological* structure of our solution must necessarily be centralized. Indeed, even the original solution by Kephart [16] describes a decentralized architecture, in which every element is an autonomous agent.

Our solution has the same intent, but takes a completely different stance. In the $\mathcal{P}i\mathcal{L}ar$ description, our solution is *apparently* centralized, in the sense that we there seems to be a set of "central" components which "manage" the adaptive behaviour, namely the updater, tester and analyzer. But actually, these are just *specialized* components which operate on certain particular set of data – they just compute a function. Therefore, though important, they are not the "center" of the adaptive behaviour.

In fact, the only suitable candidate is the *meta-level half* of the System component, which is also the only gamma in the architecture. This meta-component, conceived as a "global policy" for the architecture, would perfectly qualify for the definition of a self-adaptive center: it encapsulates the autonomic strategy which defines our autonomic behaviour. Therefore again, our approach *seems to be* centralized.

But it must be understood that our approach is *reflective*. The meta-level is not a "place" in the architecture, but a *level of description*. As explained above, the behaviour described in the meta-level is reflected onto the base components, defining an additional concurrent process which is composed in parallel to the original behaviour of *every instance*. Therefore, every instance gets transformed into an *hybrid* process, which combines base- and meta-behaviour. The result would have been essentially the same if we had defined this (autonomic) behaviour in every instance – i.e. just like every component was *autonomous*.

This means that the behaviour is *described* in a single place, but it is later replicated and distributed all through the architecture, traversing reification links, to be *executed* in multiple processes on many locations. In summary, *the reflective approach does not preclude any organization: it can define either a centralized or a decentralized pattern*, depending on the distribution policy which is chosen, and which is supported by reflective mechanisms. In the case of $\mathcal{P}i\mathcal{L}ar$, for example, it depends on the pattern of reification links which is chosen.

Therefore, our proposal can be perfectly understood as a decentralized solution, even a *self-organizing architecture*, because what is provided as a single autonomic specification, is later replicated over *every* base component. There is not a single coordinating entity, and therefore our approach is not centralized. What we have here is single coordination *strategy* – but many times replicated and distributed in many places, all throughout the architecture.

Our model does not take the form of centralized components: instead of that, the different concerns in this behaviour are modularized [9] and *reified into* the base components, as necessary. Then, the meta-level behaviour is *subsumed* inside the existing componets, using the desired distribution pattern. The resulting component exhibits a combination of their base- and meta-level features, and includes, by reflection, the autonomic behaviour. Therefore, the "autonomic subsystem", though apparently centralized when specified at the meta-level, is actually fully *distributed* within the elements of the base-level, i.e. it results in a *decentralized self-adaptive architecture*.

7   Conclusions and Future Work

This work has explored the most important structural concepts in self-adaptive systems, as the basis of an architectural approach to tackle its definition and implementation.

In the process, several properties of this approach have also been tested, including its general consistency, which is supported by compositionality.

The chapter has explored in detail the feasibility of the architectural approach to the definition of self-∗ systems – most of its first half is devoted to this. First, the features of these systems, and their modular nature, are studied in detail. Then, our results were used to create a taxonomy of elements for *self-adaptive architectures*. Finally, these elements are used to define an algebra, proving the *compositionality* of the construction. This way, we conclude that the architectural approach is indeed consistent.

Another goal of this chapter was to show that the dichotomy between self-adaptive behaviour (centralized) and self-organized structures (decentralized) can be resolved by achieving the former on top of the latter. The autonomic example from section 6 was chosen because of that, and it fits that purpose perfectly.

However, it should be noted that, even when our approach seems promising, in the sense that is is possible to apply it both to self-adaptation and self-organization, it is far for being complete with regard to the second one. A more specific examination of self-organizing architectures will probably provide several more insights, and it is very likely that our taxonomy would have to be extended to accomodate some of them. There are already some advances in this direction.

Finally, as already noted, the concept of self-∗ systems itself has a reflexive nature. Therefore, it is quite logical to suggest a reflective approach to tackle their implementation. At the same time, the architectural approach has been justified; the combination of both ideas suggests that a good solution is the specification of reflective architectures, or even more generally, of *reflective architectural descriptions*.

$\mathcal{P}i\mathcal{L}ar$, without further extensions, is capable of describing adaptive architectures – as implied above, a reflective architecture description can always support a self-adaptive system. However, probably the specification in reflective terms is not intuitive enough, specially for some applications of adaptive systems. Therefore, the definition of an specific self-adaptive syntax would probably increase the usability of the language.

Future work will explore this approach, by defining an adaptive extension of $\mathcal{P}i\mathcal{L}ar$, based on our taxonomy, to simplify the specification of self-∗ architectures.

References

1. Allen, R.J., Garlan, D.B.: A Formal Basis for Architectural Connection. ACM Transactions on Software Engineering and Methodology 6(3), 213–249 (1997)
2. Andersson, J., de Lemos, R., Malek, S., Weyns, D.: Modeling Dimensions of Self-Adaptive Software Systems. In: Cheng, B.H.C., de Lemos, R., Giese, H., Inverardi, P., Magee, J. (eds.) Software Engineering for Self-Adaptive Systems. LNCS, vol. 5525, pp. 27–47. Springer, Heidelberg (2009)
3. Babaoglu, Ö., Jelasity, M., Montresor, A., Fetzer, C., Leonardi, S., van Moorsel, A.P.A.: The Self-star Vision. In: Babaoğlu, Ö., Jelasity, M., Montresor, A., Fetzer, C., Leonardi, S., van Moorsel, A., van Steen, M. (eds.) SELF-STAR 2004. LNCS, vol. 3460, pp. 1–20. Springer, Heidelberg (2005)
4. Beer, S.: Brain Of The Firm, 2nd edn. John Wiley, Chichester (1994)
5. Conant, R.C., Ashby, W.R.: Every Good Regulator of a System Must Be a Model of that System. International Journal of Systems Science 1(2), 89–97 (1970)

6. Cuesta, C.E.: Reflection-based Dynamic Software Architecture. ProQuest/UMI, Madrid (May 2003)
7. Cuesta, C.E., de la Fuente, P., Barrio-Solórzano, M., Beato, E.: Dynamic Coordination Architecture through the use of Reflection. In: Proceedings 16th ACM Symposium on Applied Computing (SAC 2001), pp. 134–140 (March 2001)
8. Cuesta, C.E., de la Fuente, P., Barrio-Solórzano, M., Encarnación Beato, M.: Introducing Reflection in Architecture Description Languages. In: Software Architecture: System Design, Development and Maintenance, pp. 143–156. Kluwer, Dordrecht (2002)
9. Cuesta, C.E., Pilar Romay, M., de la Fuente, P., Barrio, M., Younessi, H.: Coordination in Architectural Connection: Reflective and Aspectual Introduction. L'Objet 12(1), 127–151 (2006)
10. Cuesta, C.E., Pilar Romay, M., de la Fuente, P., Barrio-Solórzano, M.: Temporal Superimposition of Aspects for Dynamic Software Architecture. In: Gorrieri, R., Wehrheim, H. (eds.) FMOODS 2006. LNCS, vol. 4037, pp. 93–107. Springer, Heidelberg (2006)
11. Galloway, A.R.: Protocol: How Control Exists after Decentralization. The MIT Press, Cambridge (2004)
12. Grace, P., Coulson, G., Blair, G.S., Porter, B.: A Distributed Architecture Meta-model for Self-Managed Middleware. In: Proceedings 5th Workshop on Adaptive and Reflective Middleware (ARM 2006), p. 3, 1–6. ACM Press, New York (2006)
13. Heylighen, F.: Principles of Systems and Cybernetics: an Evolutionary Perspective. In: Trappl, R. (ed.) Cybernetics and Systems, pp. 3–10. World Science, Singapore (1992)
14. Huang, G., Liu, T., Mei, H., Zheng, Z., Liu, Z., Fan, G.: Towards Autonomic Computing Middleware via Reflection. In: Proc. 28th Annual Intl. Computer Software and Applications Conference (COMPSAC 2004), pp. 135–140 (2004)
15. Johnson, S.: Emergence. The Connected Lives of Ants, Brains, Cities and Software. The Free Press/Simon & Schuster, New York (2001)
16. Kephart, J.O., Chess, D.M.: The Vision of Autonomic Computing. IEEE Computer 36(1), 41–50 (2003)
17. Kramer, J., Magee, J.: Self-Managed Systems: an Architectural Challenge. In: Future of Software Engineering (FOSE ICSE 2007), pp. 259–268. IEEE CS Press, Los Alamitos (2007)
18. Lin, P., MacArthur, A., Leaney, J.: Defining Autonomic Computing: A Software Engineering Perspective. In: Proc. Australian Conf. Software Engineering (ASWEC 2005), pp. 88–97. IEEE CS Press, Los Alamitos (2005)
19. Maes, P.: Concepts and Experiments in Computational Reflection. ACM SIGPLAN Notices 22(12), 147–155 (1987)
20. McKinley, P.K., Sadjadi, S.M., Kasten, E.P., Cheng, B.H.C.: Composing Adaptive Software. IEEE Computer 37(7), 56–64 (2004)
21. Morrison, R., Balasubramaniam, D., Kirby, G., Mickan, K., Warboys, B., Greenwood, M., Robertson, I., Snowdon, B.: A Framework for Supporting Dynamic Systems Co-Evolution. Automated Software Engineering 14(3), 261–292 (2007)
22. Prokopenko, M., Boschetti, F., Ryan, A.J.: An Information-Theoretic Primer On Complexity, Self-Organisation And Emergence. Complexity 15(1), 11–28 (2009)
23. Ryan, A.J.: Emergence is Coupled to Scope, Not Level. Complexity 13(1,2), 67–77 (2007)
24. Sangiorgi, D., Walker, D.: The π-calculus: A Theory of Mobile Processes. Cambridge University Press, Cambridge (2003)
25. Weyns, D., Omicini, A., Odell, J.: Environment as a First-class Abstraction in Multiagent Systems. International Journal on Autonomous Agents and Multi-Agent Systems 14(1), 5–30 (2007)

Improving Architecture-Based Self-adaptation Using Preemption

Rahul Raheja[1], Shang-Wen Cheng[2], David Garlan[1], and Bradley Schmerl[1]

[1] Carnegie Mellon University, Pittsburgh PA, 15213, USA
{rahulraheja,garlan,schmerl}@cs.cmu.edu
[2] Jet Propulsion Laboratory, 4800 Oak Grove Dr, Pasadena, CA, 91109, USA
Shang-Wen.Cheng@jpl.nasa.gov

Abstract. One common approach to self-adaptive systems is to incorporate a control layer that monitors a system, supervisorily detects problems, and applies adaptation strategies to fix problems or improve system behavior. While such approaches have been found to be quite effective, they are typically limited to carrying out a single adaptation at a time, delaying other adaptations until the current one finishes. This in turn leads to a problem in which a time-critical adaptation may have to wait for an existing long-running adaptation to complete, thereby missing a window of opportunity for that adaptation. In this paper we improve on existing practice through an approach in which adaptations can be preempted to allow for other time-critical adaptations to be scheduled. Scheduling is based on an algorithm that maximizes time-related utility for a set of concurrently executing adaptations.

Keywords: self-adaptation; preemption; utility; concurrency.

1 Introduction

Today's complex systems require considerable administrative overhead; this has led to a demand that they self-adapt at run-time to variable resource availability, loads and faults. Until recently, mechanisms for self-adaptation were largely in the form of programming language features, embedded in the code, hence limitingreusability and modifiability. Today there is an increasing trend toward *autonomic computing*, in which external control modules are used to provide adaptive capabilities that monitor and adapt the system at run time. Rainbow [1] is a framework for external control that provides capabilities for self-adaptation and provides mechanisms to balance adaptations amongst multiple stakeholder objectives [2]. It forms a closed-loop control system in which the adaptation mechanism probes, evaluates, decides, executes, and then probes again. In this respect, Rainbow is similar to other autonomic systems [3,4,5].

While such approaches have been demonstrated to be useful, they are largely limited to carrying out a single adaptation at a time; violations that occur while an adaptation is taking place will not trigger new adaptations until the current one has finished executing. Unfortunately, ignoring or delaying adaptations potentially leads to less-than-optimal adaptation, since a high-priority adaptation may have to wait for a less critical adaptation to finish. For example, it may not be desirable to delay addressing security

D. Weyns et al. (Eds.): SOAR 2009, LNCS 6090, pp. 21–37, 2010.

issues until a currently executing performance adaptation completes. By not addressing a security problem early, the system may be compromised, so that later the security adaptation will be less effective.

Ideally, whenever new adaptation conditions arise, the adaptive mechanism should be able to reconsider all objectives and determine the best course of adaptation without waiting for an existing adaptation to complete. In this paper we propose an improvement upon existing approaches that avoids delaying adaptation decisions, where the adaptation mechanism promptly considers all adaptations, including the currently executing adaptation, when new conditions arise. To make this possible, we extend Rainbow's adaptation mechanisms to support preemption of an executing adaptation strategy, reasoning amongst multiple strategies, starting new ones, and resuming preempted strategies. To facilitate this, we extended Rainbow to includean adaptation time-utility dimension for self-adaptive mechanisms that allowsreasoning about time, in addition to other concerns already considered by Rainbow, and facilitates prioritization and scheduling of adaptations at run time. The specific contributions of this paper are:

- Applying the concept of a time-utility curve (TUC) to prioritize amongst multiple adaptations taking into consideration the time it takes to complete an adaptation(see Section 3.2);
- The application of existing concepts of multi-task scheduling under the constraint of maximizing overall utility to improve current adaptation decision-making algorithms (see Sections 3.3 and 3.4);
- The use of an approximation of arely-guarantee technique to ensure consistency amongst multiple interleaving adaptation strategies (see Section 3.6); and
- The use of an architecture model of a system as a reference to provide architectural locks to ensure non-interference (see Section 3.6).

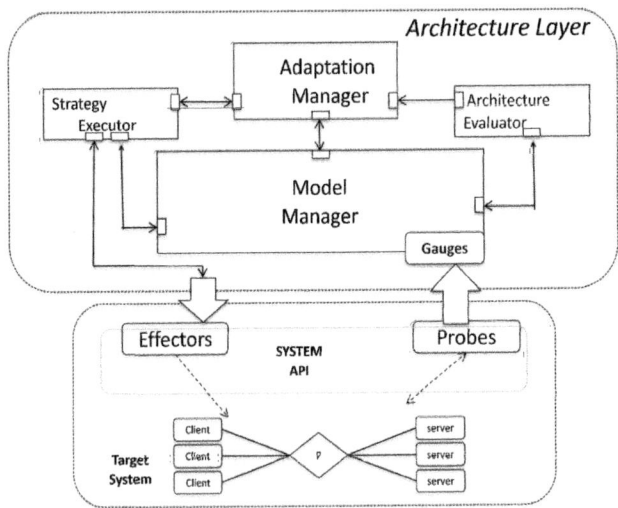

Fig. 1. Rainbow adaptation control loop components

The remainder of this paper is organized as follows: In Section 2, we give a brief overview of the existing Rainbow adaptation framework and outline a motivating scenario to illustrate how it works. In Section 3 we describeour approachand discuss its implementation. In Section 4 we presentsome experimental results. In Section 5 we discuss related work, and in Sections 6 discuss the implications of this work and outlinefuture work.

2 Background and Motivation

The Rainbow framework uses software architectures and a reusable infrastructure to support self-adaptation of software systems. Figure 1 illustrates its adaptation control loop. *Probes* are used to extract information from the target system. This information is passed to *gauges* that abstract these system observations into architecturally relevant information. Gauges then update the architectural model with this new information. The *architecture evaluator* checks constraints in the model and triggers adaptation if any violation is found. The *adaptation manager*, on receiving the adaptation trigger, chooses the best strategy to execute, and passes it to the *strategy executor*, which executes the strategy on the target system via *effectors*.

Given a particular adaptation trigger, the strategy to be executed is chosen on the basis of stakeholder utility preferences and the current state of the system as reflected in the architecture model. The underlying decision making model is based on decision theory and utility theory [6], and is discussed in detail in [2,7]; varying the utility preferences as weights on quality dimensions allows the adaptation engineer to affect which strategy is selected. Adaptations are specified using *strategies, tactics, and operators,* defined using the Stitch [7] adaptation language. A strategy captures a pattern of adaptations in which each step evaluates a set of condition-action pairs and executes an action (called a *tactic*), possibly waiting for the action to take effect. Additionally, a strategy specifies conditions of applicability that determine the contexts in which it should be involved. A tactic defines an action, packaged as a sequence of commands (operators). It specifies conditions of applicability, expected effects, and (in a separate customization step) cost-benefit attributes to relate its impact on the quality dimensions. In Rainbow, a tactic leaves the system in a consistent architectural state. This makes it possible to analyze, prior to executing the strategy, what its expected impact will be on the utility of the system, if applied. An operator is a basic command provided by the target system.

To illustrate, consider a web-based client-server system that conforms to an N-tier style ("Target System" illustrated in Figure 1). It uses a load balancer ("P") to balance requests across a pool of replicated servers, the size of which is dynamically adjusted to balance server utilization, response time, cost, etc. Assume that Rainbow can monitor the system for information such as server load, port activity on servers, etc. Assume also that we can adaptthe system at run time to add more servers to the pool, to change the fidelityof the content,orto shut down ports. Preferences on the response time experienced by clients are specified using architectural constraints. An adaptation engineer may define a strategy to deal with low response time.

Now suppose that by evaluating the architectural constraints Rainbow detects a violation of constraints associated withresponse times (i.e., response times exceed those specified for the system). This violation causes the Adaptation Manager to search through its set of repair strategies to choose the most appropriate adaptation that will fix this constraint, such as adding new servers, or reducing the quality of the content. While it is in the middle of carrying out this adaptation, probes report a port-scan activity on one of the servers and this in turn triggers a security-related adaptation. In the current implementation of Rainbow, the new trigger will be suppressedsince Rainbow is in the middle of adding servers to the server group. However, ifthe port activity isin fact an intrusion and not just a scan, and the port isnot closed (because the trigger isignored), the server will be compromised, reducing overall system utility.

Rainbow's mechanisms were deliberately built to delay adaptations in such circumstances. Rainbow considered that a new adaptation trigger in the middle of an ongoing adaptation could potentially be caused by an intermediate action of the current adaptation on the target system, and that finishing the current adaptation could potentially eliminate the need for a new one. If the condition persists, a repair will eventually be triggered. However, this assumes that the *timeliness* of a repair is not a factor. In the situation above, if a security repair that closes the port is not done in a timely fashion, it is possible that an intruder could cause further damage to the system, or have access to sensitive information, while the performance repair completes, lowering the overall system utility further.To address this, we extend Rainbow's mechanisms to be able to manage both scenarios.

3 Approach

In the previous section we motivated the need to consider the timeliness of strategies, and to pause any currently executing strategies in favor of higher priority ones. Supporting this capability necessarily requires that we develop the following:

1. A means to prioritize strategies, so that we can decide which strategies are more important than others. This prioritization needs to be sensitive to both the expectedsystem utility improvement of each strategy, as well as the timeliness for executing them.
2. A method to detect which strategies may interfere with each other, so that we can schedule strategies to avoid conflicts.
3. A policy for scheduling strategies that takes into consideration those priorities and conflicts.
4. A way to preempt strategies so that higher priority strategies may be scheduled instead, but that still leaves the system in a consistent state.

In the following sections we discuss these issues in more detail, and describe howwe address the problem described in Section 2.

3.1 Incorporating Changes in Rainbow

Before discussing the algorithms and approaches for incorporating prioritization and preemption, we discuss architecturally the components that are involved. This is to provide some context for the following sections, which in addition to describing the algorithms, also indicate what parts of Rainbow are responsible for them.

We refactor the Adaptation Manager in Rainbow to allow strategy selection *plugins*. This mechanism provides a method of changing the way in which Rainbow can

choose strategies. The default method of selecting strategies in Rainbow may be replaced with new mechanisms that may more closely meet the needs of adaptation for
particular systems. This approach also has the added benefit that it will allow us in the
future to compare alternative strategy selection approaches. The particular points of
customizability in the Adaptation Manager are prioritization knowledge, scheduling
and conflict detection.

To incorporate preemption, we add a Real Time Evaluator(*RTEvaluator*in
Figure 2) sub-component into the Adaptation Manager component. It is responsible
for ordering a set of strategies that could be executed according to customizable criteria and algorithms.

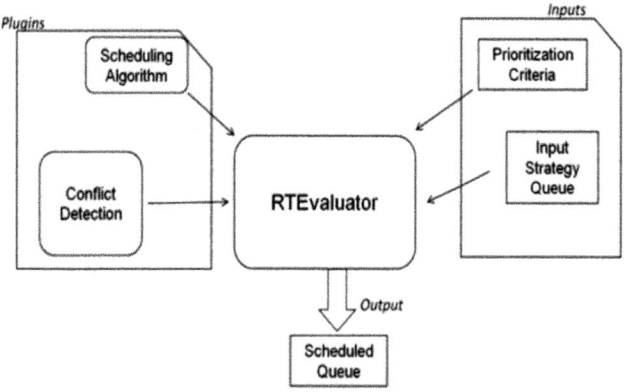

Fig. 2. Detailed design of the RTEvaluator sub-component

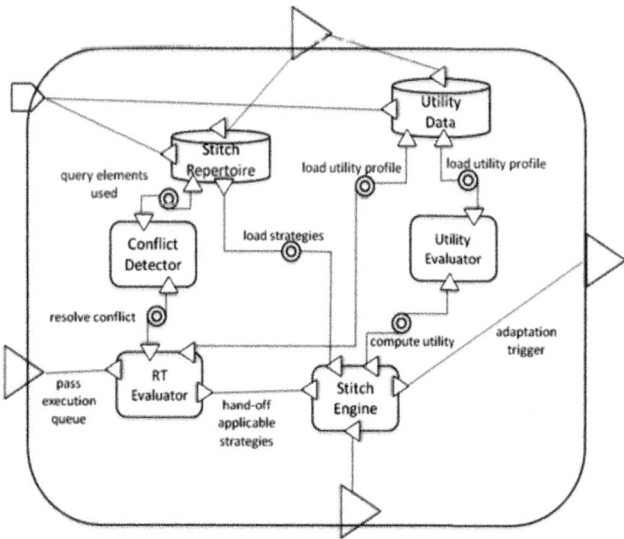

Fig. 3. Interactions of the RTEvaluator with other components in the Adaptation Manager

The RTEvaluator expects two inputs and two plug-ins. The inputs are a queue of strategies that need to be ordered for scheduling and the prioritization criteria. The plug-ins are the scheduling algorithm and the conflict detection mechanism. The evaluator, on receiving a set of strategies, uses the prioritization criteria and the scheduling algorithm, in consultation with the conflict detection module (described later), and outputs an execution order that satisfies the conflict rules and the prioritization criteria. The interactions of the RTEvaluator with other sub-components are shown in Figure 3.

When the Architecture Evaluator (in Figure 1) triggers the need for an adaptation, this information is passed to the *Stitch Engine* sub-component of the Adaptation Manager. The Stitch Engine loads the strategies to be considered from the *Stitch Repertoire* repository, and then invokes the *Utility Evaluator* to determine the set of strategies that would improve the utility of the system. Prior to the work described in this paper, the Stitch Engine would then choose the strategy resulting in the highest utility, and execute that strategy, waiting until the strategy completes before considering other violations from the Architecture Evaluator. In the improved approach, the Stitch Engine remains active: each time an adaptation is triggered, the itpreempts (pauses) any currently executing strategy[1], and invokes the RTEvaluator(after determining the set of applicable strategies) to choose the strategy to execute. The RTEvaluator calls the Conflict Detector to determine the set of applicable strategies that do not conflict with any currently preempted strategies, and then uses a scheduling algorithm to choose the most applicable strategy, which is then executed.

3.2 Prioritizing Adaptations with Time Utility Curves

To address the issue of timeliness of strategy execution, we need to be able to reason about which strategy or adaptation condition is more important given the current state of the system and user priorities. In this section, we discuss our approach of using Time Utility Curves to prioritize strategy selection, and thereby advise the Stitch Engine whether to continue any currently executing adaptation strategies, or whether to start executing a new strategy to handle a new condition.

One way to prioritize strategy selection would befor the architect of the adaptive system to specify the priorities of adaptation conditions at design time. In such a case, whenever a higher priority adaptation condition occurs, other lower priority adaptations will be aborted or preempted. This implies that if a lower priority adaptation condition occurs, then it will alwayswait for the current adaptation to finish, and, if a higher priority adaptation occurs, then it will alwaysbe serviced first.

As mentioned earlier, adaptations in self-adaptive systems are opportunities for improvement. They are aimed to increase overall system utility, which quantifies the happiness of the system stakeholders with the system. If strategies take a long time to execute over the target system, their intended utility improvement will diminish or disappear. In particular, given the static priority scheduling proposed above, a lower-priority condition strategy would *always* wait until higher-priority strategies finish executing, hence potentially losing its window of opportunity.

To avoid the above situation, we must introduce some criteria into the Adaptation Manager that will allow it to make a decision at run-time about how to prioritize

---

[1] The reasons for preempting the strategy at this point are discussed in Section 6.1.

strategies. To do this, we associate each strategy with a time utility curve(TUC) [8]. A TUC specifies the expected utility [0,1]of completing a strategy as a function of its completion time. (We discuss the use of TUCs in section 3.4).Using TUCs, we can express both hard real-time adaptations and soft real-time adaptations. TUCsalso give-one the ability to specify arbitrary utility values rather than step/linear functions(Figure 4). We can then use this information to schedule strategies according to some scheduling criteria.

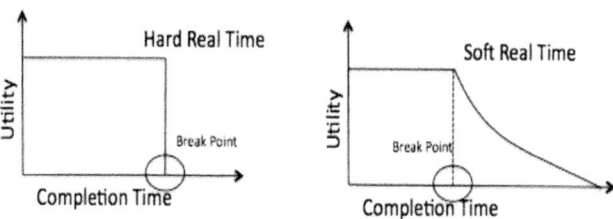

Fig. 4.Time Utility Curves for Hard/Soft Real time adaptations

3.3 Improvements to Strategy Selection Algorithm

The scheduling algorithm is encapsulated as a plug-in to the RTEvaluator, allowing different scheduling algorithms to be considered and compared in the future. The algorithm that we describe in this paper aims to maximize system utility. It is a flavor of a real-time scheduling algorithm, which tries to find the best execution order to maximize predicted system utility on remaining execution time.When the *Architecture Evaluator*triggers an adaptation, the set of applicable strategies are handed to the *RTEvaluator* by the *Stitch Engine*.The scheduler is passed the ordered queue of strategies pending execution that was chosen in prior runs of the scheduler (or empty, if this is the first time), and then executes the algorithm below:

a.   Let Q be the ordered queue of strategies pending execution; if non-empty, the head entry $Q[0]$ is the currently executing strategy, so preempt it
b.   Let P be the unordered set of Q
c.   Let A be the set of applicable strategies for the new condition
d.   Create a new empty set C of candidate strategy orderings
e.   For each strategy a in A
    a.   Pick the ordering of $a \cup P$ that has the highest Predicted System Utility (PSU) (explained in the next section)
    b.   Add this ordering and PSU to C
f.   Pick the ordering with the highest PSU from C to form Q', the new ordered queue of strategies pending execution
g.   Execute (or resume) the strategy at $Q'[0]$

Now that the Adaptation Manager no longer pauses while strategies are being executed, it is possible that the Stitch engine will be triggered by the same constraint violation numerous times, while the strategy already chosen to handle that violation is in the queue or executing. To deal with this, the Stitch Engine will only consider violations when they occur for the first time. When this happens, it remembersthe set of constraint violations that caused the adaptation to trigger. When a strategy is selected, theseconstraintsare associated with thestrategy. Ifthe Stitch Engine finds constraint violations that already have some strategy(s) associated with them, then it will not trigger a new round of strategy selection. When the strategy finishes execution, it instructs the Stitch Engine to remove its associated constraints so they can be considered from then on if they are violated.

3.4 Predicted System Utility

The Predicted System Utility (PSU) of a set of strategies gives the predicted resulting utility of the system assuming the strategies are executed in a particular order. Suppose the current system state is represented by S[], which is a vector of all utility dimensions of the system.Say,

$$S[] = [fidelity, cost, responseTime, security] = [p, q, r, t]$$

We have a set of strategies A_1, A_2 ... A_N, that will be executed in the order $[A_1, A_2 ... A_N]$.Each strategy is associated with an aggregate Attributevector[2] $agg[]_{Ai}$, and a time utility curve TUC_{Ai}

$$agg[]_{Ai} = [\Delta p, \Delta q, \Delta r, \Delta t]$$

Running A_1 first leaves the system in state $S[]_1$

$$S[]_1 = S[]_0 + TUC_{A1}(\text{exec time of } A_1)*agg[]_{A1}$$

TUC_{A1}(exec time of A_1) will give the percentage of expected change that actually happened depending on the execution time of A. Similarly running A_2, A_3 ... A_N gives

$$S[]_2 = S[]_1 + TUC_{A2}(\text{exec time of } A_1+A_2)*agg[]_{A2}$$
$$S[]_N = S[]_{N-1} + TUC_{AN}(\text{exec time of } A_1+A_2...A_N)*agg[]_{AN}$$

The state of the system at the end of executing A_1, A_2 ... $A_N$will be

$$S[]_N = [p_N, q_N, r_N, t_N]$$

We calculate utility using utility profiles and weights for the quality dimensions

$$Weight_p * UtilityCurve_p(p_N) + Weight_q * UtilityCurve_q(q_N) + Weight_r * UtilityCurve_r(r_N) + Weight_t * UtilityCurve_t(t_N) = PredictedUtility$$

This utility is the predicted system utility for the specified execution order. The strategy execution time calculation builds upon the settling time concept in Rainbow [1] by further profiling the run-time of each tactic (under serial mode with no preemption) and keeping a historical record of the mean time of execution. Rainbow is run under different configurations and the mean, standard deviation and error are calculated using multiple runs for each tactic. The execution time of the strategy is then estimated probabilistically from its tactic tree, which is then fed into the calculation of the predicted system utility. Also, to calculate the execution time of a preempted strategy, the

---

[2] The aggregate attributed vector of a strategy is the expected change to utility dimensions that it expects to provide after it finishes executing.

remaining execution time is calculated from the tactic where it was preempted and not from the beginning.Using mean time is a good start, but further evaluation is needed to completely understand the sensitivity of the adaptation outcome to variances in tactic completion time.

3.5 Granularity of Concurrency

The approach described above relies on being able to preempt any currently executing strategy, pausing it so that other higher priority adaptations may be executed instead.The granularity at which a strategy can be preempted must strike a balance between achieving maximum possible interleaving and ensuring that preemption leaves neither the target system,nor the model on which decisions are made,in an inconsistent state. In Rainbow, there are three potential points at which a strategy can be preempted: after the currently executing *strategy* has completed, after a *tactic* of the current strategy has completed, or after any *operators* comprising the tactic have finished.

First, setting preemption to only occur at the end of a strategy is the current situation in Rainbow, and so we immediately discount that.Choosing preemption to occur at the operator level, while being the finest possible level of granularity in Rainbow, has a number of drawbacks. First, it would break the assumption mentioned in Section 2 that tactics are atomic, and would require changes to the way that Rainbow calculates utility; second it would require adaptation analysts to specify conditions of applicability and timing for every operator; third, timing would incur communication overheads between Rainbow and the target to monitor and check timing. On the other hand, tactics already have conditions of applicability and some timing information associated with them, and fixing the level of granularity at this point requires no changes to the utility calculations of Rainbow. Therefore, we decided that preempting strategies after tactics have completed would be an appropriate balance.

3.6 Conflict Detection and Resolution Using Architectural Locks

Consider a strategy A that is pre-empted to execute another strategy B. Suppose that strategy A werepre-empted after executing 2 tactics, and hasyet to execute 2 more. While strategy B executes it makes some changes to parts of the target system. Now when strategy A resumes and starts executing the remaining tactics, it assumes that its target components are in thestate that they were in when the first two tactics finished. If B undid those changes, then finishing A will leave the target system in a potentially undesirable or inconsistent state.

Following the above example, we need to ensure that actions of interleaving strategies do not conflict with each other. There are a couple of ways to ensure this. First, we can leave the onus on the strategy writers to ensure that their strategies are written in ways that do not conflict with any other strategies written for this system. But there could be many strategies for a system, potentially written by different people; expecting a strategy writer to know the behavior of all other strategies would be unreasonable. Even if one assumes that the strategy writer knows about all other strategies, it could still be unreasonable to frame astrategy in a way that doesn't conflict with any other, because it degrades the reusability of strategies (it requires that each strategy have a dependency on all other strategies that could be executed).

Second, we can provide some guarantees to a strategy that other strategies will not touch the parts of the system that it acts on using an approximation of rely and guarantee reasoning [9]. Specifically, each strategy *guarantees* that it will make changes to only a subset of the system and that it *relies* on the fact that no other strategy will change this subset while it is executing. A strategy would specify the components and connectors that it is making changes to, and then the RTEvaluator could guarantee that no other strategy is selected that would make changes to any elements in that set. In this way, the architecture model of the target system and its environment[3] is used as a reference to provide virtual locks to strategies.

We encapsulate conflict detection evaluation in the Conflict Detector subcomponent of the Adaptation Manager by providing four interfaces to it. First is an interface with which a strategy that is about to be executed registers the components and connectors it will change, and second is to deregister (mark as free by removing a virtual lock). When a new strategy is to be run, it seeks permission from the Conflict Detector using the third interface. Using the fourth interface, the Detector returns a subset of a given schedule of strategies that will not conflict.

Each scheduling plugin can decide which interfaces they want to use, as well as the algorithms in the Conflict Detector. For our implementation, if a strategy about to run uses any component(s) that is also being used by some preempted strategy, it is not allowed to execute, and a new order is found that is non-conflicting and gives the next highest utility.

4 Results

We use a typical news website infrastructure, Znn.com, which is typically a three-tiered architecture, to demonstrate the preemption scenario described in Section 2. This news website infrastructure is similar to that described in Section 2 (Figure 1 "Target System"), and is in fact one of the case studies reported in [7] to show the effectiveness of Rainbow in managing multiple concerns. The servers can be adjusted to balance utilization for response time. The clients make requests for content that includes text, images, videos (static content) and templates (dynamic content).

Common objectives for a news provider are to deliver server content within a reasonable response time, while keeping the budget under control. To avoid dropping requests in high load time, the fidelity of the content is adjusted, which decreases response time, hence serving more requests in a time frame. The providers also want to ensure that all their servers are secure and no malicious activity risks the server group. In short, we short-list four quality objectives for Znn.com – cost, performance, fidelity of content and security. The number of servers in the backend group directly impacts cost analysis and hence we take the server count into consideration. For content quality, we define three levels based on type of content served – high (text, video and images), medium (text and images) and low (only text). Performance analysis comes from considering response time, bandwidth and server load. For security analysis, we associate a security Confidence Level with each server, which indicates the trust the system has regarding its security (highest value being 1.0). In case of suspicious

---

[3] The environment of a system consists of the components and connectors that are currently unused and can be added to the system. For example, unused servers that can be added to a system under high load are a part of the system's environment.

activity (e.g., on a port) this confidence level would be lowered possibly causing adaptation to be triggered.

The server activate and deactivate operators are defined to add or remove server(s) from the server group.The setFidelity operator changes fidelity of the content to be served at the specified level.TheclosePort operator closes the specified port. From these operators, we specify two pairs of tactics. One pair discharges (or enlists) a server(s) and the other shuts down the port on a ServerT type element.

4.1 Competing Scenario

The scenario that we chose to demonstrate is one in which cost adaptation competes with security adaptation. The Rainbow framework detects the cost of the system is too high. It triggers an adaptation that leads to a strategy being selected for execution – ReduceOverallCost. This will aim to reduce overall cost by discharging one or more servers. While this adaptation is in progress, a potential security breach is reported on one of the servers, and is picked up by the Evaluator. This causes the secureServer strategy to be selected. At this time Rainbow will have to make a decision.

If Rainbow is running in the serial mode, it will delay the new adaptationuntil the cost adaptation has finished.Then it will execute the security adaptation. By this time, however, the server could have been compromised. If the cost strategy is preempted, and the security adaptation is scheduled, this could prevent a server from being further compromised. The timeliness of each strategy is reflected in the time-utility curves of that are used to schedule them, as explained previously (Sections 3.2, 3.3).

We have considered only two strategies to demonstrate our case. This can easily be extended to involve more strategies by putting another quality attribute constraint violation in contention during cost strategy execution, which will chose another strategy and invoke the strategy selection algorithm to decide the optimal execution order for the new set of strategies.

4.2 System Utility

To evaluate the new approach we executed the scenario mentioned above in both the original Rainbow (*serial* mode) and with our improved scheduling (*multi* mode), for multiple runs and recorded the instantaneous utility at regular intervals. We use these values to plot the respective utility graphs represented in Figure 5; Figure 6 plots the accrued utility[4] for both runs. In both cases, but especially in serial mode, once a problem is detected it takes some time for the system to stabilize again. However, in serial mode, because the security strategy misses its deadline, the stable system utility levels out at around 0.85. When the server becomes compromised because the security strategy is not executed, its response time and its service time go high, causing the system utility to drop to almost 0.65 initially. Rainbow ends up trying to balance cost and response time until it levels out at 0.85 utility. For the multi-mode, the utility is not harmed since the cost adaptation was pre-empted and the security adaptation was carried out before a server could become compromised. This stabilizes around 0.9725. The improvement in utility of multi mode over serialonce they level out is almost 13.9%.

---

[4] Accrued utility at any instance is the sum of instantaneous utilities of all previous instants in the execution trace.

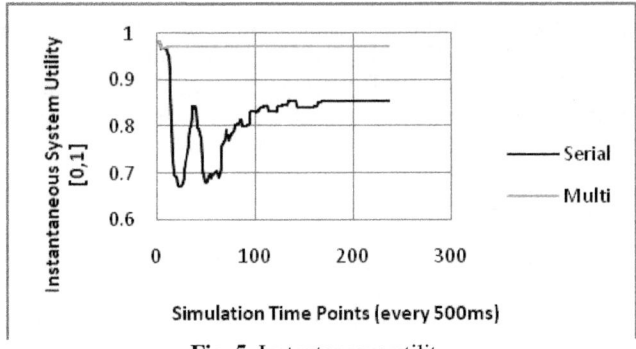

Fig. 5. Instantaneous utility

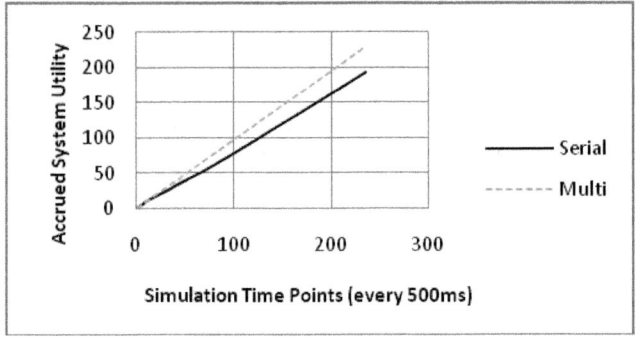

Fig. 6. Accrued utility

From the accrued utility graph, we get a measure of how well the system has been performing since it was brought online. Results show that the net improvement of accrued utility of the multi mode over the serial mode is around 20%, showing that adding preemption does in fact improve the performance of Rainbow in this case.

5   Related Work

Rainbow's architecture corresponds closely with IBM's Autonomic Computing MAPE-loopreference architecture, in which a *computing element* is managed by an *autonomic manager*, monitoring the element, analyzing it and its environment for potential problems, planning actions, and executing changes in a control loop [10].Groundbreaking work includes the dynamically reconfigurable pSeries servers and the DB2 query optimizer [11,12,13]. Our approach applies a similar control loop paradigm using an explicit architecture model combined with adaptation strategies and utility-based adaptation choice to achieve autonomic capabilities.

The technique of using rely-guarantee to ensure consistency is awell-studied concept. It has been used to ensure non-conflicting interactions for implementing fine-grain interactions on linked lists [14]. It has also been used for constructing a reasoning style for aspect-oriented programming [15]. Modifications for ease of use have been proposed as local rely and guarantee [16], and it has been combined with other concepts such as separation logic to give stronger consistency guarantees [17]. Although we have used a weak approximation of this concept, a stronger notion can be embedded into the Conflict Detector sub-component in the rainbow's Adaptation Manager.

The concept of time utility curves [8] has been explored with respect to time-critical resource management [18]. They have been used in scheduling algorithms such as GUS [19] [20]. GUS was one of the earliest proposals for scheduling under real-time and mutual exclusion constraints in the operating system domain. This algorithm, similar to Predicted System Utility as proposed in this paper, aimed to maximize accrued system utility. It produced sub-schedules that were mutually exclusive and gave maximum system accrued utility. In a similar fashion, Rainbow's RT Evaluator and Conflict Detector interact to produce schedules and sub-schedules to ensure maximum predicted utility and adherence to rely-guarantee.

The pre-emption aspect of this work could be viewed as a form of conflict resolution in strategies. While prior work [21] provided integrated support for resolving conflicts in adaptation strategies, it did not consider the case of making an adaptation decision when an existing action is already in progress. Our work explicitly tackles this problem, which combines both scheduling and resource conflicts.

6 Discussion and Future Work

6.1 Preempting on Demand

When a new adaptation is triggered during an ongoing strategy execution, the Adaptation Manager must decide whether to finish executing the current strategy before servicing the newly triggered adaptation, or preempt it and service the new condition first. The algorithm we proposed in Section 3.3 preempts the currently executing strategy and then decides the new execution order. One possible alternative would be to preempt the currently executing strategy only if the RTEvaluator chooses an order where the currently executing strategy is not the first element. The problem with this approach arises when the executing strategy is in the middle of a long tactic execution and cannot be preempted immediately. This loss of time for the new strategy would lead to a reduction in overall utility. In contrast, had we indicated the request to preempt and then started scheduling, the strategy mightbepreempted (or about to be finishing the long tactic) by the time thatthe execution order isfinalized and the new one could have been scheduled almost instantly; hence our decision to indicate preemption immediately.

6.2 Accommodating Problem Detection Time Lag

In using external control modules for self-adaptation, it takes some lag in time between when a problem occurs in the target system and when it is noticed by Rainbow (see Figure 7). Our current implementation of preemption does not include this time lag in its calculations. Some approximations can be used to accommodatethis lag, for example, by adding the execution cycle time to the constraint detection time, etc. But as of now there is no clear solution of how to negate or minimize this lag.

6.3 Where to Place Time Utility Curves

In our approach, we associate TUCs with strategies, the main rationale being that domain experts writing strategies are in the best position to state the responsiveness expected. But in Rainbow there are potentially three other places where we could associate TUCs. First, with each architectural property in the model;it could be possible that one architectural property is associated with multiple constraints, and each constraint associated with multiplestrategies. So, it would be of lesser relevance associating TUC with architectural properties. This argument similarly rules out the association of TUCs with constraint violations. Third, we can associate them with each quality dimension. In this case, the stakeholders of the system would specify the dimensions that are most important to the business. We believethat asking them to specify adaptation responsiveness for their objectives may be unreasonable.

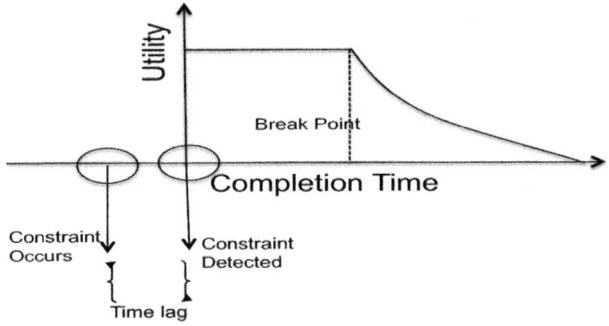

Fig. 7. Potential time lag in actual constraint violation and detection

6.4 Performance and Complexity

The scheduling algorithmsuggested in sections 3.3 and 3.4, which is used as a plug-in to the RTEvaluator, involves factorial-time computation of the number of applicable strategies to determine all possible orderings, from which the best order is selected. The computation size is bounded by the size of the Q in the algorithm in Section 3.3 (i.e., the size of the list of pending strategies). In typical systems, we expect that size to be small, and in fact the difference in the queues from one invocation to the next will be only incrementally different. Therefore, we can use well-explored techniques of caching of results. For the purposes of this paper, the computational time was negligible, but in real deployments techniques for managing the algorithm performance will need to be addressed. For this reason, we provide a plugin point for different scheduling algorithms to be easily added.

6.5 Relation to Self-organizing Systems

This paper discusses improvements to scheduling repairs in an architecture-based self-adaptive system that uses preemption. The architecture-based approach uses an architectural model that provides a global perspective of system properties against which the behavior of the system can be examined and repaired. In such an approach it is possible to show that certain changes will lead to certain properties, or that repairs will lead to improvement.An alternative approach for dynamically repairing systems is to develop systems so that they are self-organizing. In this approach, control is decentralized and the system organizes itself by addressing problems locally – global properties are emergent. Such an approach has the advantage that it is more agile to local changes in the environment and does not require central management, but it is more difficult to specify and achieve global adaptation goals.This is especially important when the trade-off between qualities leading to utility undergo change, for example when fidelity becomes more important than performance.While architecture-based approaches have mostly focused on a central architectural model, this is not a necessary limitation. In fact, we have done the work described in [21] addresses this to some extent. Coordination among different Rainbows, and in fact different self-adaptive modules, is still required.

One possible synergy between these two approaches is to design systems such that they are self-organizing along local optima, but allow Rainbow to manage global optima. This combined approach would have Rainbow discovering the architecture of the system as it self-organizes (using methods similar to those in [22]) and checking global properties. Repairs would not be direct manipulations of the system, but would be issued as guidelines for updated local optima to which the system could self-organize. Such a confluence of approaches would provide agility in responding to environmental changes, but also allow different global concerns to be traded-off and applied at different times in the systems lifetime. This synergistic approach would be similar to that discussed in [23].

It is possible to have many self-adaptive modules managing different parts of a system, as described in [21]. Taking this to the extreme, if there are many of these modules, managing different parts of the system, such a scenario means that we may need to explore issues of self-organization to manage this.

We leave exploration of these possible synergies as future work.

6.6   True Concurrency

We would like to see our work be the basis of a fully concurrent adaptive system. Having true concurrency would mean, rather than having an interleaving of multiple strategies with a single execution thread, we would have multiple threads of execution for multiple strategies, leading to true concurrency. Achieving this would depend on several factors. Firstly, we will need to have intelligent conflict detection mechanisms to ensure multiple strategies executing don't interfere with one another leaving the system in an inconsistent state. Our approach of using architectural locks described in Section 3.6 is one step towards achieving the same. Secondly, the *effectors* interface, described in Section 2, need to have the capability for executing multiple strategies over the target system. And thirdly, we need to have better decision-making mechanisms to ensure we don't address multiple problems in parallel arising from the same cause. Our scheduling algorithm described in Section 3.3 which invokes strategy selection based on new constraint violations (violations not already being addressed) could be used a stepping stone to achieve this.

7 Conclusion

Addingpreemption mechanisms in self–adaptation and allowing multiple adaptations to be considered for scheduling gives time-critical adaptations the opportunity to be scheduled promptly, and hence increase overall system utility. We also proposed a framework with which different users can customize different scheduling methodologies, conflict detection mechanisms and prioritization criteria depending on their requirements, thus making this approach flexible and promising.

Acknowledgments

This work is supported in part by the Office of Naval Research (ONR), United States Navy, N000140811223 as part of the HCSB project under OSD, the National Science Foundation (NSF) under grant CNS-0615305, and by the US Army Research Office (ARO) under grant number DAAD19-02-1-0389 ("Perpetually Available and Secure Information Systems") to Carnegie Mellon University's CyLab.

References

1. Garlan, D., Cheng, S.-W., Cheng-Huang, A., Schmerl, B., Steenkiste, P.: Rainbow: Architecture-Based Self Adaptation with Reusable Infrastructure. IEEE Computer 37(10) (2004)
2. Cheng, S.-W., Garlan, D., Schmerl, B.: Architecture-based Self-adaptation in the Presence of Multiple Objectives. In: ICSE 2006 Workshop on Software Engineering for Adaptive and Self-Managing Systems (SEAMS), Shanghai, China, May 21-22 (2006)
3. Sztajnberg, A., Loques, O.: Describing and deploying self-adaptive applications. In: Proc. 1st Latin American Autonomic Computing Symposium, July 14–20 (2006), 2.3.3
4. Vasconcelos Batista, T., Joolia, A., Coulson, G.: Managing dynamic reconfiguration in component-based systems. In: Morrison, R., Oquendo, F. (eds.) EWSA 2005. LNCS, vol. 3527, pp. 1–17. Springer, Heidelberg (2005)
5. Dashofy, E.M., Van derHoek, A., Taylor, R.N.: Towards Architecture-based Self-healing systems. In: Proceedings of the First Workshop on Self-healing Systems, pp. 21–26 (2002)
6. Bather, J.A.: Decision Theory: An Introduction to Dynamic Programming and Sequential Decisions. John Wiley and Sons, Chichester (2000)
7. Cheng, S.-W.: Rainbow: Cost-Effective Software Architecture-Based Self-Adaptation, PhD thesis, Carnegie Mellon University, Institute for Software Research Technical Report CMU-ISR-08-113 (May 2008)
8. Jensen, E.D., Locke, C.D., Tokuda, H.: A time-driven scheduling model for real-time systems. In: IEEE RTSS (1985)
9. Jones, C.B.: Tentative steps toward a development method for interfering programs. Transactions on Programming Languages and Systems 1983 5(4), 569–619 (1983)
10. Kephart, J.O., Chess, D.M.: The vision of autonomic computing. IEEE Computer 36, 1 (2003)
11. Ganak, A.G., Corbi, T.A.: The dawning of the autonomic computing era. IBM Systems Journal 42(1), 5–18 (2003)
12. Jann, J., Browning, L.M., Burugula, R.S.: Dynamic reconfiguration: Basic building blocks for autonomic computing on IBM pSeries servers. IBM Systems Journal 42(1), 29–37 (2003)

13. Markl, V., Lohman, G.M., Raman, V.: LEO: An autonomic query optimizer for DB2. IBM Systems Journal 42(1), 98–106 (2003)
14. Vafeiadis, V., Herlihy, M., Hoare, T., Shapiro, M.: Proving Correctness of Highly-Concurrent Linearisable Objects. In: Proceedings of the eleventh ACM SIGPLAN Symposium on Principles and Practice of Parallel Programming, New York, pp. 129–136 (2006)
15. Khatchadourian, R., Dovland, J., Soundarajan, N.: Enforcing Behavioral Constraints in Evolving Aspect-Oriented Programs. In: Proceedings of the 7th Workshop on Foundations of Aspect-oriented Languages, Brussels, Belgium, pp. 19–28 (2008)
16. Feng, X.: Local Rely-Guarantee Reasoning. In: Proc. 36th ACM SIGPLAN-SIGACT Symposium on Principles of Programming Languages (POPL 2009), Savannah, Georgia, USA, January 2009, pp. 315–327 (2009)
17. Vafeiadis, V., Parkinson, M.: A Marriage of Rely/Guarantee and Separation Logic. In: Caires, L., Vasconcelos, V.T. (eds.) CONCUR 2007. LNCS, vol. 4703, pp. 256–271. Springer, Heidelberg (2007)
18. Li, P., Ravindran, B., Douglas Jensen, E.: Adaptive Time-Critical Resource management Using Time-Utility Functions: Past, Present and Future. In: Proceedings of the 28th Annual International on Computer Software and Applications Conference, COMPSAC 2004, Hong Kong, September 28-30, vol. 2, pp. 12–13 (2004)
19. Li, P., Sang Wu, H., Ravindran, B., Douglas Jensen, E.: A Utility Accrual Scheduling Algorithm for Real-Time Activities with Mutual Exclusion Resource Constraints. IEEE Transactions on Computers 55(4), 454–469 (2006)
20. Chen, K., Muhlethaler, P.: A Task Scheduling algorithm for tasks described by time value function. Real-Time Systems 10, 293–312 (1996)
21. Huang, A.-C., Steenkiste, P.: Bulding Self-adaptation services using service-specific knowledge. In: Proceedings of IEEE High Performance Distributed Computing (HPDC), Research Triangle Park, NC, USA, July 2005, pp. 34–43 (2005)
22. Schmerl, B., Aldrich, J., Garlan, D., Kazman, R., Yan, H.: Discovering Architectures from Running Systems. IEEE Transactions on Software Engineering 32(7) (July 2006)
23. Georgiadis, I., Magee, J., Kramer, J.: Self-organizing software architectures for distributed systems. In: Proceedings of the Workshop on Self-healing Systems (WOSS'02), Charleston, SC, pp. 33–38 (2002)

Weaving the Fabric of the Control Loop through Aspects

Robrecht Haesevoets, Eddy Truyen, Tom Holvoet, and Wouter Joosen

DistriNet, Department of Computer Science, K.U.Leuven, Belgium
{robrecht,eddy,tom,wouter}@cs.kuleuven.be

Abstract. Self-adaptive systems are systems that are able to autonomously adapt to changing circumstances without human intervention. A number of frameworks exist that can ease the design and development of such systems by providing a generic architecture that can be reused across multiple application domains. In this paper, we study the applicability of aspect-oriented programming (AOP) to see where and how AOP can provide an interesting alternative for implementing parts of the architecture of self-adaptive frameworks. In particular, we relate two existing self-adaptive frameworks to the body of work on aspect-oriented programming techniques for self-adaptation. We present an aspect-oriented architecture for self-adaptive systems to show how AOP can be used for framework customization, event brokering and event aggregation. The potential of AOP for efficiency is shown by evaluating the architecture in a case study on decentralized traffic monitoring. Finally, we explore the potential and challenges of AOP for building scalable and decentralized self-adaptive systems.

1   Introduction

The growing complexity of today's distributed systems requires self-adaptive software that can autonomously adapt to changing user needs, changing operation environments and system failures. The well-accepted reference model for autonomic computing [20], which we will use as the foundation for studying self-adaptive frameworks, is shown in Figure 1. Self-adaptive systems typically consist of two parts: a base system that provides the normal functionalities, and a control loop system that realizes the *control loop*. The control loop often consists of several parts (see Figure 1): monitoring, analyzing, planning and execution. These parts share and use system-specific adaptation knowledge about when, where and how to adapt the system. This adaptation knowledge is typically specified by developers or administrators as high-level *adaptation policies* following the well-known *Event-Condition-Action* format.

Technologies that implement this reference model have emerged in different communities. On the one hand, the software engineering community on self-adaptive systems has developed reusable self-adaptive frameworks [12,20], but also middleware [19] that can support self-adpative concerns. On the other hand, the programming language community has investigated aspect-oriented

D. Weyns et al. (Eds.): SOAR 2009, LNCS 6090, pp. 38–65, 2010.

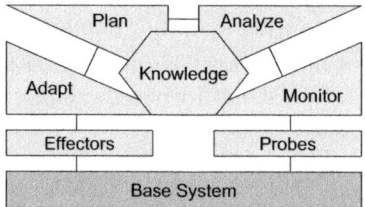

Fig. 1. A well-known reference model for self-adaptive systems [20]

programming languages [17] as key programming paradigm for building self-adaptive applications in a modular way [6,16,13,4].

Prominent examples of self-adaptive frameworks are the Rainbow framework [12] and the Autonomic Management Engine (AME) [20]. These frameworks provide support for building self-adaptive systems by offering a reusable infrastructure to realize the control loop in the base system. The infrastructure consists of components such as a constraint evaluator and action-decision components, but also standard effectors and probes. High-level domain-specific languages (DSL's) are offered by the frameworks to specify complex adaptation policies and customize the framework to a particular application. These adaptation policies are directly translated to the basic infrastructure as specific sets of configurations and rules for the different components in the infrastructure.

Aspect-oriented programming languages [17,22,27] aim to support the separation of crosscutting concerns. Cross-cutting concerns are concerns that affect the implementation of multiple modules in a system. In the context of the self-adaptive systems, as sketched above, this means that the control loop is not encoded in the base system, but in separate modules that externalize the crosscutting composition of the control loop with the system. A special compiler and associated runtime library, called the weaver, is used for knitting the aspect code back into the system. More specifically, the weaver takes as input a base program and a set of aspects, and generates at load-time a program where aspect code and base program code are partially statically composed using the compiler, and partially dynamically composed by means of the runtime library.

Several researchers have already proposed aspect-oriented programming (AOP) as a lightweight approach for implementing specific self-adaptive functionalities [6,16,13,4]. These aspect-oriented programming techniques actually provide generic support for many mechanisms found in self-adaptive frameworks, including support for monitoring, event aggregation, and dynamic adaptation.

To our knowledge, this aspect-related work has largely been performed in isolation from the above mentioned body of work on self-adaptive frameworks. The first goal of this chapter is to start a dialogue between the two research communities, in which common abstractions can be identified, and the convergence of these two bodies of work can be explored. To this end, we introduce the terminology of AOP and present a classification of different AOP-based approaches for self-adaptive software.

The second and most important goal of this chapter is to study the potential of aspects as a better alternative for implementing parts of the infrastructure of existing self-adaptive frameworks. The underlying motivation is that, in our opinion, parts of the basic infrastructure of self-adaptive frameworks can be more optimally implemented using a library of reusable aspects. This line of thought is illustrated in Figure 2. The right part shows how aspects can be used to integrate and distribute the control loop more closely with the software structure of the underlying system. Adaptation policies, formulated in a domain-specific language, are translated into a set of system-specific aspects by an aspect generator, which uses a predefined library of reusable aspects. The system-specific aspects are then woven directly into the right part of the underlying base system code by the aspect weaver at load-time. This is in contrast to existing self-adaptive frameworks, as shown in the left part of Figure 2, where a single instance of the framework is used for controlling the entire system.

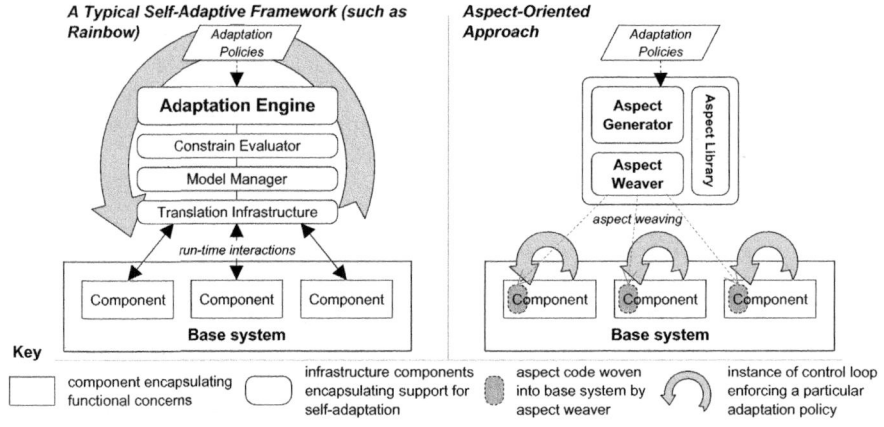

Fig. 2. Realizing the control loop in self-adaptive systems

We will explore three potential improvements that AOP could bring to the architecture and design of self-adaptive frameworks. First, we study the existing frameworks and present an aspect-oriented architecture to show the potential of AOP for *framework customization, event brokering and event aggregation*. The underlying motivation is that, as already outlined above, AOP has native language support for many constructs/mechanisms found in generic frameworks for self-adaptation. However, in contrast to most generic frameworks, AOP also offers programming support for advanced integration and customization towards existing applications.

Second, we explore the potential of AOP for achieving *more efficient execution of the control loop* with respect to time and memory. The underlying motivation is that, as shown in Figure 2, the infrastructure of existing self-adaptive frameworks consists of several heavy-weight components for evaluating adaptation policies, whereas in the aspect-oriented approach the evaluation of adaptation policies is compiled at load-time into a more efficient representation.

Third, we position the potential and challenges of AOP for scalability and decentralizing the control loop by exploring support for aspects in a distributed execution environment. As illustrated in Figure 2, a typical self-adaptive framework represents a single point of failure and introduces a potential performance bottleneck due to runtime interactions with all nodes of the system. Because aspects are woven across the different components of the system, we believe that aspects provide an appropriate abstraction for decentralizing the control loop and reducing the number of network interactions with a central point of control. We would like to stress that this third element is a vision that cannot be completely backed up by hard evidence.

In this chapter we will show how AOP is particularly useful for systems in which change is *anticipated* and adaptations are reactive instead of *time-triggered*. The adaptation cycle, however, should be *short*, because its *timeliness* cannot be guaranteed by current AOP technology.

Overview of this chapter. We start by studying two prominent examples of self-adaptive frameworks in Section 2. Next, the body of work on AOP is reviewed in Section 3 and the potential and applicability of AOP for self-adaptive frameworks is discussed. Section 4 presents our proposal for an aspect-oriented architecture for self-adaptive systems. This will show how the architectural principles from the existing self-adaptive frameworks can be used to create a library of reusable aspects for implementing the control loop and show the potential of AOP to realize framework customization, event brokering and event aggregation. Section 5 aims to demonstrate the potential that AOP holds for self-adaptive frameworks in terms of efficiency, by evaluating the proposed architecture in a case study on traffic management. The potential and challenges of using AOP to increase scalability and decentralize the control loop is explored in Section 6. Finally, Section 7 concludes with lessons learned and future challenges.

2   Study of Self-adaptive Frameworks

In this section we look at two prominent approaches for building self-adaptive frameworks in the literature, namely the Rainbow framework and the Autonomic Management Engine. We describe and compare these approaches from an architectural perspective and their potential for reuse.

2.1   The Rainbow Framework

The goal of the Rainbow framework is to offer a generic architecture for building self-adaptive systems such that the various components can be reused across a family of systems. The architecture of the Rainbow framework consists of three layers (see Fig. 3): a system-specific infrastructure layer, an architectural layer, and a translation layer.

The *system-specific infrastructure layer* offers low-level probes for measuring all kinds of properties such as response-time of connections and loads of servers,

effectors for performing change, and other infrastructural services such as resource discovery services.

At the architectural layer, the Rainbow framework includes in its run-time
system an architectural model of the executing system. This architectural model
typically represents the executing system as a set of components, connectors,
properties attached to components and connectors (e.g. response time of a connector, load of a component) and constraints (to restrict the components, connectors and properties within certain well-defined configurations). The model
manager component gives access to this architectural model. Gauges aggregate
events from the probes and update the architectural model. The constraint evaluator will periodically evaluate the constraints of the architectural model. In case
of constraint violations, adaptations will be triggered. The adaptation engine
will then determine the course of action depending on the circumstances.

The translation layer is responsible for bridging the abstraction gap between
the system layer and the architectural layer. An example is the translation of an
architectural-level change-operator to a system-specific effector mechanism.

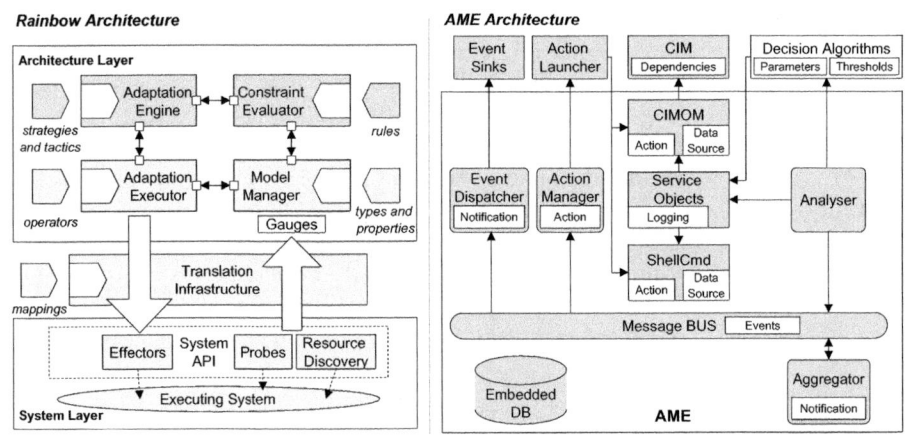

Fig. 3. Architectures of the Rainbow framework [12] and AME framework [20]

In order to apply the framework to a specific system, the framework must
be populated with specific adaptation knowledge about the system. This includes the architectural style and model for representing the system (i.e. the
component and connector types and their properties, and specific action operators to adapt the system's elements), the rules for evaluating constraints, the
adaptation strategies, and the tactics that are used in these strategies. A special
self-adaptive language, named Stitch [5], is offered for specifying that knowledge.

The authors conclude as major lesson learned that reuse of this system-specific
adaptation knowledge across a family of systems is also possible. The extent
to which this reuse is possible depends on whether these systems share the
same architectural style (e.g. client-server) and the same system concerns to be
achieved (e.g. performance, availability).

2.2   The Autonomic Management Engine

The goal of the autonomic management engine (AME) [20] is to offer a generic architecture and a complete toolkit for adding a self-adaptive control loop to existing applications. Events are represented as first-class entities and contain all kind of relevant data such as the reporting component, the affected component and the situation of the event. Events are generated by the applications themselves or are extracted from existing logs.

The architecture of the AME is depicted in Fig. 3. The AME is built around a message bus that distributes events to the interested components: event dispatcher, action manager, analyzer and aggregator. The way each component handles the event is configured in the resource model. This resource model contains the system-specific adaptation knowledge for a specific application.

The Analyzer component is responsible for monitoring the application and issues indication events to the message bus if needed. The analyzer executes the decision algorithms as defined by the resource model. The decision algorithms gather information using service objects which represent the relevant resources of the executing system.

The Aggregator component aggregates the indication events so that if an indication event has occurred a certain number of times (in consecutive cycles), an aggregated indication event is issued to the message bus. It is possible to configure holes so that, for example, an aggregated indication event is triggered if the indication event occurs two out of three times (one hole) [20].

The Action Manager calls the Action Launcher component when an aggregated event is received from the message bus. The Action Launcher component in turn uses the Service Objects to effectively perform the change.

Service Objects make use of Common Information Model (CIM) classes that to monitor and effect a particular resource [20]. A CIM class offers three kinds of reflective operations: ENUM, which allows the enumeration of all instances of a particular resource, GET for querying properties of a specific resource instance, and INVOKE for manipulating the resource in a particular way by invoking a method that performs a reconfiguration action. CIM's are declaratively defined using the Managed Object Format (MOF) language which is an IDL for defining the various CIM methods and connecting them to specific implementation classes (called ITL classes) of the underlying application. These ITL classes define hooks for plugging in specific classes of the underlying application and thus act as some sort of framework interface. CIM's are managed by the Common Information Model Object Manager component (CIMOM). Service Objects may also use standard shell commands (i.e. basic operating system services) for monitoring and manipulating resources.

2.3   High-Level Comparison

It is clear that there is a strong overlap between the architectures of the two frameworks. But there are also some interesting differences to mention. The biggest difference between the two frameworks is the way the frameworks make abstraction of the underlying application: AME takes a resource-centric approach, while the Rainbow framework takes an architecture-centric approach. In AME, the application is modelled as a set of service objects that each represent a certain resource. In Rainbow, the application is represented using an overall architectural style.

Another difference is that the AME framework also supports aggregating events whereas the Rainbow framework does not. In AME, the Analyzer component will detect the occurrence of problematic situations and trigger an indication event to indicate that a problem has occurred, while the Aggregator component keeps a historical view on the problem in development and decides when corresponding action must be taken. Planning when to adapt thus requires analysis of sequences of events and must take into account the history of previous execution traces.

3   Aspect-Oriented Solutions for Self-adaptation

This section first introduces the basic and advanced AO programming concepts that will be used throughout this paper. Subsequently the state-of-the-art on AOP techniques for self-adaptation is reviewed. Finally, we will discuss potential and applicability of AOP to realize parts of the frameworks discussed in the previous section. Support for aspects in distributed execution environment and the resulting potential will be discussed in Section 6.

3.1   An Aspect-Oriented Programming Model

Aspect-oriented programming languages aim to encapsulate crosscutting concerns which are concerns that affect the implementation of multiple modules in a system. The most well-known aspect-oriented programming model is that of AspectJ which is an extension of the OO programming language Java.

Join points, pointcut and advice. In AspectJ the implementation of a crosscutting concern is modularized in aspect modules. An *aspect* has state and methods like a normal class, but also intervenes in the control flow of other classes by attaching extra functionality at certain join points. A *join point* is a well-defined point in the execution of a program. Examples are the call or execution of a method, reading or writing an instance variable. A *pointcut* is a composition-like query for determining at run-time the join points from which the extra functionality must be invoked. Aspects can be woven into the program code at compile-time or at load-time in AspectJ. This sketches the essence of the computational model of AspectJ. This model is implemented in an optimized manner through bytecode transformation and pointcut residues [14].

Consider, for example, the following logging aspect that consists of a pointcut named `socketCreation()` that queries for the construction of new `Socket` objects in an program.

```
1  public aspect LoggingAspect {
2    pointcut socketCreation(Socket s):
3      execution(Socket.new) && this(s);
4
5    after(Socket s) returning:  socketCreation(s) {
6        System.out.println(s);
7    }
8  }
```

The crosscutting functionality that needs to be executed at the join points depicted by this pointcut, is defined in a construct called *advice* (line 5 to 7). One also has to specify if the advice has to execute before, after or around the join points through specific composition operators. Before and after advice are self-explanatory; after returning advice is the same as after with the difference that it is only executed when a method or constructor normally returns. around advice replaces the original behavior at the join point and has the ability to call that original behavior by using proceed - comparable with a super call in a method override.

If the advice needs context information from the intercepted join point, point-cut parameters must be used. These parameters behave like output parameters that get bound during pointcut evaluation, and can subsequently be used in the advice. In the above example on line 5, the newly created Socket object is passed as context info to the advice using a pointcut parameter. The this construct is used for exposing the created object. Method arguments can also be exposed through the args construct.

Event-based AOP. Event-based AOP [8] is an extension of the above aspect-oriented programming model that allows writing expressive pointcuts that can evaluate over a historical sequence of joinpoints. Various researchers have studied and created language constructs for history-based pointcuts, but the *tracematch* construct [3] has been fully integrated in a high-quality implementation of As-pectJ, called abc (the AspectBench Compiler) [2][1]. For example the following aspect uses the tracematch construct to detect when a program has created more than 100 Sockets.

```
1  public aspect MAXNumberOfSockets {
2    int MAX = 101;
3    tracematch {
4      sym socketCreation before:  call (Socket.new);
5
6      socketCreation [MAX] {
7        throw new TooManySocketsCreated();
8      }
9    }
10 }
```

Inter-type declarations. AspectJ also supports *inter-type declarations* that alter the static structure of a program, e.g. by adding new fields or methods to

---

[1] Abc is a research project that aims to make it easy to implement both extensions and optimizations of the core language. The project is maintained to a certain extent and a substantial number of other AOP language research projects have successfully prototyped their research in this compiler.

a class or interface, or extending certain inheritance hierarchies. For example, the following statements declare the class `Entity` a subtype of `Iterator` and introduces the boolean field `hasNext`:

```
1 declare parents : Entity implements Iterator;
2 private boolean Entity.hasNext;
```

Dealing with overlapping aspects. Finally, when two independent aspects share a join point, AspectJ allows defining the order in which their respective advices must be executed. In AspectJ this happens through a 'declare precedence: *TypePatternList*' statement. This signifies that if any join point has advice from two concrete aspects matched by some pattern in *TypePatternList*, then the order of execution of the advice will be the order of that list. The following example specificies that at each join point, the advice of *MAXNumberOfSockets* has precedence over the advice of *LoggingAspect*, which has precedence over any other advice:

```
1 declare precedence : MAXNumberOfSockets, LoggingAspect, *;
```

3.2   Self-adaptation as an Aspect

Many AOP researchers have advocated the use of AOP as an approach for implementing self-adaptive behavior. We use the well-accepted reference model of autonomic computing [20] for classifying these existing AO approaches (see Figure 1). This reference model proposes a self-adaptive system that implements a control loop consisting of several functionalities: monitoring, analyzing, planning and execution of adaptation. These parts share and use system-specific adaptation knowledge about when, where and how to adapt the system. This adaptation knowledge is typically specified by developers or administrators as high-level adaptation policies following the well-known 'Event-Condition-Action' (ECA) format. The control loop mechanism collects data from the system through probes and adapts the system through effectors. Using this reference model, we can now distinguish between three categories of approaches.

AOP as integration technology. A first category of approaches [34,11] uses AOP for integrating a separate adaptation engine into an existing, non-adaptive system. The adaptation engine implements the control loop but expects that the underlying system adheres to a certain component model so that it knows how to manage the system and how to deploy probes and effectors. AOP technology is used for weaving this component model into the system. For example, Yang et al [34] uses aspects for realizing an infrastructure that can deploy probes and effectors on request of the adaptation engine. Fleissner et al. [11] uses aspects for encapsulating the functionality of probes, effectors, and any others interfaces expected by their adaptation engine.

Dynamic AOP as effector technology. Another set of approaches uses dynamic aspect weaving technology for implementing effectors [28,9,13,10,7,4]. This means that aspects are deployed or removed from the running system per request of a separate control system. This type of approach is used for dynamically selecting the best implementation strategy for a certain functionality based on external or internal system conditions (e.g. different encryption algorithms, each optimized for a particular resource availability). Event-condition-action policies are expressed through a domain-specific language, but the action part typically corresponds to a plan for (un)deploying one or more aspects. In this context it has also been shown that AOP languages can be made efficient and therefore ready for execution on environments with limited processing capabilities [10,7].

AOP for monitoring, analysis, and adaptation. Another set of approaches proposes an AOP language as the primary mechanisms for expressing monitoring, analyis and adaptation policies [32,31,30,6,13,16,4]. Shomrat et al. [32] and Serban et al. [31] show how architectural principles can be automatically enforced through first-class aspects. Seiter et al. [30] proposes to use aspects for modeling and controlling the macroscopic system-wide behavior of a multi-agent system. David et al. [6] proposes a domain-specific, event-based AOP language for expressing event-condition-action policies as first-class aspects and use the Fractal component model for implementing the adaptations. Similarly, Huang et al. [16] proposes an event-based AOP mechanism for monitoring historical sequences of events in BPEL proceses. A subset of the approaches from the previous category also belong to this category: Charfi et al. [4] implements ECA policies for self-organizing BPEL processes through a pair of monitoring and adaptation aspects. Monitor aspects define pointcuts for probing the execution of the system while their associated advice triggers the dynamic deployment of one or more adaptation aspects which implement the actual adaptation. Greenwood et al. [13] takes a similar approach but proposes a domain-specific policy language for expressing the monitoring aspects.

Although this category of work explores a promising direction, these approaches have never been systematically related to the body of work on self-adaptive frameworks such as Rainbow [12] and the Autonomic Management Engine [20]. On the one hand, it would be interesting to study which parts of the architecture of self-adaptive frameworks would benefit from an aspect oriented implementation. On the other hand, the architectural principles behind the design of these self-adaptive frameworks could offer valuable architectural guidance to the aspect-oriented design of reusable self-adaptive systems.

The applicability and reusability of most of the AO approaches in this category is also rather limited because these depend on a domain-specific language or are tied to a specific component model. It would instead be better to use a general purpose AOP language (such as AspectJ) as primary implementation mechanism because it can be assumed that such general purpose language is more mainstream and can therefore more easily be made available in a practical development environment. This paper therefore explores the feasibility of employing such a general purpose AOPL in the context of existing approaches for self-adaptive frameworks.

3.3   Potential and Applicability of AOP in Self-adaptive Frameworks

This section discusses the potential and applicability of AOP as alternative implementation mechanism for parts of the existing self-adaptive frameworks. The potential for framework customization, event brokering and event aggregation will be illustrated by our aspect-oriented architecture in Section 4. The potential for efficiency will be further explored in Section 5 by evaluating the proposed architecture in a concrete case study. The potential for scalability will be discussed in Section 6, by exploring distributed aspects for decentralizing the control loop.

Framework customization. AME and Rainbow offer hooks for specializing the framework towards the underlying application. For example, in the AME framework, these hooks are represented by the ILT interfaces. These ILT interfaces are implemented by automatically generated classes that wrap around existing classes of the system. It is well-known, however, that in the pure object-oriented programming model, wrappers lead to various problems such as object identity hell and problems with callbacks [21].

These problems can be resolved by using aspects, as demonstrated by the first category of AO-based approaches in Section 3.2. The architectural model expected by framework can be directly woven in the application through inter-type declarations of AspectJ. Existing classes are extended in place with additional interfaces and corresponding method bodies for querying the internal state of the system.

Event brokering and aggregation. Instead of using the message bus in AME, AOP can be used as underlying communication mechanism. As indicated by Hiltunen et al. [15], event-based communication can be simulated by AOP in various ways.

The Event Aggregator component in the AME framework is implemented as a user library that contains functionality for aggregating events. As demonstrated by the third category of AO approaches in Section 3.2, event-based AOP [8] languages already support this functionality through well-defined language constructs. Using event-based AOP, pointcuts can be written that match with a historical sequence of events. This ability seems an interesting way for subsuming a user library, especially if the latter is cumbersome or difficult to use.

Scalability. Rainbow and AME choose central control as their main approach. This enables a single point for decision-making and simplifies the adaptation process. Combined with the use of an architectural model, Rainbow is even able to take end-to-end system conditions into consideration and to observe emerging behavior. Unfortunately, central control also introduces a single-point-of-failure. When one or more of the components in these frameworks fail, they need to be restarted. Otherwise, the framework stops operating. Luckily, most of the components store minimal internal states and thus can be quickly restarted upon failure.

In Rainbow, the model manager component, however, is a bottleneck which leads to potential scalability issues. As the model manager stores all the data of the architectural representation of the underlying system, additional fault tolerance techniques must be applied: when restarting the model manager component after failure, the entire internal model must be reconstructed to ensure consistency with the state of running system [5]. In AME, the collected data is distributed across Service Objects. This makes the AME approach more scalable with respect to dealing with failures. When a Service Object fails it can be quickly restarted after which it queries the state of the underlying resource.

We consider the problems with the Model Manager component in Rainbow a good use case for AOP. As demonstrated by the first category of AO-based approaches in Section 3.2, the representation of the architectural model can be easily distributed by weaving the necessary data structures across different components of the system itself. In other words, the existing system software can be transparently remodularized to implement the abstractions of the architectural model.

Efficiency. A number of efficiency issues are reported by the authors of the Rainbow approach. The adaptation cycle – from probes to adaptation framework and back to effectors – may incur an inherent time delay. According to reported performance tests [5], this time delay is hundreds of milliseconds at best, seconds on average, and minutes at worst. This time scale limits the applicability of Rainbow to most embedded and real-time systems, which usually operate at sub-millisecond time-scale.

We consider AOP a potential candidate technology for executing the control loop in a more efficient way. As demonstrated by the third category of AO approaches in Section 3.2, adaptation policies implemented as aspects can be directly injected at the relevant locations in the code of the running system. As such certain analysis and reasoning can be performed within the system and even some corrective actions can be taken from within the system. Potentially, this helps to reduce the time for executing an adaptation cycle, provided that the aspects can efficiently gain access to the data of architectural model.

Limitations and Applicability of AOP. Current AOP technology introduces a number of limitations to the applicability of AOP in self-adaptive frameworks and aspect-based architectures for self-adaptive systems. We first discuss the applicability of AOP according to a subset of the modeling dimensions for self-adaptive systems as proposed by Andersson et al. [1], more specifically, the dimensions of *Change* and *Mechanisms*.

According to the dimension of *Change*, an aspect-based architecture can be considered less suitable for *unanticipated* changes, because the architectural model fixes the type of changes that can be monitored. According to the dimension of *Mechanisms*, an aspect-based architecture can be considered less suitable for *time-triggered* adaptations, due to the inherent nature of AOP. Furthermore, if the control loop and application interact synchronously, only *short* adaptation cycles can be tolerated. The *timeliness* of the adaptation cycle cannot be guaranteed as the current AOP technology does not provide any means for this.

Another limitation of AOP is the potential for creating a domain-specific language (DSL). Although AOP does allow the reification of a system's architectural model, it is in essence still a technology at language-level. Existing frameworks, such as Rainbow, however, can take an inherently architectural perspective, for example, by defining adaptation policies in the context of architectural styles. We believe that current AOP technology does not have the potential to create a DSL itself, for formulating high-level adaptation policies, that is as expressive as specialized languages such as Stitch [5]. This is reflected in our vision of the aspect-based approach (cfr. Figure 2). One of the key ideas behind this vision is the translation of high-level adaptation policies, formulated in a separate DSL, into a set of system-specific aspects by an aspect generator, which uses a predefined library of parameterized reusable aspects.

4   Aspect-Based Architecture for Self-adaptative Systems

This section presents our proposal for an aspect-oriented architecture for implementing self-adaptive systems. This architecture aims to decouple the control loop system from the underlying application through an explicit architectural model similar to the Rainbow approach. Furthermore, event aggregation functionality (such at the one of AME) is supported by using the `tracematch` construct. Figure 4 gives an overview of the architecture. The architecture consists of three parts: (1) An architectural model that specifies all the architectural abstractions that are needed to formulate and execute the system-specific adaption policies within a certain family of self-adaptive systems; (2) An application-specific binding that connects a particular application to the architectural model; (3) The control loop system that consists of a monitoring, analysis and adaptation module for implementing each phase of the control loop. Each module consists of a set of reusable aspects. In general, the aspects interact with each other by means of the event-based AOP mechanism.

In this section we outline how the aspect-oriented architecture can be implemented in the context of the abc implementation of the AspectJ programming language. The illustrative running example for illustrative purposes is based on a slightly adapted application scenario from the Rainbow project [12].

4.1   Architectural Model

The architectural model provides an abstraction layer that reflects only information relevant for the self-adaptive framework. The architectural model is similar to the architectural model of the Rainbow framework in that it specifies all that is needed to formulate system-specific adaption policies. Similar to the Rainbow framework, the architectural model views an application as a set of components and connectors and both component, and connectors can have properties and adaptation operators attached to them. The properties describe the kind of information that is needed to determine when and what change is necessary. The operators facilitate the basic primitives that are needed to implement various adaptation strategies.

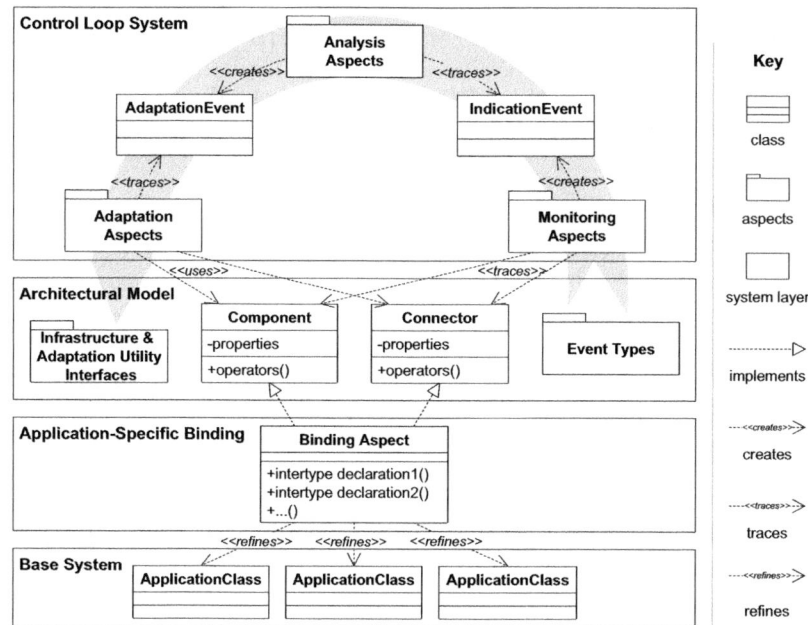

Fig. 4. An aspect-oriented architecture for self-adaptive systems

We propose to extend Rainbow's approach by also declaring various event types in the architectural model. These event types represent semantic information about problematic situations that may require change.

Consider the following concrete application, a web-based, client-server system: "*The system consists of a set of clients that each make stateless requests to one of separate server groups. Clients connected to a server group send requests to the group's shared request queue, and servers that belong to the group grab requests from the queue. The system concerns focus primarily on performance and availability. Specifically, the response time and availability as experienced by the clients. A queuing theory analysis of the system identifies that the server load and available bandwidth are two properties that affect the response time*" [12]. We extend this case study by also including server failures as an important type of event. Failures are important information, because not immediately dealing with a server failure may seriously affect the availability and responsiveness of the overall server group.

Based on this information, the developer defines the architectural model for the system. For example, the architectural model can consist of at least the following elements:

– Components and Connectors: *Server, Client, ServerGroup, Link*
– Properties: *Server.load, Client.responseTime, ServerGroup.load, Link. bandwidth*
– Adaptation Operators: *Client.move(), ServerGroup.addServer()*

- Event types: *ServerFailure, LinkResponseFailure, ResponseTimeExceeded*
- Infrastructural operators: *findBestServer(), findBestServerGroup*

In AspectJ the architectural model can be implemented as a set of interfaces. These interfaces can be embedded in a top-level abstract class or in a package. Each interface introduces a particular component type, connector type, or event type:

```
1    public interface Component {}
2    public interface Connector {}
3    public interface AdaptiveEvent {}
4    public interface IndicationEvent {}
```

Note that two categories of events are identified. Indication events represent the occurrence of a constraint violation or any other problematic situation which does not require immediate change, but may do so in the future. An adaptive event is triggered to communicate that adaptation is required now. Finally, the architectural model can optionally also declare various operations that represent common infrastructure services such as resource discovery services.

The concrete architectural model for the client-server architectural style, as introduced above, is then implemented as illustrated below:

```
1    //Components and connectors
2    public interface Client extends Component {
3      static long MAX_TRESHOLD = 5000;
4      public long responseTime ();
5      public void setResponseTime (long milli );
6      public void move(ServerGroup from , ServerGroup to );
7      public Link getLink ();
8    }
9    public interface Link extends Connector {
10     public int bandwidth ();
11   }
12   public interface Server extends Component {...}
13   public interface ServerGroup extends Component {...}
14
15   //Events
16   public class ResponseTimeExceeded implements IndicationEvent {..}
17   public class LinkResponseFailure implements AdaptiveEvent {..}
18   public class ServerFailure implements AdaptiveEvent {..}
19   public class Events {
20     pointcut client_request (Client c, Server s):
21       call (* Server.*(..)) & this(c) & target(s);
22     pointcut responseTimeExceededEvent(Client c, Server s}:
23       call (ReponseTimeExceeded.new( Client , Server )) & args(c,s );
24     pointcut serverFailureEvent(ServerFailure f ):
25       execution (ServerFailure.new()) & this(f );
26     pointcut linkResponseFailureEvent(LinkResponseFailure f ):
27       execution (LinkResponseFailure.new()) & this(f );
28     ...
29   }
30
31   //Infrastructure
32   public class Infrastructure {
33     abstract Server findBestServer(Client [] cl );
34     ...
35   }
```

Note that the architectural model also contains a set of primitive pointcuts (line 20 to 27) that make explicit the code conventions that must be used by the aspects and applications for signaling events. In this case we have opted to

impose the convention of creating a new object of the appropriate event class.
For example, a failure detection component that detects server crashes can signal
these failures as follows:

```
1  //detection of failure using internal mechanism
2  //collection of information about failure
3  new ServerFailure(<collected information>);
```

4.2 Application-Specific Binding

The architectural model must be bound to a concrete system through an application-specific binding. We subdivide this binding into four parts. (1) It first
maps software elements from the underlying system to component and connector
types of the architectural model. (2) It determines how to compute the necessary
properties of components and connectors by accessing state from the application classes or by using the underlying probes. (3) It implemens the adaptation
operators by exploiting the interfaces of the software elements of the system or
by means of a reflective API that supports implementing dynamic adaptations
(through either dynamic aspect weaving or a component-based reconfiguration
technique). (4) Finally, it implements the common utility interfaces such as the
infrastructure services for resource discovery.

When developing an application from scratch and the architectural model
has already been defined, a good strategy is to let the application implement
the interfaces of architectural model directly. When starting from an existing
application, the implementation of the architectural model must be woven into
the existing application code. AspectJ offers basic support for this integration
in the form of inter-type declarations. To illustrate this, the application-specific
binding for the Link interface of the architectural model is shown below:

```
1  public aspect ClientServerBinding {
2    declare parents: WebClient implements Client;
3    private Link WebClient.link;
4    public Link WebClient.getLink {
5      return link;
6    }
7
8  //[Pointcuts for creating link wrappers]
9    pointcut start(WebClient c). call(WebClient.main)) && target(c),
10
11   pointcut linkCreation(Socket s, WebClient c):
12     execution(Socket.new) && this(s) && cflowbelow(start(c));
13
14   after (Socket s, WebClient c): returning linkCreation(s,c) {
15     ((Client)c).link = new LinkWrapper(c,s);
16     ...
17   }
18  }
19
20  public class LinkWrapper implements Link {
21     private WebClient c;
22     private Socket s;
23
24     public int bandwidth() {
25       //connect to underlying probe monitoring socket s.
26     }
27  }
```

4.3   Control Loop System

The control loop system consists of a monitoring, analysis and adaptation module. Each module consists of a set of reusable aspects. Monitor aspects aim to detect certain (sequences of) events or constraint violations and pass such violations to the analysis aspects by creating an `IndicationEvent`. Analysis aspects evaluate over (sequences of) such indication events to determine when action is really needed. If so, an `AdaptiveEvent` is passed to the Adaptation aspects. Adaptation aspects then offer an appropriate strategy for dealing with each type of adaptive event. This strategy is implemented as part of the advice that depends on the adaptation operators of the architectural model.

In general, an event-condition-action adaptation policy is thus expressed in AspectJ as a three-part structure consisting of a monitoring, analysis and adaptation aspect. Note this three-part structure is not a fixed template for defining event-condition-action policies. It is also possible to express an event-condition-action policy in a single adaptation aspect. Suppose, for example, when the above `ServerFailure` event is created, it should be directly treated by an adaptation aspect without interception by an analysis aspect. The aspect below encodes the following adaptation policy: *when a server crashes in a certain server group, a new server must be added to the server group; this new server must suit the needs of the clients that are currently connected to that server group*:

```
1  public aspect FailureAdaption {
2    after(ServerFailure f) returning:Events.serverFailureEvent(f) {
3      Server s= Infrastructure.instance().findBestServer(f.affectedClients());
4      (f.failedServerGrp()).addServer(s);
5    }
6  }
```

Note that the control loop system aspects depend on the interfaces of the architectural model. These interfaces should define stable abstractions in the scope of a well-defined family of applications. More specifically, the concrete architectural model shown above corresponds to a family of applications that share the same architectural style (client-server) and that target the same system concerns (responsiveness and availability) [12]. As a result, the control loop system is reusable for every application within that family of applications.

The remainder of this section will now illustrate the implementation of a more complex adaptation policy as a three-part structure. The adaptation policy states that *when a link fails to provide the appropriate response time, move the client to another server group. A link is considered to fail after the client experiences a slow response during 10 consecutive requests.*

First, the monitoring aspect creates the `ReponseTimeExceeded` indication event when the response time observed by a particular client exceeds a well-defined threshold. The notion of client request has been defined as part of the architectural model using the pointcut `client_request`.

```
1  public aspect ResponsetimeMonitor {
2
3    around(Client c, Server s): client_request(c,s) {
4      long time 1 = System.currentTimeMillis();
5      Object response = proceed();
```

```
 6        long time 2 = System.currentTimeMillis();
 7        c.setResponseTime(time2-time1);
 8        if (c.responseTime() > c.MAX_TRESHOLD)
 9           new ResponseTimeExceeded(c,s);
10        return response;
11     }
12  }
```

Secondly, the analysis aspect looks for certain sequences of indication events by means of the `tracematch` construct. The following tracematch states that the `LinkResponseFailure` adaptive event should be created after 10 occurrences of the `ResponseTimeExceeded` event.

```
1  tracematch (Client c, Server s) {
2     sym responseTimeExceeded after: responseTimeExceededEvent(c,s);
3
4     reponseTimeExceeded[10] {
5        new LinkResponseFailure(c,s);
6     }
7  }
```

Finally, the `LinkFailureAdaptation` aspect will move the client to the best server group when it receives a `LinkResponseFailure` event:

```
1  public aspect LinkFailureAdaption {
2     after(LinkResponseFailure f) returning: Events.linkResponseFailureEvent(f) {
3        Client c = f.affectedClient;
4        c.move((f.server()).getServerGroup(),
5        Infrastructure.instance().findBestServerGroup(new Client[]{c}));
6     }
7  }
```

4.4   Interaction between Control Loop System and Application

The problem of overlapping aspects on one join point can be addressed by using the `declare predence` statement, as explained in Section 3.1. This statement allows to assign relative priorities to different aspects, such as multiple adaptation aspects that are triggered by the same AdaptiveEvent. In addition, our architecture uses events to clearly define the join points throughout the application code. This allows us to control where aspects will be triggered by the application code.

In order to prevent nondeterministic behavior due to race conditions between parallel executing adaptation aspects, the execution of adaptation aspects must be serialized. Consider for instance the undesired situation that one adaptation aspect is triggered to move a component, while another adaptation aspect is still in the progress of replacing that component[2]. In our architecture, the underlying language run-time has full control over the scheduling of the aspect behavior and therefore aspects need to be programmed with appropriate synchronization code.

The most simple programming strategy is to add no additional synchronization code. This will make that the control loop system and the application interact synchronously by default. A disadvantage is that the execution of the

---

[2] This issue does not appear for monitoring and analysis aspects because these are expected to only inspect the behavior and state of the application.

application is suspended each time an adaptation cycle is performed, but parallel adaptation cycles can also not be triggered provided that the underlying application executes sequentially. A disadvantage is of course that the application is temporarily blocked each time an adaptation cycle is processed. When a synchronous approach cannot be tolerated, an asynchronous approach must be used in the monitoring aspects by executing their advice in a separate thread. Race conditions can then be prevented by using a synchronized advice. In order to serialize multiple adaptation aspects, one must synchronize on a lock object that is available to the adaptation aspects only:

```
1 before () : pc () {
2   synchronized (sharedLock) {
3     // advice code here
4   }
5 }
```

5 Efficiency of the Aspect-Based Architecture: A Traffic Monitoring Case Study

The purpose of this section is to show the potential of AOP for efficient implementation of self-adaptive systems, by evaluating the aspect-based architecture in a traffic monitoring case. We do not aim to directly compare the aspect-based architecture with existing self-adaptive frameworks, but use a dedicated object-oriented (OO) solution as reference for evaluation.

The motivation behind this is as follows. The design of existing self-adaptive systems can be classified into 2 categories that clearly illustrate the trade-off between reusability and efficiency: reusable frameworks versus dedicated solutions. Frameworks provide a generic software architecture that can be reused for multiple applications. Advantages include improved development productivity and software quality, and a relatively well separation of non-adaptive concerns that improves the maintainability and modifiability of the system software. Dedicated solutions implement tailored support for self-adaptation to the specific needs of an application, often in procedural or OO programming languages. Such solutions are typically preferred when the system requires very specific algorithms, needs to execute on top of a domain-specific platform, or an adaptation time window at the scale of sub-milliseconds is required. By mixing self-adaptive and functional concerns in an optimal way, dedicated solutions can easily outperform frameworks in terms of efficiency. Such efficiency gains, however, come at the cost of a strong coupling between adaptive and non-adaptive concerns, reducing the overall reusability.

By using a dedicated (OO) solution as reference point, we can show that we can get similar results in terms of performance, scalability and memory usage with the aspect-based architecture, but score better in terms of modularity.

In the rest of this section, we first introduce the traffic monitoring case study. Next, we illustrate the aspect-based and dedicated OO architecture for this case study. Finally, we compare the different implementations of these architectures and measure the trade-off between efficiency and modularity.

5.1   Case Study: Decentralized Traffic Monitoring

Within an applied research track on multi-agent system [33], we have been using a traffic monitoring case to study agent organizations. In this case, a number of cameras are distributed over a road network. A software agent is deployed on each camera. Because there is no central control and because each camera has a limited view, camera agents have to collaborate to observe larger phenomena such as traffic jams. To deal with the dynamic operating environment of cameras, in which camera sensors and communication hardware can fail at any time, a reliable control loop system is required. This control loop system should allow stakeholders to specify adaptation policies that define how collaborations between agents should be adapted as traffic jams occur and cameras fail.

5.2   An Aspect-Based Architecture for Decentralized Traffic Monitoring

This section gives a short overview of the aspect-based architecture for the traffic monitoring case, illustrating the architectural model and control loop system, and showing how tracematches can be used to deal with fluctuations. The application-specific binding was realized by letting the application directly implement the interfaces of architectural model, a strategy discussed in Section 4.2.

The architectural model. The design of the architectural model is based on an extensive domain analysis within the original research project of the traffic monitoring case [33]. A subset of the model is shown in Figure 5, containing the key abstractions for managing collaborations between agents in a traffic monitoring system at a high-level.

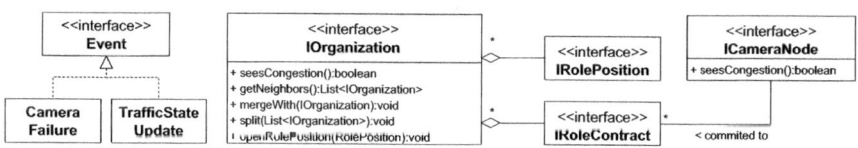

Fig. 5. A Subset of the architectural model

Control loop system. In this case study, the aspects of the control loop system are not generated by an aspect generator but are custom designed, based on the adaptation policies. An example is shown in Figure 6, illustrating the realization of a policy to adapt collaborations among agents.

The monitoring aspect intercepts `TrafficStateUpdate` events, defined in the architectural model, and triggers the analysis aspect. `TrafficStateUpdate` events are generated on a regular basis by the underlying system whenever the traffic state observed by a camera is refreshed. The analysis aspect checks whether collaborations have to be adapted. To do this, the aspect makes use

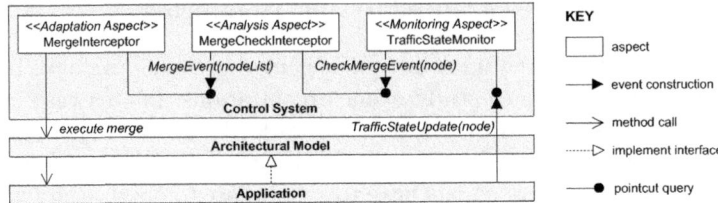

Fig. 6. Different aspects realizing an adaptation policy for agent collaborations

of the interfaces defined in the architectural model. Allowing it to retrieve additional data. If adaptation is required, the analysis aspect will create a new `MergeEvent`, triggering the corresponding adaptation aspect. Finally, the adaptation aspect will execute the actual adaptation, by adapting the corresponding collaborations. All the required operations are defined in the architectural model.

Dealing with fluctuations. Rapid fluctuations in traffic state may cause undesired fluctuations in the collaborations between agents. To deal with this problem, an event aggregation mechanism can be used in the analysis aspect of Figure 6. One way to realize this mechanism is the use of a tracematch construct to filter out possible fluctuations in `TrafficStateUpdate` events, before triggering the corresponding analysis aspect. An example is shown in Figure 7, where the upper string of events contains no matching trace of five subsequent congestion events, while the lower string does. An alternative is to use a custom event aggregation mechanism that aggregates the same number of events.

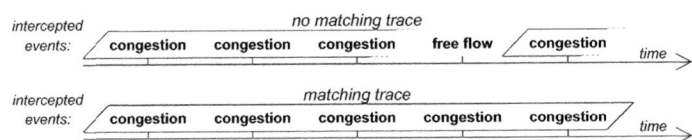

Fig. 7. An example of two strings of intercepted events and the matching traces

5.3   A Dedicated OO Architecture as Reference Point

An overview of the dedicated OO architecture, used as reference for evaluation, is shown in Figure 8. The architecture resembles the aspect-based architecture, but uses a dedicated event bus and specialized components to trigger and connect the different monitoring, analysis and adaption phases. To handle fluctuations, it uses a custom aggregation mechanism. This mechanism can also be used in the aspect-based architecture as an alternative to tracematches.

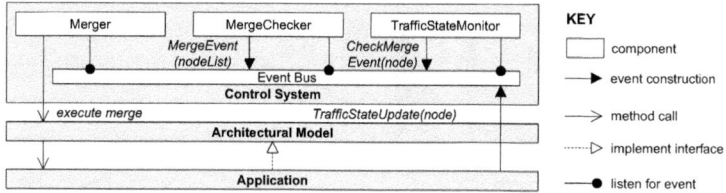

Fig. 8. Architecture of the dedicated OO solution using a custom event bus

5.4 Evaluation

In this section we compare[3] three different implementations of the self-adaptive system for the traffic monitoring case. A first implementation uses the aspect-based architecture with custom event aggregation (compiled with aspectj v.1.6.4), a second implementation uses the aspect-based architecture with tracematches (compiled with abc v.1.3.0), and a third implementation uses the dedicated OO architecture (compiled with java 1.6). The implementations are non-distributed and all experiments are performed on a single computer[4].

The evaluation focuses on two experiments, the overhead of a single adaptation cycle and the scalability in a complete adaptation scenario. The scenario in these experiments consists of introducing an overall traffic jam in the viewing range of all cameras. As a result, the control loop system has to adapt the collaboration between all cameras.

Overhead of a single adaptation cycle. An adaptation cycle is defined as the execution of a single control loop[5], consisting of the execution of the monitoring, analysis and adaptation aspect. The experiment is repeated over 10,000 adaptation cycles for each implementation, measured over different runs of the scenario:

implementation	OO	AOP without tracematches	AOP with tracematches
adaptation cycle (in milliseconds)	0.172	0.181	0.571

Scalability with respect to a complete adaptation scenario. A complete adaptation scenario is defined as the time between the introduction of the overall traffic jam and the final adaptation of the overall collaboration. The experiment is repeated for a growing number of cameras. The results are shown in Figure 9(a). Each datapoint represents the average of 1,000 runs of the scenario with a specific number of cameras.

---

[3] Measurements are performed using a hard-coded logging mechanism. This mechanism uses the standard `System.currentTimeMillis()` and `System.nanoTime()` methods and writes results to an external file.

[4] A computer running Windows Vista SP2 and java 1.6, with a 2.20 Ghz Core 2 Duo processor and 2 GB ram.

[5] Not all adaptation cycles will trigger an analysis and adaptation. The experiment only considers those cycles that contain all three phases.

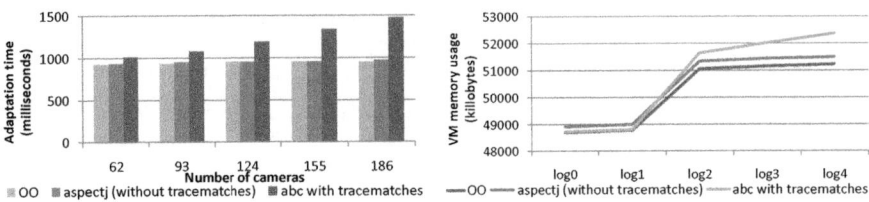

(a) Overhead of a complete adaptation (b) Memory usage of virtual machine for scenario for the different implementations. the different implementations.

Fig. 9.

Memory usage. Figure 9(b) shows the memory usage[6] of the java virtual machine process that is measured at 5 separate data points during the scenario. The traffic jam is introduced between log point 1 and 2, resulting in a memory usage rise in all configurations.

Modularity Comparison. Based on the more expressive compositional power of AOP, it can be expected that the aspect-based implementations exposes a better modularity than the event-based OO implementations. We have used the specifically designed metrics suite [29] for comparing the aspect-oriented version with tracematches versus the object-oriented implementations in terms of separation of concerns, coupling and size. We have applied this metrics suite to the implementation of the adaptation policies within the control loop system layer. For separation of concerns, the Concern Diffusion over Operations (CDO) and the Concern Diffusion over Components (CDC) metrics are used for measuring the number of operations and components that contribute to the implementation of each of the adaptation policies. The Coupling Between Components (CBC) metric is used for measuring the number of references among the implementation modules for each adaptation policy within the control loop system. Finally, the Lines Of Code (LOC) metric is used to measure the lines of code for implementing the adaptation policies. The results are summarized below, by adding up the numbers we got for each adaptation policy:

	CDC	CDO	CBC	LOC
Aspect-Based	38	42	42	370
OO	49	101	56	604

Based on this, we can conclude that in the aspect-based architecture each adaptation policy can be implemented more concisely (lower LOC), the resulting implementation code is better localized (lower CDO & CDC), and there is less coupling among the different aspects among the implementation modules.

Interpretation of the Results. The experiments showed that an implementation of the aspect-based architecture can achieve similar results in terms of

---

[6] Measured using Windows's Reliability and Performance Monitor tool.

performance, scalability and memory usage as a dedicated OO implementation, which can be considered the most optimal solution for a specific application in terms of efficiency. Both in terms of overhead and scalability, the aspectj implementation achieves similar results as the dedicated OO implementation. The tracematch implementation does score a bit less, but still very well within acceptable limits. This was to be expected as the tracematches is a generic mechanism while the custom aggregation mechanisms in the other implementations are specifically targeted for the purpose of dealing with fluctuations in the forming of organizations. For memory usage, all three implementations achieve similar results. However, in terms of modularity, the aspect-based implementations clearly outperforms the dedicated OO solution.

These results provide a clear indication that the aspect-based architecture is a valid and realistic solution for implementing self-adaptive systems. It scores well, both on efficiency and modularity. How the aspect-based architecture relates to existing frameworks for self-adaptive systems, however, is part of future work.

6   Decentralizing the Control Loop: Opportunities and Challenges

Most existing self-adaptive frameworks are based on a centralized architecture. They maintain a central architectural model or rely on a central adaptation engine to enforce the adaptation policies. Such an architecture suffers from potential scalability problems, performance bottlenecks and single-points of failure. Additionally, a number of application domains require an inherently decentralized solution, due to the lack of central authority. Examples are open platforms, such as a traffic monitoring system where cameras have different owners.

In this section, we first introduce the notion of distributed aspects and then discuss their potential and challenges for decentralizing the control loop.

Distributed aspects. A body of work exists on languages and aspect-oriented middleware to support the creation, deployment and execution of distributed aspects [24,26,18,25,23]. Distributed aspects can be dynamically deployed on different nodes. The key concepts are remote pointcuts and remote advice. Remote pointcuts extend the notion of joinpoints for distributed systems. They can evaluate on calls and executions of remote method invocations, and refer to distributed context information, such as information on components running on a different Java virtual machines. Distributed tracematches can also be supported [23]. Remote advice can be transparently executed, both synchronously and asynchronously, in remote environments.

Potential of aspects for decentralizing the control loop. Distributed aspects provide a number of key features to realize the control loop in a decentralized way. In fact, we believe that distributed aspects allow to integrate and distribute the control loop more closely with the software structure of the underlying system, while increasing the reliability and scalability of the system.

First, remote pointcuts and advice allow to easily distribute the representation of the architectural model by weaving the necessary data structures across different components of the system itself. In other words, the existing system software can be transparently remodularized to implement the abstractions of the architectural model. This allows the system to be used as a model of itself, instead of maintaining an explicit architectural model of the system. Doing so, can increase the availability of the model and reduce the risk of a single point of failure, because the failure of some part of the system no longer affects the model of the rest of the system.

Secondly, distributed aspects reduce the need for a dedicated or centralized infrastructure to distribute events and enforce adaptation policies. Remote pointcuts can monitor remote events locally and remote advice can be used to enforce adaptation policies, without relying on an explicit distributed infrastructure for effecting the adaptations.

Finally, aspects can be used to dynamically update adaptation policies, probes and effectors distributed across the system at runtime. Using dynamic weaving, aspects can be dynamically deployed within a distributed application.

Challenges ahead. Although distributed aspects provide some promising potential to realize a decentralized control loop, there are also a number of challenges. A first challenge is the need for a suitable coordination mechanism. Decentralizing the control loop implies that there will no longer be central control. As a result, a decentralized coordination mechanism is required. This mechanism should provide the necessary synchronization and allow adaptation policies to be evaluated and enforced on a local basis. Both AOP and the current self-adaptive frameworks do not provide such coordination mechanisms.

A second challenge is to realize the full synchronization of the architectural model. Remote pointcuts can be defined to propagate information through a distributed environment, making the information locally accessible to aspects without relying on a centralized infrastructure. However, this information is limited to the execution context of the corresponding join points. A typical control loop can require additional data for analysis and planning that is outside the scope of this context. As a result, there is still need for some distribution infrastructure in the advice in order to fully synchronize the architectural model and get access to all of the required information. This infrastructure could be realized in the control loop itself, or in the application specific binding.

7   Conclusions and Remaining Challenges

In this chapter, we presented the vision of using aspects as key abstraction to distribute and integrate the control loop in self-adaptive systems more closely with the software structure of the underlying system, improving the overall efficiency, scalability, and reusability. We support this vision with a study of existing self-adaptive frameworks, such as Rainbow [12] and AME [20], a proposal and evaluation of an aspect-oriented architecture for self-adaptive systems, and an exploration of using distributed aspects to decentralize the control loop.

The proposed architecture is based on the architectural principles of Rainbow and AME, and shows where aspects can provide a promising alternative to implement parts of these frameworks. In particular, it shows how framework customization, event brokering and event aggregation can be realized using basic constructs provided by the aspect-based languages. Constructs such as inter-type declaration allow the existing system software to be transparently remodularized to implement the abstractions of the architectural model, similar to the model used in Rainbow. Event aggregation, such as in AME, is supported more concisely using the `tracematch` construct of the AspectBench compiler for AspectJ.

The applicability of the proposed architecture is the most promising in self-adaptive systems where change is anticipated, and adaptations are short and reactive, instead of time-triggered. The potential for creating a domain-specific language (DSL) based on the AOP programming concepts is limited however.

An evaluation of the proposed architecture shows that similar results, in terms of efficiency, can be achieved as a dedicated OO solution, tailored to the specific needs of the application in an optimal way. The aspect-based architecture, however, provides a better trade-off between efficiency and modularity.

We would like to stress that our vision cannot be completely backed up by hard evidence and still presents several challenges. To further test our hypothesis, an actual comparison is needed of the aspect-based architecture and the existing self-adaptive frameworks. One of the major challenges of realizing our vision, will be the design of an aspect generator for translating the adaptation policies written in a separate DSL, such as Stitch, into a set of application-specific monitoring, analysis and adaptation aspects.

References

1. Andersson, J., de Lemos, R., Malek, S., Weyns, D.: Modeling dimensions of self-adaptive software systems. In: Cheng, B.H.C., de Lemos, R., Giese, H., Inverardi, P., Magee, J. (eds.) Software Engineering for Self-Adaptive Systems. LNCS, vol. 5525, pp. 27–47. Springer, Heidelberg (2009)
2. abc, The AspectBench Compiler for Aspect/J, http://aspectbench.org
3. Allan, C., et al.: Adding trace matching with free variables to AspectJ. In: OOP-SLA 2005 (2005)
4. Charfi, A., Dinkelaker, T., Mezini, M.: A plug-in architecture for self-adaptive web service compositions. In: IEEE International Conference on Web Services (ICWS 2009), pp. 35–42. IEEE, Los Alamitos (2009)
5. Cheng, S.-W.: Rainbow: Cost-Effective Software Architecture-Based Self-Adaptation. PhD thesis, School of Computer Science, Carnegie Mellon University, CMU-ISR-08-113 (2008); Chapter: Limitations and Issues
6. David, P.-C., Ledoux, T.: An aspect-oriented approach for developing self-adaptive fractal components. In: Löwe, W., Südholt, M. (eds.) SC 2006. LNCS, vol. 4089, pp. 82–97. Springer, Heidelberg (2006)
7. Douence, R., Fritz, T., Loriant, N., Menaud, J.-M., Ségura-Devillechaise, M., Südholt, M.: An expressive aspect language for system applications with arachne. In: 4th International Conference on Aspect-Oriented Software Development (AOSD 2005), pp. 27–38. ACM, New York (2005)

64   R. Haesevoets et al.

8. Douence, R., Motelet, O., Südholt, M.: A formal definition of crosscuts. In: Yonezawa, A., Matsuoka, S. (eds.) REFLECTION 2001. LNCS, vol. 2192, pp. 170–186. Springer, Heidelberg (2001)
9. Duzan, G., Loyall, J.P., Schantz, R.E., Shapiro, R., Zinky, J.A.: Building adaptive distributed applications with middleware and aspects. In: Proceedings of the 3rd International Conference on Aspect-Oriented Software Development (AOSD 2004), pp. 66–73. ACM, New York (2004)
10. Engel, M., Freisleben, B.: Supporting autonomic computing functionality via dynamic operating system kernel aspects. In: 4th International Conference on Aspect-Oriented Software Development (AOSD 2005), pp. 51–62. ACM, New York (2005)
11. Fleissner, S., Baniassad, E.L.A.: Epi-aspects: aspect-oriented conscientious software. In: Proceedings of the 22nd Annual ACM SIGPLAN Conference on Object-Oriented Programming, Systems, Languages, and Applications (OOPSLA 2007), ONWARD! session, pp. 659–674. ACM, New York (2007)
12. Garlan, D., Cheng, S.-W., Huang, A.-C., Schmerl, B.R., Steenkiste, P.: Rainbow: Architecture-based self-adaptation with reusable infrastructure. IEEE Computer 37(10), 46–54 (2004)
13. Greenwood, P., Blair, L.: A framework for policy driven auto-adaptive systems using dynamic framed aspects. In: Rashid, A., Aksit, M. (eds.) Transactions on Aspect-Oriented Software Development II. LNCS, vol. 4242, pp. 30–65. Springer, Heidelberg (2006)
14. Hilsdale, E., Hugunin, J.: Advice weaving in aspectj. In: 3rd International Conference on Aspect-Oriented Software Development (AOSD 2004), pp. 26–35. ACM, New York (2004)
15. Hiltunen, M.A., Taïani, F., Schlichting, R.D.: Reflections on aspects and configurable protocols. In: 5th International Conference on Aspect-Oriented Software Development (AOSD 2006), pp. 87–98. ACM, New York (2006)
16. Huang, T., Wu, G.-Q., Wei, J.: Runtime monitoring composite web services through stateful aspect extension. Journal of Computer Science and Technology 24(3) (2009)
17. Kiczales, G., Lamping, J., Mendhekar, A., Maeda, C., Lopes, C.V., Loingtier, J.-M., Irwin, J.: Aspect-oriented programming. In: Aksit, M., Matsuoka, S. (eds.) ECOOP 1997. LNCS, vol. 1241, pp. 220–242. Springer, Heidelberg (1997)
18. Lagaisse, B., Joosen, W.: True and transparent distributed composition of aspect-components. In: van Steen, M., Henning, M. (eds.) Middleware 2006. LNCS, vol. 4290, pp. 42–61. Springer, Heidelberg (2006)
19. Malek, S., Mikic-Rakic, M., Medvidovic, N.: A style-aware architectural middleware for resource-constrained, distributed systems. IEEE Transactions on Software Engineering 31(3), 256–272 (2005)
20. Manoel, E., Nielsen, M.J., Salahshour, A., Sampath, S., Sudarshanan, S.: Problem Determination Using Self-Managing Autonomic Technology. IBM redbooks (June 2005)
21. Mezini, M., Ostermann, K.: Integrating independent components with on-demand remodularization. In: OOPSLA, pp. 52–67 (2002)
22. Mezini, M., Ostermann, K.: Conquering aspects with caesar. In: AOSD, pp. 90–99 (2003)
23. Navarro, L.D.B., Südholt, M., Vanderperren, W., De Fraine, B., Suvée, D.: Explicitly distributed AOP using AWED. In: Proceedings of the 5th International Conference on Aspect-Oriented Software Development, p. 62. ACM, New York (2006)

24. Nishizawa, M., Chiba, S., Tatsubori, M.: Remote pointcut: a language construct for distributed AOP. In: Proceedings of the 3rd International Conference on Aspect-Oriented Software Development, p. 15. ACM, New York (2004)
25. Pawlak, R., Duchien, L., Florin, G., Seinturier, L.: JAC: A flexible solution for aspect-oriented programming in Java. In: Yonezawa, A., Matsuoka, S. (eds.) Reflection 2001. LNCS, vol. 2192, pp. 1–24. Springer, Heidelberg (2001)
26. Pinto, M., Fuentes, L., Troya, J.M.: A dynamic component and aspect-oriented platform. The Computer Journal 48(4), 401 (2005)
27. Rajan, H., Leavens, G.T.: Ptolemy: A language with quantified, typed events. In: Vitek, J. (ed.) ECOOP 2008. LNCS, vol. 5142, pp. 155–179. Springer, Heidelberg (2008)
28. Rouvoy, R., Eliassen, F., Beauvois, M.: Dynamic planning and weaving of dependability concerns for self-adaptive ubiquitous services. In: Proceedings of the 2009 ACM Symposium on Applied Computing, SAC 2009, pp. 1021–1028. ACM, New York (2009)
29. Sant'anna, C., Garcia, A., Chavez, C., Lucena, C., von Staa, A.V.: On the reuse and maintenance of aspect-oriented software: An assessment framework. In: The XVII Brazilian Symposium on Software Engineering (2003)
30. Seiter, L., Palmer, D.W., Kirschenbaum, M.: An aspect-oriented approach for modeling self-organizing emergent structures. In: International Workshop on Software Engineering for Large-Scale Multi-Agent Systems (SELMAS 2006), pp. 59–66. ACM, New York (2006)
31. Serban, C., Tyszberowicz, S., Feldman, Y.A., Minsky, N.: Developing law-governed systems using aspects. Journal of Object Technology 7(9), 25–46 (2008)
32. Shomrat, M., Yehudai, A.: Obvious or not?: regulating architectural decisions using aspect-oriented programming. In: AOSD, pp. 3–9 (2002)
33. Weyns, D., Haesevoets, R., Helleboogh, A., Holvoet, T., Joosen, W.: The MACODO Middleware for Context-Driven Dynamic Agent Organzations. ACM Transaction on Autonomous and Adaptive Systems (2010)
34. Yang, Z., Cheng, B.H.C., Kurt Stirewalt, R.E., Sowell, J., Sadjadi, S.M., McKinley, P.K.: An aspect-oriented approach to dynamic adaptation. In: Proceedings of the First Workshop on Self-Healing Systems (WOSS 2002), pp. 85–92. ACM, New York (2002)

Self-organisation for Survival in Complex Computer Architectures

Fiona A.C. Polack*

Department of Computer Science, University of York, UK YO10 5DD
fiona@cs.york.ac.uk

Abstract. This chapter steps back from specific self-organisational architectural solutions to consider what self-organisation means in the context of complex systems. Drawing on insights in complexity and self-organisation, the chapter explores how the natural propensity of complex systems (such as the mammalian immune system) to self-organise could be exploited as a mechanism for adaptation in complex computer architectures. Drawing on experience of immune-inspired fault-tolerant swarm robotics, the chapter speculates on how complex systems such as global information systems could adopt self-organised survivability.

1 Introduction

In the context of computer architecture, the self-organisation concept supports goals such as self-adaptation, self-configuration, self-healing, self-optimisation and self-protection [9,24]. These goals are ideals that would permit complex computer systems to autonomously adapt to change. Much research on self-organising computer architectures focuses on explicit solutions for specific goals. Approaches may provide well-engineered solutions to change in requirements, environment or performance. However, self-organising computer architecture solutions tend to be fragile to unexpected change, and the self-organisation mechanisms require significant extra processing and storage.

This chapter steps back from the current research on self-organising computer architectures to seek other potential inspirations. Self-organisation in relation to computer architecture is understood here to be the ability of a system to adapt to change, whether or not that change is anticipated by the designer. The computer architectures to which the chapter applies are complex computer systems with high levels of possibly-ad-hoc inter-connectivity, such as enterprise information systems, cloud or grid-based applications and service-oriented architectures. The desired self-organised property is labelled *dynamic homeostasis*, the ability continue to operate whilst continuously adapting to change. The chapter assumes

---
* The input and advice of Prof. Jon Timmis (Department of Computer Science and Department of Electronics, University of York) is gratefully acknowledged, particularly in relation to the SYMBRION project, www.symbrion.eu. This chapter is inspired by work on EPSRC-funded projects, TUNA (grant EP/C516966/1) and CoSMoS (grants EP/E053505/1 and EP/E049419/1, www.cosmosresearch.org/).

D. Weyns et al. (Eds.): SOAR 2009, LNCS 6090, pp. 66–83, 2010.

that an approach that provides dynamic homeostasis is an effective approach to achieving many of the goals of a self-organising architecture.

Information-theoretic research (section 1.2) suggests that self-organisation should be achievable at a lower cost than that of current self-organisation mechanisms. However, the problem of engineering a system that organises itself in a desirable way is non-trivial. The chapter considers the nature of change (section 2), then presents evidence that lightweight self-organisation to address change is feasible, with illustrations from work on fault-tolerant swarm robotics (section 3). Returning to the context of computer architectures, the chapter explores how the ideas behind immune-inspired lightweight self-organisation might enhance survivability of a complex information system (section 4). The chapter concludes with a brief discussion of validation issues, a significant problem in any area of complex systems engineering.

1.1 Complexity and Systems Architectures

Complexity is unavoidable in large-scale computer architectures. Complexity can be exploited to make systems more robust to change. There are many definitions of a complex system. The one that is used in this chapter is based on the author's work on engineering emergent systems [28,30,29].

> A complex system is a system that is made up of many components (which are themselves systems). The components interact continuously, in ways that are not always those designed or intended, with each other and with the environment. At least some behaviour at the system level cannot be described or predicted directly from the behaviours of the interacting component systems.

This definition is consistent with the definitions of "system" in computer architectures and software engineering, except that it makes explicit that the component-level interactions and the high-level behaviour of a complex system may be unpredictable (the non-linearity that is fundamental to complexity in science). The traditional approach in engineered systems has been to try to contain, mitigate, or eliminate the effects of complexity. However, in natural systems, the freedom to interact allows exactly the sort of adaptation to changing circumstances that is sought for engineered systems.

Traditional engineering assumes that component interaction can be defined and controlled. In systems-theory terms, given all possible inputs, the engineer designs the functions that convert inputs into required system outputs [11,14]. However, in a complex-systems view, the designer cannot know all the inputs to the system. For instance, it is possible that things other than those considered in the design can use the identified input channels: an example would be SQL injection via a text field in a web-form interface to an information system [16]. Also, systems can interact through channels other than those explicitly defined by the designers: some unintended interactions with an information system might be addition or changes to stored data at or below the operating-system level, or the use of resource-usage monitoring in a security attack. This means that there will be situations in which system outputs are not those intended by the designer.

Undesigned interaction is the traditional focus of security and safety engineering. However, undesigned interaction is a perhaps-inevitable consequence of computer systems with high levels of interaction or inter-connectivity (such as enterprise information systems, cloud or grid-based applications, service-oriented architectures). If we accept that these systems are complex (in the scientific sense, above), then we can turn to complexity research to seek inspiration for self-organising architectures.

1.2   Complexity and Self-organisation

The term *self-organisation* was coined by Ashby in 1947 [7]. Ashby states that the current state of organisation is functionally dependent on the past state and some inputs; in self-organisation, the system's dynamics (rather than external inputs) are responsible for organisational state change. Ashby's definition is a reminder that a self-organising architecture is a continuum. It is also a warning that if a system moves away from its intended behaviour, self-organisation is likely to make the situation worse as it is to return the system "normal".

Shalizi [34,35] elaborates Ashby's insight from an information-theoretic perspective, showing that a necessary condition of self-organisation is an increase in statistical complexity at the component level. (A corollary is that emergence represents a decrease in statistical complexity, at a higher level.) In order to self-organise, extra information is needed, because a component's next state is a function not just of its past state, but also of the states of components with which it interacts. This is consistent with self-organising approaches such as reflective architectures: for example, Andersson et al. [3] consider an architecture in which a system is represented in a meta-computation, and the system domain is represented in a metamodel. Self-repairability in the base system is supported by comparison with the ideal, or blueprint, held in the meta-levels. The system is able to adapt to changes in the environment, because changes in the domain, are represented in the metamodel.

Natural self-organising systems do require a lot of sophistication of extra information. One reason for the significant extra information in designed self-organisation is that, unlike natural systems, reflective systems (and other self-organising architectures in general) guide self-organisation in ways that the designers (or system maintainers) consider desirable. There is an implicit assumption that we know what a system should do, and that we want it to continue to do that. In nature there is adaptation to changing circumstances, but no pre-defined intent; engineered systems have a specific purpose, which determines that some adaptations are desirable, some are acceptable, but many are deprecated. Self-organisation or self-adaptation in engineered systems has an overhead in introducing the complexity necessary to adapt, but a much more significant overhead is the cost of directing adaptation.

2   Change in Natural and Engineered Complex Systems

Self-organising architectures seek to produce appropriate responses to change in requirements or internal malfunction. The response is to something that is identified as sub-optimal in the current situation. Taking inspiration from natural complex systems, this chapter does not distinguish among sorts of change.

To illustrate adaptation in natural complex systems, consider the immune system that maintains dynamic homeostasis in mammals. An immune system adapts through response to changes such as new pathogens (viruses etc.), changes in the host (aging processes, amputation etc.), modification of resources (change of quantity or type of food and drink). Normally, the immune system can normally adapt to gradual changes (aging, varying concentrations of particular pathogens over time, regular medical intervention), but may be significantly challenged by abrupt changes (amputation, a virulent new virus, a sudden change in diet or medical intervention). In the face of several acute changes (e.g. serious injury to a malnourished mammal with new viral infection), it is unlikely that the immune system will be able to continue to maintain the mammal.

Turning to an example of the need for adaptation in a computer architecture, consider enterprise information systems [37]. Enterprise information systems should ideally accommodate addition of new structures and applications, and the removal or replacement of old applications. The systems should be robust to server down-time and other (predictable or accidental) reliability issues. The systems should also support changes in requirements such as those due to changes in the strategy and goals of the enterprise. With high levels of integration or inter-connectivity, many changes could arise simultaneously. The ideal is to adapt to detected change, within reasonable bounds.

The detection of change is referred to as *anomaly detection*. An *anomaly* is something that is different from expectation. Expectation may refer to any characteristic: behaviour, state, input, output, patterns of interaction etc. The mismatch could be due to a "problem" with the system or a "problem" with the expectation. In self-organisation, anomalies initiate organisational state change. Anomalies are not always symptoms of unwanted system behaviours: some anomalies are indicative of the changing operational context (or requirements) of a system; others are temporary and do not merit a system response.

Internally-detectable anomalies can be variously characterised: Timmis et al. [38] identify: theoretically-impossible states (Type 0), things which are inconsistent with the recent behaviour of the system (Type 1), and longer term divergence (Type 2). These types of anomaly can be related to self-organising architectures. For example, in a reflective self-organising architectures [3], a type 0 anomaly would arise when the actual system diverges from the reflective model of the system: a component fails, or the model of the system is updated; the self-organising architecture should respond by realigning itself to the model. Type 1 anomalies [38] are based on the assumption that recent history is a good predictor of future system behaviour; any self-organising architecture that collects and analyses short-term time-series data (audit logs) can potentially detect when new data is unacceptably far from the current trend.

Anomaly detection provides the stimulus for self-organised responses, but response requires self-diagnosis. A problem may be survivable, either in the short term or the long term, or may require instant action. A change in the environment may be benign, may require evolution to new operational boundaries, or may be a symptom of a fault, hazard or threat to the system. An anomaly may be symptomatic of one or many problems, whilst a problem may cause one or many anomalies (at a time or over time). Diagnosis and selection of appropriate responses is harder than detection.

Approaches to diagnosis and response often turn to nature for inspiration. However, whilst natural systems can provide useful insights and inspire engineered mechanisms, the natural inspiration is often somewhat vague. For example, self-* research often implies that self-awareness is a characteristic of robust natural systems[1]. In reality, natural complex systems are robust not because they are self-aware, but because they evolve to changing environmental conditions by subtle ongoing change. Survivors tend to have evolved to be robust to removal or restriction of some components. They may also survive significant alteration of their operational environment, so long as it is slow enough to allow adaptation and does not take the natural system outside its operational envelope.

Finally, natural systems use degeneracy rather than redundancy to cope with unexpected failure. A degenerate system such as the immune system, has structurally-different components that produce similar outputs under normal conditions (like redundant components). However, in abnormal situations, components may adopt new behaviours and produce different outputs [39,15,5]. A degenerate system is adaptable to unpredictable changes in circumstances and output requirements. Tononi et al. provide a convincing information-theoretic analysis of advantages of degeneracy over redundancy, and state, "It is no accident that selectional processes such as those that underlie evolution and the immune system show widespread evidence of degeneracy."[39]. In immune systems, degeneracy has been shown to be a more powerful mechanism for fault tolerance than the traditional engineering solution of redundancy [26,12,15].

The following section explores immune-inspiration further, considering self-organising robot organisms. The section describes features of artificial immune systems used for identifying and responding to various forms of change. In the subsequent section, immune inspiration is speculatively applied to a hypothetical complex information system.

3   Adapting to Change: Using Immune Inspiration

The motivation for turning to immune systems for inspiration is the observation that mammals are complex systems in dynamic homeostasis, which is maintained by the interaction of neural, endocrine and immune systems. Roboticists and others use immune analogies in the design of many self-organising behaviours

---

[1] Some researchers imply that natural systems such as cells have self-awareness that allows adaptation and self-repair (e.g. [4]); in other cases, a belief in natural self-awareness is implied by analogy (e.g. [20]) or terminology (e.g. [41]).

(e.g. learning [42] and foraging [40]). Of particular relevance here is work on fault tolerance and problem solving: the focus of, for example, the SYMBRION project (www.symbrion.eu). In this context, Owens et al. [26] review key immunological theories, and justify the immune inspiration with reference to Cohen's thesis [12] that the immune system behaves as if it has an intended purpose, namely the maintenance of the organism in a viable state. (Whether the "purpose" is there by accident or design depends on your theological standpoint.)

The immune system operates on many temporal and spatial scales. The *innate* immune system provides a programmed response to known pathogens on an evolutionary time scale. The somatic, or *adaptive*, immune systems respond to new, unforeseen problems. The systems interact at many levels, to maximise the effectiveness of detection. In addition, the immune system has layered mechanisms and defences that can destroy, contain or repair, depending on circumstances. These features are exploited in immune inspired robotics, and might be used more generally in self-organising architectures.

In the SYMBRION project, Kernbach et al. [21] describe the conceptual architecture of an adaptable, self-organising robot swarm that can form and maintain a higher robotic organism. The individual robots each have an artificial immune system (AIS) that can determine when its operation is moving outside normal limits. In the higher robot organism, the immunological (performance) information is shared across an "immune network" that emerges naturally through the interactions of robots [21]. Kernbach et al. [21] describe how the collection of robots moves from acting as a loose swarm, to a self-organised adaptive organism that can evolve solutions to problems of the environment (ranging from navigational issues and overcoming obstacles to dealing with malfunctions in some of the robots). To support adaptivity, Kernbach et al. [21] reward information acquisition from the environment. The information is used in maintenance of the robot organism, as the AIS of individual robots make adjustments to maintain the state of the higher organism (to maintain dynamic homeostasis).

3.1 Comparing Swarm Robots and Complex Computer Architectures

A complex system (section 1) is made up of interacting components that are themselves systems. In swarm robotics, the components are individual robots. In a complex computer architecture, the components are the servers, processors, applications etc. that make up the system.

From systems theory, any system can be represented as a set of inputs, a set of outputs, and mappings from inputs and outputs (which may use resources internal to the system). A robot takes inputs by monitoring its environment, and the outputs are movement and other external behaviours. Similarly an architectural component can be represented as its input and output data and events. In both situations, the designed and unintended interactions among components represent a flexible network. If the computer architecture is truly complex then, like the swarm robots, we can envisage the connected system as a dynamic network that requires the same sort of adaptive responses as robot organism maintenance. As in the robotic analogy, the complex computer architecture should be able adapt, to repair or replace components, and to respond to new components, new challenges or changed requirements.

In the case of complex computer architectures, the novel problem is formulating appropriate solutions: in the robotic organism [21], the robots determine what they need to do by analysing their environment, which essentially means recognising different sorts of obstacles or different sorts of malfunction in themselves or other robots; in computer architectures, a wider range of challenges is anticipated. However, the multi-level approach development and maintenance of dynamic homeostasis described by Kernbach et al. [21] is flexible enough to cope with a wider range of problems. Before pursuing the analogies further, a specific self-organising adaptive aspect of the swarm robots is considered in more detail: anomaly detection and fault tolerance.

3.2   Immune-Inspired Fault Tolerance in Swarm Robots

Timmis et al.[38] detail an immune-inspired approach to fault tolerance[2] for the SYMBRION robots and robot organism described by Kernbach et al. [21], in which each robot AIS provides potentially useful information concerning the state of the system.

The robot fault tolerance mechanisms are inspired by the role of lymphnodes in the innate and adaptive mammalian immune system [38,5,26]. In simple terms, *lymphnodes* are organs that host the immune response, which is based on the current state of the immune system and signals generated by cells. The lymphnodes act on immune system cells, white blood cells collectively referred to as *lymphocytes*. The *innate immune system* is the inherent protection against *pathogens*: potentially damaging invaders. Innate immunity also has important roles in detection and in initiation of more sophisticated immune responses. The immune cells involved in innate immunity bind to the characteristic molecular patterns of micro-organisms and present them to the *T cells* that are the key to most immune responses. Innate immunity is based on a simple model of the "normal" organism; change to what is "normal" can occur, but only on an evolutionary time scale. The *adaptive immune system* comprises interacting systems that drive acquired immunity, adapting to the changing circumstances. The combined behaviour of the innate and adaptive immune systems is to detect and respond to *antigens*. Antigens are toxins or enzymes emitted by micro-organisms; some are responses to pathogens and represent danger, but others are emitted in normal situations and must be ignored. The circumstances of both normal and dangerous situations change over the life of the host mammal. A simple summary of how adaptive response is achieved starts when lymphocytes are activated by antigen binding and proliferate (called clonal selection); clones produce *antibodies* that are specific to the identified antigen. Some clones become memory cells

---

[2] A fault is any non-specific sub-optimality in a system; a fault can only be detected through "observable" errors – changes of behaviour. Diagnosis is not exact, since a fault may give rise to multiple or no errors, whilst the same observed errors may be due to different (combinations of) faults.

– thus memory is enforced by subsequent exposure to the same antigen. Other clones produce large amounts of antigen to attract the cells that destroy the antigen source.

In the SYMBRION robots, anomalies represent pathogens or antigens; anomaly detection takes place both in component systems (robots) and in the whole robot organism [38]. The different capabilities of the immune-inspired system tackle the different types of anomaly (summarised in section 2, above). Type 0 anomalies are detectable by the innate immune system, which holds the model of "normal" – this is not a full definition of the operational state of the system, but a simple representation of the *operational boundaries* of the components. Detection of a type 0 anomaly occurs when a parameter value is outside the defined normal operational boundaries, and the diagnosis is simply a record of which parameters are out-of-range. Detection of type 1 and type 2 anomalies uses the combined innate and adaptive immune systems, and is analogous to identifying a new antigen. The processes rely on continuous analysis of logged performance and audit data, with comparison both against what is expected, and against what has been seen in the past.

The artificial innate response is not just a response to single parameter variation, but a means of correlating the various input sources over a time window, to identify the likely source(s) of detected anomalies; the analysis can indicate the severity of the anomalous situation (should the component be shut down, or can it just be ignored, for instance). Furthermore, because the artificial innate immune system uses time-window data, it can evolve slowly as the normal operating conditions change. A classic dendritic cell algorithm [17] can be used, whilst Mokhtar et al. [25] present a minimal version of the algorithm for fault tolerance in limited resource situations. Timmis et al. [38] illustrate the approach using the weighted sum of signals from three sensors, with graded thresholds relating to the certainty that an anomaly has, in fact, occurred. The algorithms detect type 1 anomalies and diagnose the robot sensor(s) responsible for the anomaly.

An important aspect of the AIS described by Timmis et al. [38] and Mokhtar et al. [25], is the lymphocyte's *self-detector*: an array of values that express the state of a robot as it is performing a particular task. An instance of the self-detector includes the resource-characteristics of the task, the state of each of the components during this task, and a number of measured *actuation outputs* for the runnings of the task (describing the values – exact or fuzzy – of various system parameters). Finally, each self-detector includes a two-part *health measure* that records the apparent health of the whole system and a measure of whether the self-detector is a candidate for (probabilistic) removal from the self-detector set [38]. Evolution is supported by introduction of new self-detector instances that are slight variations on existing instances (allowing new anomalies to be detected), and a process of removing instances that are no longer useful. The approach uses *gene libraries* [10], which constrain generated instances to realistic profiles. The algorithm detects type 1 anomalies, and provides a potential contribution to detection of type 2 anomalies [38,25]. The T-cell inspired part of the artificial immune system uses a system of *generalised receptors* that has been shown to reliably detect type 1 and type 2 anomalies [38,27].

Like many other anomaly detection approaches, the AIS described here are essentially ways of learning or encoding the probability that an observation lies within a particular region of the input space. Timmis et al [38] show, for instance, that the T-cell approach is akin to other density estimation techniques, whilst the clonal selection approach is akin to a *k*-nearest-neighbour approach.

For organism-level anomaly detection, Timmis et al. [38] describe two approaches to sharing of immunological information. The first requires robots to pass generated immune information (danger signal values, clones, receptors) to other robots. The second (which has been used in practice for error diagnosis in ATMs [13]), maintains a central set of *network detectors* derived from the immune information of individual components. The cells, instances, signals etc are subject to the same evolutionary processes and maintenance criteria to ensure relevance and freshness.

This brief review of the immune-inspired fault detection shows that adaptive anomaly detection and diagnosis is possible. Because the approaches are designed for simple robots, the mechanisms are compact and efficient: Mokhtar et al. [25] note that the modified dendritic cell algorithm fits comfortably within the 256KB flash memory of a robot micro-controller.

3.3   Tolerance and Survivability vs. Correction

The work on swarms and robot organisms is concerned with adaptability and fault tolerance – the ability to respond to new challenges, and to maintain dynamic homeostasis in the organism despite the presence of malfunctioning individuals. Timmis et al. [38] explicitly aim at *fault tolerance* rather than correction: that is, the robot organism can continue to achieve system-level behaviours when some of the individual robots have failed and others have non-fatal faults. In many situations, fault tolerance is an acceptable solution. Systems continue to operate, with some degradation of service.

Fault tolerance is also recommended in work on critical software architectures. Knight and Strunk [22] recommend *survivability*, in which systems can continue to provide some functionality in the presence of faults, generating diagnostic information. A system may be operating below its full potential but maintain critical functionality. Knight and Strunk [22] are interested in what it means to specify (and to meet a specification of) survivability. This refocuses design attention on to what is needed to maintain the most critical functionality of systems. This uses a simple form of degeneracy, in which existing architectural components can adopt a survival mode in appropriate contexts.

The notion of criticality depends on the context or environmental situation, and varies over time. Knight and Strunk [22] give the example of a failure of the automatic landing subsystem of an aircraft control system. Whilst this is a critical function, its loss is survivable, but the pilots need to be alerted. If the plane is not engaged in landing, a simple notification is sufficient. However, when the automatic landing system is controlling the landing process, survivability would require a graceful degrade process that safely transfers control back to the pilots. The contextual aspect of survivability fits with the features of the immune-inspired anomaly detection and tolerance – for instance, the continuous monitoring of operational boundaries can be used to determine the state and context of the system when a fault leads to loss of functionality. Furthermore, the mechanisms such as the AIS self-detectors relate *health* to *task*, allowing the identification of tasks that are still viable in reduced operating situations.

Focusing on survivability is appropriate when external intervention is possible. The challenge of co-ordinating self-organising maintenance mechanisms and human interventions applies in many contexts. For example, the PerAda work on pervasive adaptation [19] considers both human intervention in a system, and human intervention to extend and change the repertoire of responses.

The following section brings together some of the ideas of immune-inspired fault tolerance and survivable software architectures. The example is from Knight and Strunk [22]: a complex financial payment system that displays some limited, programmed ability to adapt to anomalies such as component failure and security attacks. The discussion proposes ways in which the immune ideas could enhance adaptability to cope with multiple failures as well as evolution of what is "normal" through maintenance of dynamic homeostasis.

4   A Hypothetical Example: Financial Information

One of the system examples used by Knight and Strunk [22] to illustrate survivability is a critical information system supporting a financial payment system. Their analysis predates much of the rise in Internet banking, and the so-called banking crisis of 2009.

Banking and financial payment systems, whilst complex and critical, are accessible and highly regulated. There is potential for human intervention in fault correction, but systems must continue to operate whilst intervention is prepared. Survivability therefore requires timely identification of critical errors with appropriate notification to human operators, and maintenance of appropriate services whilst or until human intervention takes place.

The business-criticality of financial information systems means that the human and financial resource required to engineer high-quality systems should be available. The system described by Knight and Strunk [22] is highly distributed; the presence of humans in the system and the need to survive partial failure increases the system complexity. The components of financial information systems typically support large-scale data storage and high-volume transaction processing, so computing resource is not an issue (in contrast to the robot organism). Furthermore, there is a design-acceptance in financial information systems of the need to use system resource for system maintenance and security.

The financial information system [22] supports a typical banking hierarchy, from branches up to central banks, with financial services from domestic banking through retail and commercial banking to management of the international money supply. Hazards relate to local or regional failures (systems, communications, power); security threats include compromised services and co-ordinated attacks. Changes in requirements are not considered [22]. The authors identify five example levels of tolerable service: *primary, industry/government, financial markets, foreign transfers*, and *government bonds* [22]. The primary, or preferred, level of service is the normal expectation. The different services of the banking organisation have different temporal and volume characteristics. For example, domestic markets are closed at night; stock, bond and commodity market services must continue to function under exceptional and unexpectedly high volumes of trading. Some contextual issues are noted: for example, exceptional levels of trading in government bonds is associated with political crises [22].

Knight and Strunk [22] show how monitoring functions and context can be used to direct the parts of the financial payment system that should be supported when a failure, such as complete loss of a server, occurs at different times of day and in different political and trading contexts. Their analysis also shows how a sequence of failures – their example is a major server failure followed by a co-ordinated security attack – changes the survivability option. The pattern of survivability is determined by a response to a single anomaly, or a sequence of anomalies, but not to concurrent anomalies, and the response is the same regardless of the type or severity of fault. To support survivability, the monitoring functions analyse time series data: the illustration requires two bytes of extra information, comprising three 1-bit state variables, five 2-bit state variables and six priority levels [22]. Some simple self-organisation capability is supported, allowing the system to continue operating whilst anomalies are investigated.

It would not be unreasonable to expect that the banking sector could afford more than two bytes of information, and a few basic operations, to support self-organising maintenance of its information systems. With further information on the nature of the anomaly and its effects, a more sophisticated survivability response could support better targeted adaptation to concurrent anomalies and to changing operating contexts (requirements).

4.1   Potential Enhancement of Survivability with Immune Fault Tolerance

Natural complex systems are simply evolutionary survivors, whereas engineered systems are produced for a purpose. There are few systematic approaches to nature-inspired systems development. Stepney et al. [36] propose a principled approach which has been used experimentally to derive immune algorithms [6]. Stepney et al. [36] also identify five *meta-probes*, or general criteria, under which natural systems can be analysed: openness, diversity, interaction, structure, and scale. Azmi et al. [8] use the meta-probes to analyse and compare the characteristics of an application (adaptive information filtering) and aspects of the mammalian immune system, in order to extract natural idioms for adaptation. This systematic approach could be used to design self-organising mechanisms by matching self-* requirements to the characteristics of the natural systems from which inspiration is sought. For now, however, a more general overview of how immune mechanisms might enhance survivability is presented. How might the survivability of such a system be enhanced by immune-inspired self-organising dynamic homeostasis? What is needed to provide immune-inspired self-organising dynamic homeostasis?

The immune-inspired approach outlined in section 3 would require that any component system of the architecture has its own immune-inspired anomaly detection, and that information can be shared with other component systems.

The innate and adaptive immune functions require that the normal operating boundaries of the individual system, and of the whole financial payment system, are analysed and defined. Whereas in the robot systems these boundaries related to sensor information, motor output etc., in the case of an information system, the operational boundaries relate to normal usage and normal performance criteria. In a critical information system, the performance criteria should be determined in the design process, and much of the required monitoring information would be routinely logged for audit trails. However, determining the most effective information to store and the most useful way to represent it is a task that requires careful analysis and design. The analysis task is closely related to existing safety and security analysis.

Turning to the complete financial payment system, the connectivity of the integrated systems is akin to the ad hoc dynamic communication network of an immune system, and of the robot organism [38]. The component systems can communicate information about operational state alongside the regular system interactions, enabling the same emergent immune network support for system-wide error detection and fault tolerance. As an illustration, considering the case of a major server failure during a normal working day. The component AIS might be able to anticipate the failure through the three-pronged anomaly detection of the innate and adaptive systems. Even if the failure of the server was inevitable, interacting components would have been alerted, allowing time to degrade services and inform to users, and to pass diagnostic information to engineers. As in the robot organism, the expectation is that the response would be an evolved survivability strategy of the system, so that it could make best use of resources: a failing server might, for instance, be automatically backed-up and might delegate critical operation to servers elsewhere in the network.

A significant challenge in many AIS designs is to determine the most efficient data and data representation. However, the approach inspired by the innate and adaptive immune systems has been shown to be robust to sub-optimal data representation. Kernbach et al. [21] recommend collecting a wide range of monitoring data; this allows the AIS to self-organise to new operating situations, exploiting the data that are most effective at any point.

Knight and Strunk [22] do not explore survivability in the context of requirements change. The recent history of the banking sector gives some pointers to the scale of changes to which such a critical information system might be expected to adapt. For instance, consider what happens to Knight and Strunk's survivability strategies [22] in truly-global financial markets. In a global financial market, simple maintenance of foreign transfer services in the financial payment system [22] is too crude a strategy: anomalies might affect some transfers but not all. Better information coverage, coupled with the ability to track the source of faults from collections of detected anomalies using self adaptation (as demonstrated in [25]), might allow the AIS mechanisms to identify whether an anomaly affected all foreign transfers, and to determine a targeted survivability strategy. For example, if only Hong Kong trading is affected, all other foreign trading could continue normally (if that was financially appropriate) whilst the fault was attended to; furthermore, the AIS might be able to use the emergent immune network to evolve re-routing strategies that would allow Hong Kong transfers to continue by an indirect route. The issue of maintaining communications across fragile networks is, of course, an essential feature of global Internet provision; the example is used here to demonstrate that holding information about which operational boundaries are breached allows a targeted survival strategy.

A radical change such as the introduction of Internet banking would be a significant challenge to the conventional survivability approaches. Indeed, the way domestic banking is conducted breaks one of the fundamental assumptions of [22]: that domestic markets are shut at night. New survivability analyses would have been required, of existing as well as new services. Potential failure scenarios would have to be developed, and appropriate responses identified.

However, the introduction of all-night domestic payment activity would be typical of the long-term change that can be accommodated by a self-organising AIS. Whilst the first few domestic transactions at midnight would be detected as type 0 anomalies (behaviour outside the normal operational boundaries), the continued night-time activity of the domestic payment market would gradually appear as a type 2 anomaly, and when it became apparent that this was the new norm, the innate immune system would evolve to recognise the operational boundaries of 24-hour banking as the normal pattern of behaviour.

The radical change problem is akin to the problem of initialising an immune system component on a complex critical system: there is a bootstrap period, which may have to be off-line or based on simulated observations, before the immune system can be allowed to interfere directly with the running of the host system. One immune analogy is the mammalian foetus's developing immune system before exposure to the outside world, or the clean-room isolation of a patient whose immune system has been compromised by drug therapies, to allow their own immune system to re-establish itself. An example of AIS bootstrapping is provided by the immune-inspired detection system for ATMs [13]: an off-line process creates an initial population of detectors, which are then continuously updated by the on-line system. Another immune analogy would be immuno-suppressant drug treatments for transplant patients. In the model described by Timmis et al. [38] and Mokhtar et al. [25], safe antigen signals or self detectors corresponding to the new service tasks could be inserted in to the existing repertoire of the AIS, to ensure that the new system tasks were not mistaken for, for example, personation or middleman attacks.

The web services and security services required for trustworthy Internet banking is a significant development exercise. If immune-inspired self-organising dynamic homeostasis mechanisms had already been part of the existing financial systems, then the additional work is in analysing and implementing component-level immune protection to the new system components, extending (but not radically changing) the security analyses that are required for new Internet banking systems. At the whole-system level, the operational boundaries of the enhanced financial payment system would also have to be reviewed, with possible changes or additions. In both these cases, most new information requirements are likely to have existing solutions, but some new operational boundaries and representational problems may need addressing.

Since the new financial services have interactions with the existing services, they would naturally support immune interaction, but the immune algorithms need to be robust to the addition of new services. In principle, this is no different to the situation where a service has been unavailable for a period and is then re-started. An extension phase with some human intervention in the immune system would facilitate the smooth transition to new banking modes.

5   Validation and Assurance

In work on self-organising architectures, validation is crucial. Whilst there has been a reluctance to include non-deterministic or heuristic solutions, there is accumulating evidence that such solutions can be incorporated in critical systems. For example, Kurd et al. have investigated the use of neural networks in critical systems [23].

The underlying system which would be subject to immune-inspired self-organisation mechanisms is, of course, subject to conventional validation. In the financial payment example explored in section 4, the inputs to the immune-inspired fault detection system are part of the underlying system, or simple derivations from these: the operational boundaries are taken form the system specification and design, whilst the monitoring data comes from system variables and audit-trail logging. Much of the analysis exploits conventional critical systems analysis needed for dependability and safety or security assurance. It should be stressed that high-quality engineering, of the functional and non-functional parts of the composite systems, is a prerequisite of the validation of such a system augmented by immune-inspired fault tolerance and survivability.

In considering how to validate immune-inspired dynamic homeostasis, one way to explore the dynamic behaviour of a complex system is simulation. Alexander and others show that agent-based simulation can be used in analysis of a complex system, in the context of safety [2,1]. A common feature of the work on critical system validation and bio-inspired or heuristic techniques is its reliance on arguments of assurance. Polack et al. [29] propose argumentation for validation of agent-based simulations, complementing traditional engineering of simulations (see, for instance, the work of Sargent [33,32]). If a simulation can be argued to be a valid model of a dynamic system, then it can be used to provide evidence of the presumed behaviour of that system in different contexts.

Mokhtar et al. [25] summarise results of simulating the effect of running a modified dendritic cell algorithm on the self-organising robot organism, and of running the algorithm on the real robots. They have the advantage of working with a simulator that is explicitly designed for the robots that they use. However, the principle of simulating the effects of self-organising fault tolerant or survival behaviours can be explored for other system architectures.

How could simulation be used in the validation of immune-inspired dynamic homeostasis? Essentially, we have two complex systems to consider: the host architecture, and the AIS. In simulations of immune systems (natural or algorithmic), it is common to simplify the model of the host system so that only those parts that interact directly with the immune system are included (see, for example [31,18]). In engineering terms, the abstract model of the host system is the test harness for the immune algorithms. In the case of a engineered system such as the hypothetical financial payment system in section 4, conventional design approaches start from abstract models of the system, and simple representations of component interaction should be sufficient to create a simulation of how the systems interact. Simulation can be derived from the abstract design models of the system, using operational and log data, to express transaction volumes and variation etc. On this simple-as-possible simulation of the host system, the immune-inspired system can be run directly, in the same way that code modules are tested on a test harness. The design of tests for the AIS running on the abstract simulation of the host system should follow guidelines for conventional testing. For instance, testing should seek to establish sufficient coverage of the domain (the possible anomalies, faults, and survivability scenarios). Cases of expected and unexpected evolution should also be explored as far as possible.

6   Conclusion

This chapter has developed a case for treating the problem of self-organising computer architecture from the perspective of a complex system maintained in dynamic homeostasis. Self-organisation requires additional information and freedom to adapt to unforeseen situations (anomalies).

The viability of such a solution is demonstrated by the existing work on fault tolerance in robot swarm organisms, which provides a lightweight implementation of a sophisticated immune-inspired fault tolerance system. The robot fault tolerance is inherently similar to the survivability approach taken in some critical systems work.

The robot organism's immune-inspired dynamic homeostasis approach uses an interacting set of algorithms based on innate and adaptive immunity. Self-organising computer architectures of the complexity considered here would have the resources to support AIS mechanisms, offering the potential to extend and develop evolvable survivability into critical information systems. Immune-inspired solutions have the additional advantage that much of the required basis exists in systems design and conventional audit-trail data from system logs. The analysis required to design survivability is similar to conventional security analysis, using design information about the intended operational boundaries of the system and its components.

Whilst validation of non-deterministic approaches such as AIS is often seen as problematic, the conventional validation of the host system could be augmented by simulation of the immune-inspired survivability mechanisms, as demonstrated in critical systems assurance.

The AIS mechanisms, and indeed the focus on survivability rather than repair, lack the predictability of self-organising approaches such as reflectivity. However, the relatively lightweight nature of AIS fault tolerance, and the robustness to new operating circumstances, along with the scope for better targeting of survival strategies, makes the approach considered here potentially attractive for constantly-developing large scale complex systems such as enterprise information systems, cloud or grid-based applications and service-oriented architectures

References

1. Alexander, R., Kazakov, D., Kelly, T.: System of systems hazard analysis using simulation and machine learning. In: Górski, J. (ed.) SAFECOMP 2006. LNCS, vol. 4166, pp. 1–14. Springer, Heidelberg (2006)
2. Alexander, R., Kelly, T.: Simulation and prediction in safety case evidence. In: International System Safety Conference (2008)
3. Andersson, J., de Lemos, R., Malek, S., Weyns, D.: Reflecting on self-adaptive software systems. In: SEAMS, pp. 38–47. IEEE, Los Alamitos (2009)
4. Andras, P., Charlton, B.G.: Self-aware software – will it become a reality? In: Babaoğlu, Ö., Jelasity, M., Montresor, A., Fetzer, C., Leonardi, S., van Moorsel, A., van Steen, M. (eds.) SELF-STAR 2004. LNCS, vol. 3460, pp. 229–259. Springer, Heidelberg (2005)
5. Andrews, P.S.: An investigation of a methodology for the development of artificial immune systems: a case study in immune receptor degeneracy. PhD thesis, Department of Computer Science, University of York, YCST-2008-17 (2008)
6. Andrews, P.S., Timmis, J.: Adaptable lymphocytes for artificial immune systems. In: Bentley, P.J., Lee, D., Jung, S. (eds.) ICARIS 2008. LNCS, vol. 5132, pp. 376–386. Springer, Heidelberg (2008)
7. Ashby, W.R.: Principles of the self-organizing dynamic system. General Psychology 37, 125–128 (1947)
8. Mohd Azmi, N.F., Timmis, J., Polack, F.A.C.: Profile adaptation in adaptive information filtering: An immune inspired approach. In: SoCPaR, pp. 420–429. IEEE Press, Los Alamitos (2009)
9. Babaoğlu, Ö., Jelasity, M., Montresor, A., Fetzer, C., Leonardi, S., van Moorsel, A.P.A., van Steen, M. (eds.): SELF-STAR 2004. LNCS, vol. 3460. Springer, Heidelberg (2005)
10. Cayzer, S., Smith, J.: Gene libraries: Coverage, efficiency and diversity. In: Bersini, H., Carneiro, J. (eds.) ICARIS 2006. LNCS, vol. 4163, pp. 136–149. Springer, Heidelberg (2006)
11. Checkland, P.: Systems thinking, systems practice. Wiley, Chichester (1981)
12. Cohen, I.R.: Tending Adam's Garden. Evolving the Cognitive Immune Self. Elsevier, Amsterdam (2000)
13. de Lemos, R., Timmis, J., Forrest, S., Ayara, M.: Immune-inspired adaptable error detection for automated teller machines. IEEE SMC Part C: Applications and Reviews 37(5), 873–886 (2007)
14. Dyer, M.: The Cleanroom approach to quality software development. Wiley, Chichester (1992)

15. Edelman, G.M., Gally, J.A.: Degeneracy and complexity in biological systems. PNAS 98(24), 13763–13768 (2001)

16. Ge, X., Polack, F., Laleau, R.: Secure databases: an analysis of Clark-Wilson model in a database environment. In: Persson, A., Stirna, J. (eds.) CAiSE 2004. LNCS, vol. 3084, pp. 234–247. Springer, Heidelberg (2004)

17. Greensmith, J., Twycross, J., Aickelin, U.: Dendritic cells for anomaly detection. In: Congress on Evolutionary Computation, pp. 664–671. IEEE Press, Los Alamitos (2006)

18. Harel, D., Setty, Y., Efroni, S., Swerdlin, N., Cohen, I.R.: Concurrency in biological modeling: Behavior, execution and visualization. In: FBTC 2007. ENTCS, vol. 194(3), pp. 119–131. Elsevier, Amsterdam (2007)

19. Herrmann, K., Timmis, J., Fairclough, S.: Challenges in pervasive adaptation (under review, 2009)

20. Hinchey, M.G., Sterritt, R.: Self-managing software. IEEE Computer 39(2), 107–109 (2006)

21. Kernbach, S., Hamann, H., Stradner, J., Thenius, R., Schmickl, T., van Rossum, A.C., Sebag, M., Bredeche, N., Yao, Y., Baele, G., Van de Peer, Y., Timmis, J., Mohktar, M., Tyrrell, A., Eiben, A.E., McKibbin, S.P., Liu, W., Winfield, A.F.T.: On adaptive self-organization in artificial robot organisms. In: Adaptive and Self-adaptive Systems and Applications. IEEE Press, Los Alamitos (2009)

22. Knight, J.C., Strunk, E.A.: Achieving critical system survivability through software architectures. In: de Lemos, R., Gacek, C., Romanovsky, A. (eds.) Architecting Dependable Systems II. LNCS, vol. 3069, pp. 51–78. Springer, Heidelberg (2004)

23. Kurd, Z., Kelly, T., Austin, J.: Developing artificial neural networks for safety critical systems. Neural Computing and Applications 16(1), 11–19 (2006)

24. Miller, B.: The autonomic computing edge: Can you chop up autonomic computing? IBM Technical Library (March 2008), http://www.ibm.com/developerworks/library/ac-edge4/index.html

25. Mokhtar, M., Bi, R., Timmis, J., Tyrrell, A.M.: A modified dendritic cell algorithm for on-line error detection in robotic systems. In: Congress on Evolutionary Computation, pp. 2055–2062. IEEE Press, Los Alamitos (2009)

26. Owens, N.D., Timmis, J., Greensted, A.J., Tyrrell, A.M.: On immune inspired homeostasis for electronic systems. In: de Castro, L.N., Von Zuben, F.J., Knidel, H. (eds.) ICARIS 2007. LNCS, vol. 4628, pp. 216–227. Springer, Heidelberg (2007)

27. Owens, N.D.L., Greensted, A.J., Timmis, J., Tyrrell, A.M.: T cell receptor signalling inspired kernel density estimation and anomaly detection. In: Andrews, P.S. (ed.) ICARIS 2009. LNCS, vol. 5666, pp. 122–135. Springer, Heidelberg (2009)

28. Polack, F., Stepney, S., Turner, H., Welch, P., Barnes, F.: An architecture for modelling emergence in CA-like systems. In: Capcarrère, M.S., Freitas, A.A., Bentley, P.J., Johnson, C.G., Timmis, J. (eds.) ECAL 2005. LNCS (LNAI), vol. 3630, pp. 433–442. Springer, Heidelberg (2005)

29. Polack, F.A.C., Andrews, P.S., Sampson, A.T.: The engineering of concurrent simulations of complex systems. In: Congress on Evolutionary Computation, pp. 217–224. IEEE Press, Los Alamitos (2009)

30. Polack, F.A.C., Hoverd, T., Sampson, A.T., Stepney, S., Timmis, J.: Complex systems models: Engineering simulations. In: ALife XI. MIT Press, Cambridge (2008)

31. Read, M., Andrews, P.S., Timmis, J., Kumar, V.: A domain model of Experimental Autoimmune Encephalomyelitis. In: Workshop on Complex Systems Modelling and Simulation, pp. 9–44. Luniver Press (2009)

32. Sargent, R.G.: The use of graphical models in model validation. In: 18th Winter Simulation Conference, pp. 237–241. ACM, New York (1986)
33. Sargent, R.G.: Verification and validation of simulation models. In: 37th Winter Simulation Conference, pp. 130–143. ACM, New York (2005)
34. Shalizi, C.R.: Causal Architecture, Complexity, and Self-Organization in Time Series and Cellular Automata. PhD thesis, Physics Department, University of Wisconsin at Madison (2001)
35. Shalizi, C.R., Shalizi, K.L., Haslinger, R.: Quantifying self-organization with optimal predictors. Physical Review Letters 93(11), 118701 (2004)
36. Stepney, S., Smith, R.E., Timmis, J., Tyrrell, A.M.: Towards a conceptual framework for artificial immune systems. In: Nicosia, G., Cutello, V., Bentley, P.J., Timmis, J. (eds.) ICARIS 2004. LNCS, vol. 3239, pp. 53–64. Springer, Heidelberg (2004)
37. Tabatabaie, M., Paige, R.F., Kimble, C.: Exploring enterprise information systems. In: Cruz-Cunha, M. (ed.) Social, Managerial, and Organizational Dimensions of Enterprise Information Systems, pp. 415–433. IGI Global (2010)
38. Timmis, J., Tyrrell, A., Mokhtar, M., Ismail, A., Owens, N., Bi, R.: An artificial immune system for robot organisms. In: Levi, P., Kernbach, S. (eds.) Symbiotic Multi-Robot Organisms: Reliability, Adaptability, Evolution, pp. 268–288. Springer, Heidelberg (April 2010) (to appear)
39. Tononi, G., Sporns, O., Edelman, G.M.: Measures of degeneracy and redundancy in biological networks. PNAS 96(6), 3257–3262 (1999)
40. Tsankova, D., Georgieva, V., Zezulka, F., Bradac, Z.: Immune network control for stigmergy based foraging behaviour of autonomous mobile robots. International Journal of Adaptive Control and Signal Processing 21(2), 265–286 (2006)
41. Wenkstern, R.Z., Steel, T., Leask, G.: A self-organizing architecture for traffic management. In: Self-Organizing Architectures (2009)
42. Whitbrook, A., Aickelin, U., Garibaldi, J.: An idiotypic immune network as a short-term learning architecture for mobile robots. In: Bentley, P.J., Lee, D., Jung, S. (eds.) ICARIS 2008. LNCS, vol. 5132, pp. 266–278. Springer, Heidelberg (2008)

Self-organising Sensors for Wide Area Surveillance Using the Max-sum Algorithm

Alex Rogers[1], Alessandro Farinelli[2], and Nicholas R. Jennings[1]

[1] School of Electronics and Computer Science
University of Southampton, Southampton, SO17 1BJ, UK
{acr,nrj}@ecs.soton.ac.uk
[2] Department of Computer Science
University of Verona, Verona, Italy
alessandro.farinelli@univr.it

Abstract. In this paper, we consider the self-organisation of sensors within a network deployed for wide area surveillance. We present a decentralised coordination algorithm based upon the max-sum algorithm and demonstrate how self-organisation can be achieved within a setting where sensors are deployed with no *a priori* information regarding their local environment. These energy-constrained sensors first learn how their actions interact with those of neighbouring sensors, and then use the max-sum algorithm to coordinate their sense/sleep schedules in order to maximise the effectiveness of the sensor network as a whole. In a simulation we show that this approach yields a 30% reduction in the number of vehicles that the sensor network fails to detect (compared to an uncoordinated network), and this performance is close to that achieved by a benchmark centralised optimisation algorithm (simulated annealing).

1 Introduction

The vision of computational systems composed of multiple autonomous interacting agents has been a research aim for at least two decades now, and is increasingly being realised in the real-world by the deployment of wireless sensor networks [1]. Such networks have found application in wide-area surveillance, animal tracking, and for monitoring environmental phenomena in remote locations, and a fundamental challenge within all such applications arises due to the fact that the sensors within these networks are often deployed in an ad-hoc manner (e.g. dropped from an aircraft or ground vehicle within a military surveillance application). In this case, the local environment of each sensor, and hence the exact configuration of the network, cannot be determined prior to deployment, and thus, the sensors themselves must be equipped with capability to *self-organise* sometime after deployment once the local environment in which they (and their neighbours) find themselves has been determined. Examples of such self-organisation include determining the most energy-efficient communication paths within the network once the actual reliability of communication links between individual sensors can be measured in situ, determining the optimal

D. Weyns et al. (Eds.): SOAR 2009, LNCS 6090, pp. 84–100, 2010.

orientation of sensors to track multiple moving targets that move through the sensor network, and in the application that we consider in detail in this paper, coordinating the sense/sleep schedules (or duty cycles) of power constrained sensors deployed in a wide-area surveillance task once the degree of overlap of the sensing fields of nearby sensors has been determined.

A common feature of these self-organisation problems is that the sensors must typically choose between a small number of possible states (e.g. which neighbouring sensor to transmit data to, or which sense/sleep schedule to adopt), and the effectiveness of the sensor network as a whole depends not only on the individual choices of state made by each sensor, but on the joint choices of interacting sensors. Thus, to maximise the overall effectiveness of the sensor network, the sensors within the network must typically make coordinated, rather than independent, decisions. Furthermore, this coordinated decision must be performed despite the specific constraints of each individual device (such as limited power, communication and computational resources), and the fact that each device can typically only communicate with the few other devices in its local neighbourhood (due to the use of low-power wireless transceivers, the small form factor of the device and antenna, and the hostile environments in which they are deployed). Additional challenges arise through the need to perform such coordination in a decentralised manner such that there is no central point of failure and no communication bottleneck, and to ensure that the deployed solution scales well as the number of devices within the network increases.

Problems of this nature are often described within the multi-agent systems literature as *Distributed Constraint Optimisation Problems* (DCOPs), and are often represented by graphs in which the nodes represent the agents (in this case the sensors), and edges represent real valued constraints that arise between the agents depending on their combined choice of state. A number of distributed complete algorithms have been proposed that allow the agents within the graph to make coordinated decisions over their states (communicating only with other agents who are connected to themselves through a constraint) such that the sum over all of the constraints is maximised (or minimised). An example of such an algorithm is DPOP which preprocess the constraint graph, arranging it into a *Depth First Search* (DFS) tree, before exchanging messages over this tree [2]. However, complete algorithms calculate the globally optimal solution, and such optimality demands that some aspect of the algorithm grows exponential in size. For example, within DPOP, the size of the messages is exponential in the width of the tree. Such exponential relationships are simply unacceptable for the constrained devices deployed within typical sensor networks.

In contrast, a large number of approximate algorithms have also been proposed for solving DCOPs. These algorithms are typically based upon entirely local computation, whereby each agent updates its state based only on the communicated (or observed) states of those local neighbours that influence its utility. As such, these approaches are well suited for large scale distributed applications, and in this context, the Distributed Stochastic Algorithm (DSA), is one of the most promising having been proposed for decentralised coordination within sensor networks [3]. However, algorithms of this type often converge to poor quality solutions since the agents do not explicitly communicate their utility for being in any particular state, but only communicate their preferred state (i.e. the one that will maximise their own utility) based on the current preferred state of their neighbours. Furthermore, DSA introduces a parameter (which represents the probablity that an agent should actually update its state if the states of its neighbours indicate that this would be approapriate) whose value must be determined through trial and error prior to deployment of the algorithm.

Thus, against this background, there is a clear need for decentralised coordination algorithms that make efficient use of the constrained computational and communication resources found within wireless networks sensor systems, and yet are able to effectively represent and communicate complex utility relationships between sensors. To address this shortcoming, in this paper, we describe an approximate decentralised coordination algorithm that can be applied to the general problem of maximising the overall effectiveness of a sensor network (or any other decentralised system) in which the utility of any individual sensor is dependent not only on its own state, but also on the state of a number of interacting neighbours (as is the case in the example applications we have discussed previously). Our solution is based upon a novel representation of the problem as a cyclic bipartite factor graph and exploits message passing techniques that are frequently used in the context of information theory to decompose complex computations on single processors. We have previously demonstrated the effectiveness of this approach on a number of benchmark graph-colouring problems [11], and shown that it's performance is in line with the extensive evidence that demonstrates that the max-sum algorithm generates good approximate solutions when applied to cyclic graphs (in the context of approximate inference through 'loopy' belief propagation on Bayesian networks [4], iterative decoding of practical error correcting codes [5], and solving large scale K-SAT problems involving thousands of variables [6]). This algorithm effectively propagates information around the network such that it converge to a *neighborhood maximum*, rather than a simple local maximum [4].

Hence, having described a generic decentralised coordination algorithm, we then apply it within a specific problem concerning the self-organisation of sensors within a wireless network deployed within an urban environment to detect vehicle movements on a road network. Within this setting, energy management is a key challenge, and such sensors will typically control their duty cycle (effectively switching between active sensing modes and low-power sleep modes) in order to operate in an *energy neutral* mode, and hence, exhibit an indefinite lifetime [7]. However, since the sensing ranges of the sensors within the network will typically overlap with one another, the overall effectiveness of the sensor network depends not only on the sensors' individual choice of duty cycles, but also on the combined choice of neighbouring sensors whose sensing ranges overlap. With an ad-hoc sensor deployment, these interactions are not known prior to deployment, and thus, we describe how the sensors may self-organise by first learning the interactions between their neighbours (i.e. how much their neighbours' sensing fields overlap with their own), and then coordinating their sense/sleep schedules (using the decentralised coordination algorithm described above) in order to address the system-wide performance goal of maximising the probability that a vehicle is detected. We show that by self-organising in this way, we can achieve a 30% reduction in the number of vehicles that the sensor network fails to detect (compared to an uncoordinated network), and this performance is shown to be close to that achieved by a benchmark centralised optimisation algorithm (simulated annealing).

The remainder of this paper is structured as following. In section 2 we present our factor graph representation and max-sum decentralised coordination algorithm. In section 3 we describe how it can be applied within self-organising sensors in our wide-area surveillance scenario amd we present our empirical evaluation. Finally, we conclude and discuss future work in section 4.

2 The Max-sum Approach to Coordination

The max-sum algorithm is a specific instance of a general message passing algorithm that exploits the *general distributive law* in order to decompose a complex calculation by factorising it (i.e. representing it as the sum or product of a number of simpler factors) [8]. Such algorithms are frequently used in fields such as information theory, artificial intelligence and statistics. In our case, they represent an ideal combination of the best features of optimal algorithms and approximate stochastic algorithms. They can make efficient use of constrained computational and communication resources, and yet are able to effectively represent and communicate complex utility relationships through the network and attain close to optimal solutions. In the following, we first provide a formal description of the generic coordination problem we address in this paper, and then detail the max-sum algorithm itself.

2.1 Problem Description

We initially consider the general case in which there are M sensors, and the state of each sensor may be described by a discrete variable x_m. Each sensor interacts locally with a number of other sensors such that the utility of an individual sensor, $U_m(\mathbf{x}_m)$, is dependent on its own state and the states of these other sensors (defined by the set $\mathbf{x}_m$). For example, in the wide area surveillance problem that we consider in this paper, the state of the sensor represents the sense/sleep schedule that it has adopted, the interacting sensors are those whose sensing areas overlap with its own, and the utility describes the probability of detecting an event within the sensor's sensing range. However, our approach at this stage is generic, and thus, we make no specific assumptions regarding the structure of the individual utility functions.

Within this setting, we wish to find the state of each sensor, $\mathbf{x}^*$, such that the sum of the individual sensors' utilities (commonly referred to as social welfare within the multi-agent systems literature) is maximised:

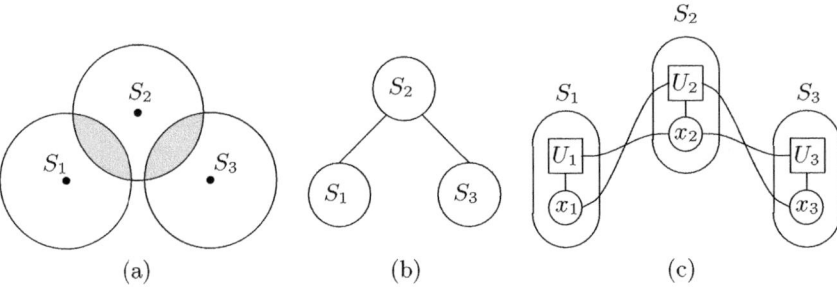

Fig. 1. Diagram showing (a) the position of three sensors in the environment whose sensing ranges overlap, and (b) the resulting factor graph with sensors decomposed into function and variable nodes

$$\mathbf{x}^* = \arg\max_{\mathbf{x}} \sum_{i=1}^{M} U_i(\mathbf{x}_i) \tag{1}$$

Furthermore, in order to enforce a truly decentralised solution, we assume that each sensor only has knowledge of, and can directly communicate with, the few neighbouring agents on whose state its own utility depends. In this way, the complexity of the calculation that the sensor performs depends only on the number of neighbours that it has (and not the total size of the network), and thus, we can achieve solutions that scale well.

2.2 Factor Graph Representation

In order to apply the max-sum algorithm, we represent the optimisation problem described in equation 1 above, as a bipartite factor graph. To this end, we decompose each sensor into a variable node that represents its state, and a function node that represents its utility. The function node of each sensor is connected to its own variable node (since its utility depends on its own state), and also to the variable nodes of other sensors whose states its utility depends on. For example, we show in figure 1a an example in which three sensors, $\{S_1, S_2, S_3\}$, interact with their immediate neighbours through the overlap of their sensing areas. In figure 1b we show the resulting constraint graph often used within the optimal algorithms, such as DPOP, discussed earlier. In figure 1c we show the resulting bipartite factor graph in which the sensors are decomposed into function nodes, $\{U_1, U_2, U_3\}$, and variable node, $\{x_1, x_2, x_3\}$. The overall function represented by this factor graph is given by $U = U_1(x_1, x_2) + U_2(x_1, x_2, x_3) + U_3(x_2, x_3)$.

2.3 Message Content of the Max-sum Algorithm

The max-sum algorithm operates directly on the factor graph representation described above. When this graph is cycle free, the algorithm is guaranteed to converge to the global optimal solution such that it finds the combination of states that maximises the sum of the sensors' utilities. When applied to cyclic graphs (as is the case here), there is no guarantee of convergence but extensive empirical evidence demonstrates that such family of algorithms generate good approximate solutions [9,10].

The max-sum algorithm solves this problem in a decentralised manner by specifying messages that should be passed from variable to function nodes, and from function nodes to variable nodes. These messages are defined as:

- **From variable to function**

$$q_{i \to j}(x_i) = \alpha_{ij} + \sum_{k \in \mathcal{M}_i \setminus j} r_{k \to i}(x_i) \tag{2}$$

where $\mathcal{M}_i$ is a vector of function indexes, indicating which function nodes are connected to variable node i, and α_{ij} is a scaler chosen such that $\sum_{x_i} q_{i \to j}(x_i) = 0$, in order to normalise the message and prevent them increasing endless in the cyclic graphs.

- **From function to variable**

$$r_{j \to i}(x_i) = \max_{\mathbf{x}_j \setminus i} \left[U_j(\mathbf{x}_j) + \sum_{k \in \mathcal{N}_j \setminus i} q_{k \to j}(x_k) \right] \tag{3}$$

where, $\mathcal{N}_j$ is a vector of variable indexes, indicating which variable nodes are connected to function node j and $\mathbf{x}_j \setminus i \equiv \{x_k : k \in \mathcal{N}_j \setminus i\}$.

Note that these message definitions are somewhat cyclic, and reflect the iterative way in which the messages are updated. The messages flowing into and out of the variable nodes within the factor graph are functions that represent the total utility of the network for each of the possible states of the variable. At any time during the propagation of these messages, sensor i is able to determine which state it should adopt such that the sum over all the sensors' utilities is maximised. This is done by locally calculating the function, $z_i(x_i)$, from the messages flowing into i's variable node:

$$z_i(x_i) = \sum_{j \in \mathcal{N}_i} r_{j \to i}(x_i) \tag{4}$$

and hence finding $\arg\max_{x_i} z_i(x_i)$.

Previous applications of the max-sum algorithm have applied it as an efficient iterative algorithm for centralised problems such as decoding error correcting codes [5]. Here, the factor graph is actually physically divided among the sensors within the network, and thus the computation of the system-wide global utility function is now carried out through a distributed computation involving message passing between sensors. Thus although the max-sum algorithm is approximating the solution to a global optimisation problem it involves only local communication and computation.

2.4   Convergence and Performance

The messages described above may be randomly initialised, and then updated whenever a sensor receives an updated message from a neighbouring sensor; there is no need for a strict ordering or synchronisation of the messages. In addition, the calculation of the marginal function shown in equation 4 can be performed at any time (using the most recent messages received), and thus, sensors have a continuously updated estimate of their optimum state. When the underlying factor graph contains cycles there is no guarantee that the max-sum algorithm will converge; nor that if it does converge it will find the optimal solution. However, extensive empirical evaluation on a number of benchmark coordination problems, including graph colouring, indicates that it does in fact produce better quality solutions than other state of the art approximate algorithms such as DSA, but at significantly lower computation and communication cost compared to complete algorithms such as DPOP [11].

2.5   Architecture

Note that the formulation of the problem as a factor graph does not actually limit where any of the computation is actually performed. At its most decentralised, each factor may reside on a single agent, and this agent may reside on a separate sensor. In this case, messages between factors and variables (and between variables and factors), represent messages exchanged between the sensors, across the communication network. However, it is also possible to use exactly the same factor graph representation as a computationally efficient means to perform the coordination even within a system where the elements are not actually distributed across different computational entities. Indeed, it is exactly this approach that is employed in the context of approximate inference through 'loopy' belief propagation on Bayesian networks [4], iterative decoding of practical error correcting codes [5], and when solving large scale K-SAT problems involving thousands of variables [6].

3   Wide Area Surveillance Problem

Having presented our max-sum coordination algorithm we now focus on its application within the self-organisation of a sensor network deployed for a wide area surveillance task. To this end, we consider a wide area surveillance problem based upon a simulation of an urban settings (using the Robocup Rescue Simulation Environment — see http://www.robocuprescue.org/). We assume that multiple wireless sensors are randomly deployed within the environment, and these sensors are tasked with detecting vehicles that travel along the roads. We assume that the sensors have no *a priori* knowledge of the road network, and do not know their own location within it. The sensors detect seismic or acoustic signals in order to indicate the binary presence, or absence, of vehicles within their sensing fields. We make no assumptions regarding the shape or range of these

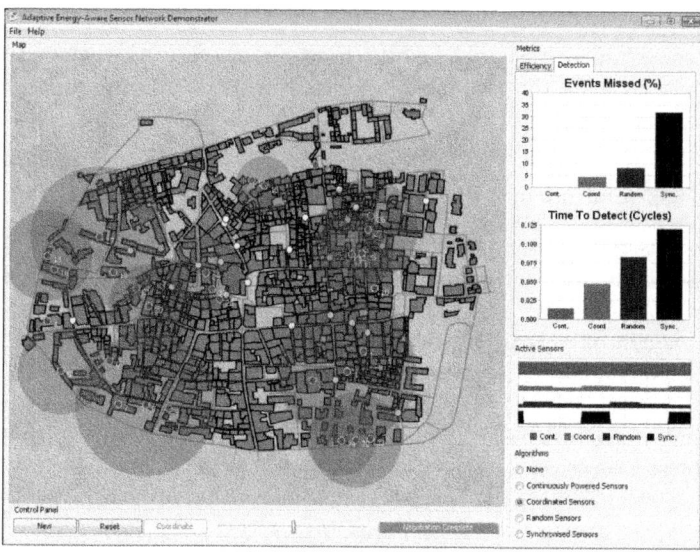

Fig. 2. Simulation of a wide area surveillance scenario (based on the Robocup Rescue Simulation Environment)

sensing fields (although for ease of simulating the setting, within our simulation we model these as circular fields with randomly assigned radii). Figure 2 shows this simulation environment in operation. The area sensed by active sensors is shown in red, and moving vehicles are shown as white markers on the roads.

We assume that the sensors are able to harvest energy from their local environment, but at a rate that is insufficient to allow them to be powered continually. Thus at any time a sensor can be in one of two states: either sensing or sleeping. In the sensing state the sensor consumes energy at a constant rate, and is able to interact with the surrounding environment (e.g. it can detect events within its sensing field and communicate with other sensors)[1]. In the sleep state the sensor can not interact with the environment but it consumes negligible energy. To maintain energy-neutral operation, and thus exhibit an indefinite lifetime, sensors adopt a duty cycle whereby within discrete time slots they switch between these two states according to a fixed schedule of length L. We denote the schedule of sensor i by a vector $\mathbf{s}_i = \{s_0^i, \ldots, s_{L-1}^i\}$ where $s_k^i \in \{0,1\}$, and $s_k^i = 1$ indicates that sensor i is in its active sensing state during time slot k (and conversely, it is sleeping when $s_k^i = 0$). We assume that this schedule is repeated indefinitely, and in this paper, we specifically consider schedules in which the sensor is in its sense state for one time slot, and in a sleep state for all $L - 1$

---

[1] Note that we assume that the energy consumed by activating the sensor is much more significant than that used in communication. This is generally true for sensors that require continuous signal processing such as the acoustic or seismic sensors considered here.

other time slots (i.e. $\sum_{k=0}^{L-1} s_k^i = 1$). This represents the simplest description of a power constrained sensing schedule, however, we note that the max-sum coordination algorithm that we have presented in the last section, can be applied for any discrete schedule.

3.1   The Coordination Problem

Figure 3 illustrates the coordination problem that results from this scenario. In this specific example, three sensors, $\{S_1, S_2, S_3\}$, are randomly deployed and exhibit overlapping sensing fields. In order to maintain energy neutral operation, each sensor can only actively sense for half of the time (i.e. $L = 2$), and thus, each sensor has a choice from two sensing schedules: either $\{1, 0\}$ or $\{0, 1\}$.

The system-wide goal is to maximise the probability that events are detected by the sensor network as a whole. This is achieved by ensuring that the area covered by the three sensors is actively sensed by at least one sensor at any time. However, with the sensing schedules available, it is clearly not possible to ensure that area $S_1 \cap S_2$, area $S_2 \cap S_3$ and area $S_1 \cap S_3$ are all sensed continually. Thus, the sensors must coordinate to ensure that the minimal area possible exhibits the minimal periods during which no sensor is actively sensing it. In this case, the optimal solution is the one shown where $s_1 = \{0, 1\}$, $s_2 = \{1, 0\}$ and $s_3 = \{0, 1\}$. Note that this leads to areas $A_{\{1,2\}}$, $A_{\{2,3\}}$ and $A_{\{1,2,3\}}$ being sensed continually, and the smallest area, $A_{\{1,3\}}$, and of course the three non-overlapping areas, exhibiting intermittent sensing.

In a larger sensor deployment, each of these three sensors is also likely to overlap with other sensors. Thus, finding the appropriate sensing schedule of each sensors, such that probability of detecting an event is maximised, is a combinatorial optimisation problem. As such, this problem is similar to the graph colouring benchmark used in the evaluation of the max-sum algorithm described earlier [11]. However, an important difference is that in our sensor scheduling problem we can have interactions between multiple sensors (as is the case in the example shown in figure 3), rather than interaction between just pairs of sensors (as is the case in the standard graph colouring problem).

3.2   Applying the Max-sum Algorithm

To apply the max-sum coordination algorithm to this problem it is necessary to first decompose the system-wide goal that we face (that of maximising the probability that an event is detected) into individual sensor utility functions. As shown above, the utility of each sensor is determined by its own sense/sleep schedule, and by those of sensors whose sensing fields overlap with its own. Thus, we define $\mathbf{N}_i$ to be a set of indexes indicating which other sensors' sensing fields overlap with that of sensor i and $\mathbf{k}$ is any subset of $\mathbf{N}_i$ (including the empty set). $A_{\{i\} \cup \mathbf{k}}$ is the area that is overlapped only by sensor i and those sensors in $\mathbf{k}$. For example, with respect to figure 3, the area $A_{\{1,2\}}$ is the area that is sensed only by sensors 1 and 2. In a slight abuse of notation, we represent the entire sensing area of sensor S_1 as S_1, and thus, note that the area $A_{\{1,2\}}$ is different from

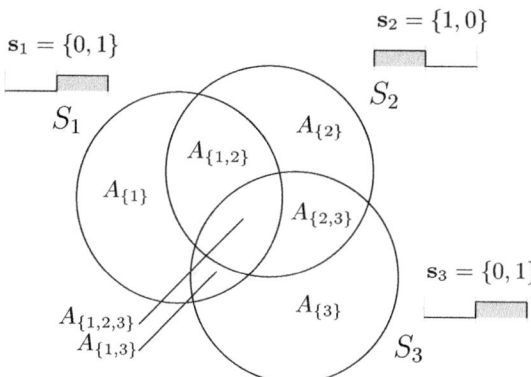

Fig. 3. Example coordination problem in which three sensors, $\{S_1, S_2, S_3\}$, have sensing fields that overlap

$S_1 \cap S_2$ because the area $S_1 \cap S_2$ would include also the sub area $S_1 \cap S_2 \cap S_3$. In general, we have:

$$A_{\{i\}\cup \mathbf{k}} = \bigcap_{j \in (\{i\}\cup \mathbf{k})} S_j \setminus \bigcup_{l \notin (\{i\}\cup \mathbf{k})} S_l$$

We define a function $G : 2^{\mathbf{X}} \to \mathcal{S}$ and $G(\mathbf{x}_{\{i\}\cup \mathbf{k}})$ is the combined sensing schedule of sensor i and those sensors in $\mathbf{k}$ (calculated through the logical 'OR' of each individual schedule). Now, assuming that events occur uniformly over the environment, then the utility of sensor i is given by:

$$U_i(\mathbf{x}_i) = \sum_{\mathbf{k} \subseteq N_i} \frac{A_{\{i\}\cup \mathbf{k}}}{|\{i\} \cup \mathbf{k}|} \times P(\text{detection}|\lambda_d, G(\mathbf{x}_{\{i\}\cup \mathbf{k}})) \qquad (5)$$

where $P(\text{detection}|\lambda_d, G(\mathbf{x}_{\{i\}\cup \mathbf{k}}))$ is the probability of detecting an event given the combined sensing schedules of the overlapping sensors and a parameter, λ_d, that describes the typical duration of an event. We model this duration as a Poisson process such that the probability of an event lasting time t is given by $\lambda_d e^{-\lambda_d t}$. Note, that we scale the area by the number of sensors who can sense it to avoid double-counting areas which are represented by multiple sensors. Also, note that when the set $\mathbf{k}$ is empty we consider the area covered only by the single sensor. For example, the utility of sensor S_2 shown in figure 3, is calculated by considering the areas $A_{\{2\}}$, $A_{\{1,2\}}$, $A_{\{2,3\}}$ and $A_{\{1,2,3\}}$.

3.3   Learning the Mutual Interaction of Sensing Fields

The utility function presented in equation 5 assumes that the sensors are able to determine the area of overlap of their own and neighbouring sensors' sensing fields, and that they know the parameter λ_d that describes the detectable duration of events. In reality, sensors may have highly irregular and obscured sensing

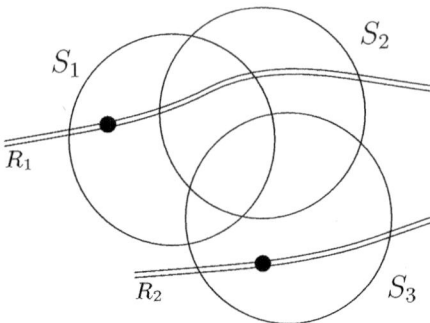

Fig. 4. Example showing the paths of two vehicles on roads, $\{R_1, R_2\}$, crossing the sensing fields of three overlapping sensors S_1, S_2 and S_3

areas, they may not be able to determine the exact position of themselves, let alone neighbouring sensors, and events may be known to be more likely to occur in some areas than others. Thus, we now relax these constraints, and describe how the sensors may learn these relationships in order to make a coordinated decision regarding their sense/sleep schedules. In more detail, we implement the following scheme:

Calibration Phase
We assume that after deployment, all sensors initially select the same sensing schedule, and thus, the sensors are all active and sense simultaneously[2]. At regular intervals during this phase sensors exchange information regarding the events that they have detected, and they keep track of (i) the number of events that they observe individually, O_i, and (ii) the number of events that are both detected by themselves and a subset of their neighbours, $O_{\{i\}\cup k}$. The exact form that this exchange of information takes depends on the nature of the sensors used, and the events that they are detecting. Within our wide area surveillance scenario, we assume that sensors are able to time stamp the appearance and disappearance of vehicles within their sensing fields. Comparison of the time stamps of observations of other sensors then allow the sensor to identify whether vehicles are detected by multiple sensors as they cross its own sensing field.

For example, consider figure 4 in which the two vehicles crossing three overlapping sensing fields, and assume that sensor S_1 time stamps the appearance and disappearance of a vehicle at times 09:02:12 and 09:02:21 respectively, sensor S_2 time stamps the appearance and disappearance of a vehicle at times 09:02:15 and 09:02:24 respectively, and finally, sensor S_3 time stamps the appearance and disappearance of a vehicle at times 09:02:27 and 09:02:33 respectively. In this case, the intersection of the time stamps of sensors S_1 and S_2 lead these two

---
[2] Note that the performance of the network is somewhat degraded throughout this calibration phase since all the sensors select the same sensing schedule and are synchronised. We intend to address this within future work (see section 4).

sensors to conclude that $O_{\{1\}} = 1$, $O_{\{1,2\}} = 1$, $O_{\{2\}} = 1$, while the non-intersection of the time stamps of sensor S_3 leads it to conclude that $O_{\{3\}} = 1$. Note that in general, more complex techniques may be required to differentiate events when they occur concurrently. This will typically require some additional information such as the position of the event, or some recognisable characteristic of the event, and within the data fusion and tracking literature, this problem is commonly known as data or track association. Here, we assume that events are uniquely identified, since data association is not the focus of this paper. However, we note that this data association need not be error-free and the performance of the network degrades slowly if errors do occur.

Finally, the duration of the events is used to calculate an estimate of λ_d. This is easily done, since the maximum likelihood estimate of λ_d is simply the mean of the observed event durations.

Coordination Phase
The numbers of events observed in the calibration phase now acts as a proxy for the unknown areas of overlap between neighbouring sensors. Furthermore, it also captures the fact that events will not occur evenly over the entire area, but are restricted to certain areas (i.e. the roads in our case). Hence, the sensors now calculate their utility based on a modification of equation 5 given by:

$$U_i(\mathbf{x}_i) = \sum_{\mathbf{k} \subseteq \mathbf{N}_i} \frac{O_{\{i\} \cup \mathbf{k}}}{|\{i\} \cup \mathbf{k}|} \times P(\text{detection} | \lambda_d, G(\mathbf{x}_{\{i\} \cup \mathbf{k}})) \tag{6}$$

The sensors can now use the max-sum coordination algorithm presented earlier to coordinate their choice of sense/sleep schedule such that the utility of the overall sensor network is maximised, and hence, the probability of detection of a vehicle traveling within the area covered by the sensor network is maximised.

Operational Phase
Finally, the operational phase proceeds as before, sensors simply follow the sense/sleep schedule determined in the previous coordination phase. If during this phase a sensor fails, then the coordination algorithm above may simply be re-run to coordinate over the smaller sensor network. Likewise, should the position of sensors change, or new sensors be added, both the calibration phase and the coordination phase can be re-run to coordinate over the new environment in which the sensors find themselves. In section 4 we shall describe our future work developing a more principled approach that allows for continuous self-adaption of the sensor network as the state of the environment, or the sensors themselves, changes over time.

To validate this approach we now perform an empirical evaluation within our simulation environment.

3.4   Empirical Evaluation

To this end, we simulated the above three phases using random deployments of the sensors whose sensing ranges are assumed to be circular discs with radius drawn uniformly between $0.05d$ and $0.15d$ (where d is the maximum dimension of the area in which the sensors are deployed). During the calibration phase we simulated the movement of 500 vehicles between random start and end points, and the sensors exchanged observations with one another regarding their observations during this time. During the coordination phase, the sensors use the max-sum algorithm over a fixed number of cycles, in order to coordinate their sensing schedules. Finally, during the operational phase the sensors use the sensing schedules determined in the negotiation phase, and we again simulate the movement of 500 vehicles between random start and end points.

We measure the operational effectiveness of the sensor network by calculating the percentage of vehicles that are missed by the sensor network (i.e. vehicles that move between their start and end point without ever being within the sensing field of an actively sensing sensor) and for those vehicles that are detected, we measure the time taken for the first detection (i.e. the time at which the network first becomes aware of the presence of the vehicle after it leaves its start point). We repeat the experiments 100 times for three different length sensing schedules ($L = 2$, 3 and 4) and we investigate three different ranges of sensor number such that the effective number of sensors (given by N/L) remained constant. In this way, each deployment had the same effective sensing capability. We compare results for four different coordination mechanisms:

1. **Randomly Coordinated Sensors**
   As before, the choice of each sensors' sense/sleep schedule is made randomly by each individual sensor with no coordination.

2. **DSA Coordinated Sensors**
   Using the results of the calibration phase, the sensors use the DSA algorithm (as discussed earlier) to coordinate their sense/sleep schedules. The probability that any agent actually updates its preferred state is 0.4 (determined through empirical optimisation).

3. **Max-sum Coordinated Sensors**
   Using the results of the calibration phase, the sensors use the max-sum algorithm to coordinate their sense/sleep schedules.

4. **Simulated Annealing Coordinated Sensors**
   We use an offline centralised optimisation algorithm to calculate an upper bound on the performance that we can expect from our decentralised max-sum approach. This solution cannot be used in practice to coordinate the sense/sleep schedules of the real sensors since it is centralised and assumes full knowledge of the topology of the network, however, it provides an upper bound on the performance of a coordinated solution.

The results of these experiments are shown in figures 5 and 6, where the error bars represent the standard error of the mean in the repeated experiments. In more detail, figure 5 shows the percentage of vehicles that fail to be detected by

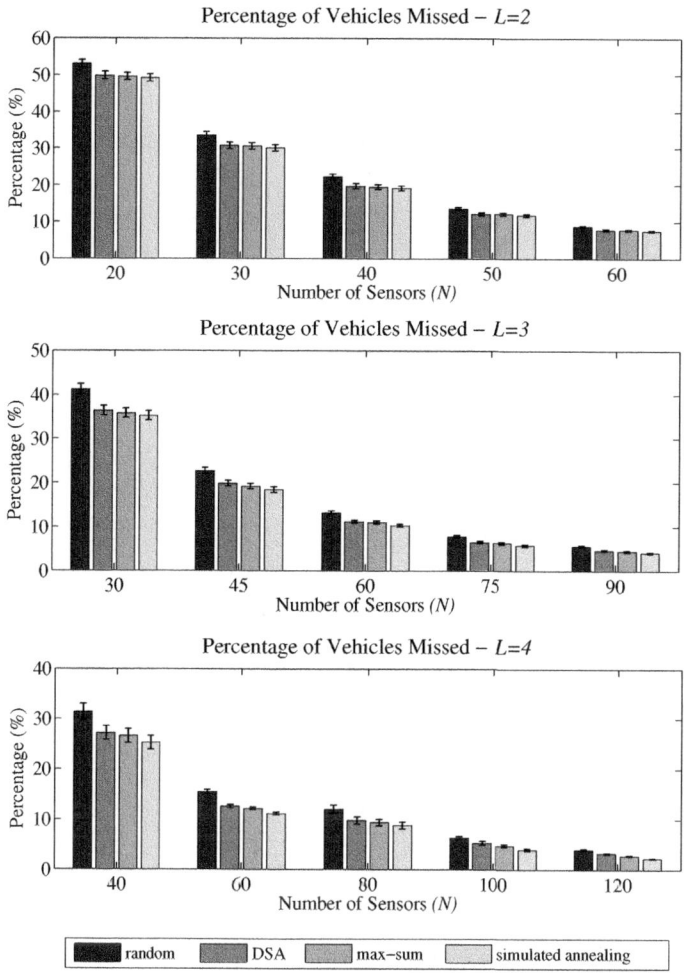

Fig. 5. Comparison of simulation results reporting the percentage of missed vehicles, for a sensor network using random, DSA, max-sum, and centralised simulated annealing coordination algorithms plotted against the number of deployed sensors

the sensor network; our main metric for the performance of the network. Figure 6 shows the time that it took the sensor network to first detect each vehicle; a metric that we do not actively seek to minimise. Note that in all cases, the randomly coordinated sensor network performs the worst (failing to detect more vehicles and taking a longer time to detect them), and that the centralised simulated annealing approach provides the best solutions. In each case, the max-sum approach out performs the DSA algorithm, and does so without the need to empirically tune a parameter. The difference between the algorithms increases as both the number of sensors within the network and the length of

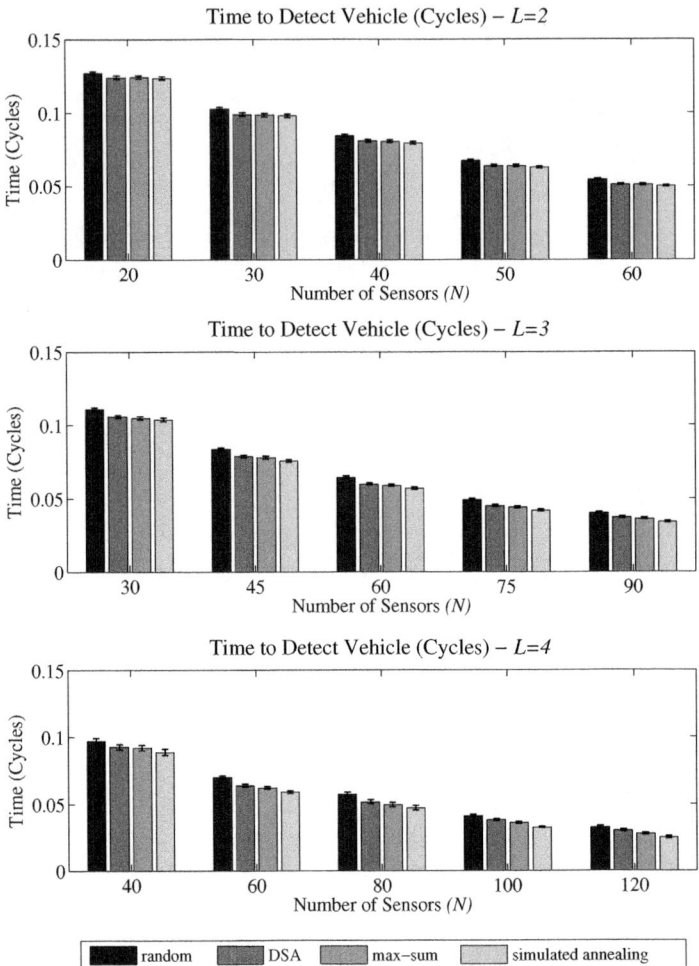

Fig. 6. Comparison of simulation results reporting the mean time to first detect a vehicle, for a sensor network using random, DSA, max-sum, and centralised simulated annealing coordination algorithms plotted against the number of deployed sensors

sensing schedules increase. This trend is expected as the combinatorial coordination problem becomes harder as both these factors increase.

Table 1 shows the results for both of these metrics for the case when $L = 4$ and $N = 120$. In this case, by using our approach, we achieve a 30% reduction in the number of missed vehicles (compared to the uncoordinated network), and this performance is shown to be close to that achieved by the benchmark centralised optimisation algorithm (simulated annealing) and significantly better than DSA (a state of the art decentralised approach).

Table 1. Comparison of percentage of vehicles missed and time to detect vehicles for each coordination algorithm when $L = 4$ and $N = 120$

Coordination Algorithm	Percentage of Vehicles Missed (%)	Time to Detect Vehicle (Cycles)
random	4.0±[0.4]	0.033±[0.002]
DSA	3.2±[0.2]	0.030±[0.002]
max-sum	2.7±[0.2]	0.028±[0.002]
simulated annealing	2.2±[0.2]	0.025±[0.002]

4   Conclusions

In this paper, we have considered the self-organisation of sensors within a network deployed for wide area surveillance. We have presented a decentralised coordination algorithm based upon the max-sum algorithm and demonstrated how self-organisation can be achieved within a setting where sensors are deployed with no *a priori* information regarding their local environment. We showed the sensors can learn how their actions interact with those of neighbouring sensors, and then use the max-sum algorithm to coordinate their sense/sleep schedules in order to maximise the effectiveness of the sensor network as a whole. In a software simulation we showed that this approach yields significant improvements in performance over the case of random coordination, and closely approaches that of a benchmark centralised optimisation algorithm.

Our future work consists of extending these results in order to relax the requirement of a separate calibration phase prior to the negotiation phase. The synchronised schedules of the sensors during the calibration phase corresponds to a period of poor system-wide performance that is offset by improved system-wide performance during the operational phase. However, it is also possible to learn about the occurrence of events, and hence the overlap of sensors' sensing fields, during this operational phase. Thus, we would like to investigate online algorithms that can trade-off between exploratory behaviour (synchronising with neighbouring sensors to learn about the occurrence of events), and exploitative behaviour (using relationships already learnt to coordinate the sensors). Principled Bayesian heuristics exist for performing such trade-offs [12], and applying these approaches within this setting would remove the requirement for three distinct phases. Rather, the sensors would continuously self-organise and self-adapt, changing sense/sleep schedules continuously to trade-off between exploration and exploitation. Such an approach would also naturally apply within dynamic settings where sensors' utilities may change at any time, sensors may fail, or additional sensors may be deployed. The max-sum coordination algorithm used in this paper already supports this continual behaviour since utility messages can be communicated, and sensors can estimate their optimal state, at anytime, and thus, it would appear to be a solid base on which to develop this more advanced self-organising behaviour.

References

1. Rogers, A., Corkill, D.D., Jennings, N.R.: Agent technologies for sensor networks. IEEE Intelligent Systems 24(2), 13–17 (2009)
2. Petcu, A., Faltings, B.: DPOP: A scalable method for multiagent constraint optimization. In: Proceedings of the 19th International Joint Conference on Artificial Intelligence, pp. 266–271 (2005)
3. Fitzpatrick, S., Meetrens, L.: Distributed Coordination through Anarchic Optimization. In: Distributed Sensor Networks A Multiagent Perspective, pp. 257–293. Kluwer Academic, Dordrecht (2003)
4. Weiss, Y., Freeman, W.T.: On the optimality of solutions of the max-product belief propagation algorithm in arbitrary graphs. IEEE Transactions on Information Theory 47(2), 723–735 (2001)
5. MacKay, D.: Good error-correcting codes based on very sparse matrices. IEEE Transactions on Information Theory 45(2), 399–431 (1999)
6. Mezard, M., Parisi, G., Zecchina, R.: Analytic and algorithmic solution of random satisfiability problems. Science 297(5582), 812–815 (2002)
7. Kansal, A., Hsu, J., Zahedi, S., Srivastava, M.B.: Power management in energy harvesting sensor networks. ACM Transactions on Embedded Computing Systems 6(4) (2007)
8. Aji, S., McEliece, R.: The generalized distributive law. IEEE Transactions on Information Theory 46(2), 325–343 (2000)
9. Kschischang, F.R., Frey, B.J., Loeliger, H.A.: Factor graphs and the sum-product algorithm. IEEE Transactions on Information Theory 42(2), 498–519 (2001)
10. MacKay, D.J.C.: Information Theory, Inference, and Learning Algorithms. Cambridge University Press, Cambridge (2003)
11. Farinelli, A., Rogers, A., Petcu, A., Jennings, N.R.: Decentralised coordination of low-power embedded devices using the max-sum algorithm. In: Proceedings of the 7th International Conference on Autonomous Agents and Multiagent Systems, pp. 639–646 (2008)
12. Dearden, R., Friedman, N., Andre, D.: Model based Bayesian Exploration. In: Proceedings of the 15th Conference on Uncertainty in Artificial Intelligence, pp. 150–159 (1999)

Multi-policy Optimization in Self-organizing Systems

Ivana Dusparic and Vinny Cahill

Lero – The Irish Software Engineering Research Centre
Distributed Systems Group
School of Computer Science and Statistics
Trinity College Dublin
{ivana.dusparic,vinny.cahill}@scss.tcd.ie

Abstract. Self-organizing systems are often implemented as collections of collaborating agents. Such agents may need to optimize their own performance according to multiple policies as well as contribute to the optimization of overall system performance towards a potentially different set of policies. These policies can be heterogeneous, i.e., be implemented on different sets of agents, be active at different times and have different levels of priority, leading to the heterogeneity of the agents of which the system is composed. Numerous biologically-inspired techniques as well as techniques from artificial intelligence have been used to implement such self-organizing systems. In this paper we review the most commonly used techniques for multi-policy optimization in such systems, specifically, those based on ant colony optimization, evolutionary algorithms, particle swarm optimization and reinforcement learning (RL). We analyze the characteristics and existing applications of the reviewed algorithms, assessing their suitability for particular types of optimization problems, based on the environment and policy characteristics. We focus on RL, as it is considered particularly suitable for large-scale self-organizing systems due to its ability to take into account the long-term consequences of the actions executed. Therefore, RL enables the system to learn not only the immediate payoffs of its actions, but also the best actions for the long-term performance of the system. Existing RL implementations mostly focus on optimization towards a single system policy, while most multi-policy RL-based optimization techniques have so far been implemented only on a single agent. We argue that, in order to be more widely utilized as a technique for self-optimization, RL needs to address both multiple policies and multiple agents simultaneously, and analyze the challenges associated with extending existing or developing new RL optimization techniques.

1 Self-adaptive Systems

As computing systems become increasingly large-scale and geographically dispersed, traditional centralized and hierarchical management of such systems becomes intractable. Such large-scale systems also need to be able to adapt to

D. Weyns et al. (Eds.): SOAR 2009, LNCS 6090, pp. 101–126, 2010.

changing operating conditions, not all of which can be anticipated at design time. Moreover, systems might be required to optimize their performance according to multiple, potentially conflicting, heterogeneous policies (e.g., policies with different spatial and temporal scopes, or of different priority).

For the purposes of this paper, we define *policy* as "any type of formal behavioural guide" [43], while also using the word *goal* more informally to denote "an aim or desired result" [54] of a system's behaviour, where policy is a formal expression of a goal. Additionally, we use the word *objective* synonymously with the word goal, when retaining the original terminology of the specific work discussed.

Examples of large-scale systems that are required to meet multiple performance policies include large-scale critical infrastructures (e.g., transportation networks, electricity, gas, and water supply), as well as numerous other applications that involve, for example, scheduling, task allocation, routing, or load balancing, such as global supply chains. We believe multi-agent approaches based on self-organizing principles [18] are a suitable mechanism for decentralized management of these large-scale systems. In multi-agent systems (MAS), autonomous agents can learn their behaviour for given environmental conditions and coordinate their actions with immediate neighbours allowing global system behaviour to emerge from local learning, actions and interactions. Agents can learn from their interaction with the environment, without requiring detailed models of the environment (i.e., without the detailed knowledge of "how the world works" [65]), which are potentially time-consuming and complex to construct in large-scale or complex systems. A number of biologically-inspired agent-based techniques are used to support self-organization in large-scale decentralized systems, such as ant colony optimization, particle swarm optimization, evolutionary algorithms, and artificial neural networks, as well as reinforcement learning (RL). RL, originally a single-agent machine learning technique, is considered particularly suitable for learning in large-scale systems, as it enables agents to learn the actions that are the most suitable for their long-term performance, rather than just those with the most suitable immediate consequences [77]. In this paper we review self-organizing multi-policy optimization techniques, and particularly focus on RL by providing details of existing multi-agent single-policy and single-agent multi-policy techniques and identifying issues in extending these into multi-policy multi-agent techniques.

The remainder of the paper is organized as follows. Section 2 reviews the most commonly used self-organizing algorithms and analyzes the types of problems for which they are suitable. Section 3 focuses solely on the use of RL, reviewing multi-policy RL approaches and multi-agent RL approaches, as the number of RL-based optimization techniques that are both multi-policy and multi-agent is very limited. Section 4 summarizes the findings of this review and concludes the paper.

2   MAS Techniques

In the following sections we review some of the most commonly-used algorithms that exhibit self-organizing behaviours. We provide a basic introduction to each technique, both in their single-policy and multi-policy forms, and provide pointers to existing survey material where available. We refer to sample applications of each algorithm and based on algorithm characteristics analyze the type of problems for which they are suitable.

2.1   Ant Colony Optimization

Ant colony optimization refers to a family of algorithms inspired by the behaviour of ants in an ant colony. When searching for a food source, ants in a colony converge to moving over the shortest path, among different available paths, when moving between their nest and the food source (see Figure 1).

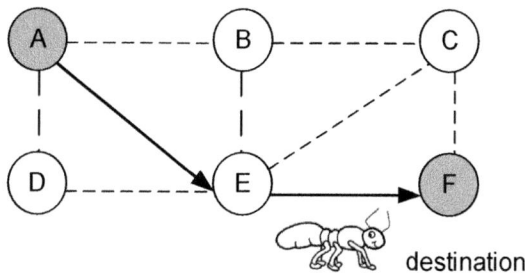

Fig. 1. An ant searching for the shortest path [20]

This behaviour is enabled by stigmergy, a form of indirect communication between ants through the environment, realized by depositing a substance called a pheromone on the path. Shorter paths get completed more quickly, causing more trips to be made on those routes and therefore more pheromone to be deposited on them. Stronger pheromone trails attract more ants, further increasing the number of ants following the route and increasing the amount of pheromone [20,48]. Figure 1 illustrates this process. Ants following route A $\rightarrow$ E$\rightarrow$ F will make a round trip start$\rightarrow$ destination$\rightarrow$ start quicker than ants following A$\rightarrow$ B$\rightarrow$ C$\rightarrow$ F route. The path selection of further ants leaving the starting position is therefore biased, as ants deposit pheromone on that route quicker than ants that are taking a longer trip.

In agent-based systems ants are mapped to individual agents searching for the solution to an optimization problem. Agents leave feedback about the effectiveness of their solution for other agents, similar to a pheromone trail, where the amount of pheromone deposited is proportional to the quality of the solution.

In Multi-Objective Ant Colony Optimization (MOACO) [1], multiple objectives are assigned with different weights (reflecting the different degrees of importance or priority of the goals to an agent) either a priori or dynamically during the search. Quality of the solution is determined by combining weighted qualities of that solution towards multiple objectives. MOACO algorithms can use a single colony of ants that store pheromone values into a single pheromone matrix, or can use a separate colony of ants for each objective and maintain multiple pheromone matrices, one for each objective. A single pheromone matrix is generally used if the relative weights of the objectives can be specified a priori, as weights specified are used to combine the quality of the solution towards multiple objectives into a single pheromone value to be stored. If the relative weights of solutions change dynamically, a separate colony of ants is assigned to each objective, and separate pheromone matrices are maintained. Information from the matrices is combined at solution construction time using objective weightings at that particular time. If separate ant colonies are used for each objective, each colony will find the optimal solution for its own objective, leaving compromise solutions undiscovered. To address this issue, "spy" ants can be introduced [19], that occasionally combine information gathered by all of the colonies to find trade-off solutions. Detailed reviews of existing MOACO techniques and comparisons of their performance are provided in [1] and [30].

2.2   Evolutionary Algorithms

Evolutionary algorithms [25] are a family of optimization algorithms inspired by biological evolution. The initial population of solutions is created randomly, and through the evolutionary processes of selection, crossover and mutation, the most suitable solutions are found after a number of generations. A fitness function is used to quantify the quality of the solution against the required optimization objective, and to select high-quality solutions that should pass on their genetic material to the next generation. Selected solutions crossover to create offspring, consisting of the genetic material of both parents. Additionally, mutations, or random changes to the genetic material, are also occasionally introduced, to generate solutions that differ from any other solution currently present in the population. If the new mutation is beneficial, it will survive through selection and crossover, otherwise it will die out.

Evolutionary algorithms are extensively used in multi-objective optimization [74], due to their ability to find multiple optimal solutions [81] in a single run [13]. Evolutionary algorithms can use a number of approaches to combine multiple objectives to be optimized. A weighted sum approach can be used to combine the objectives into a single fitness function, using weighting coefficients that reflect the importance of the objectives. The goal programming approach requires a designer to specify target goals for each of the objectives, and an evolutionary algorithm is used to minimize the absolute deviations from the targets specified. Using the ϵ-Constraint method, a single objective optimization is carried out for the most important objective, while other objective functions are considered as constraints bound by allowable level ϵ. Comprehensive reviews of evolutionary algorithms used in multi-objective optimization can be found in [13], [81], [40] and [74].

2.3   Particle Swarm Optimization

Particle swarm optimization [42] is a self-organizing optimization technique inspired by the flocking behaviour of birds. Each particle (a bird) in a swarm (a flock, a population) captures a potential solution, and moves through the multidimensional problem search space (possible set of solutions) seeking an optimal solution. Solutions are evaluated against a fitness function that represents the optimization objective. Particles broadcast their current position (i.e. the quality of their current solution) to their neighbours. Each particle then accelerates its movement towards a function of the best position it has found so far and the best position found by its neighbours [7]. For example, consider Figure 2, depicting a five-particle PSO population. During the optimization process, particle 3 receives broadcasts from its neighbours 2 and 4. As particle 2 currently has the best position in 3's neighbourhood, 3 will accelerate towards a function of its own previous best position and 2's current position. As the swarm iterates,

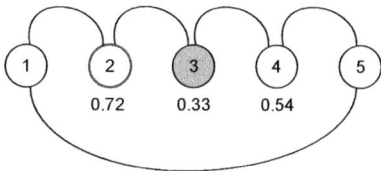

Fig. 2. Particle swarm optimization [42]

the fitness of the solutions increases, and particles focus on the area of the search space with high-quality solutions. Therefore, optimal solutions to the optimization problem arise in a self-organizing way, where each particle only needs to communicate with its neighbours to move towards regions with higher fitness.

Particle swarm optimization can be applied to multi-objective optimization problems by combining all objectives into a single weighted function, by using importance ordering of objectives specified by users, by assigning a population of swarms to each objective and recombining solutions, or by considering a particle's performance towards multiple objectives when selecting a leader towards which to accelerate. Reviews and comparisons of these techniques are provided in [56] and [62].

2.4 Artificial Neural Networks

Artificial neural networks are adaptive learning systems based on biological neural networks. They consist of an interconnected group of artificial neurons where connections between neurons change during the learning process. An artificial neural network is provided with a set of inputs, which, through a series of hidden layers, lead to one of the outputs, as shown in Figure 3. Each of the circles in Figure 3 represents a neuron, while lines connecting them represent connections with associated weights. During the training phase, connections between neurons strengthen each time a set of inputs generates a given output. In this way, an artificial neural network learns, based on a given input, to select the output that has the highest probability (connection strength) of being an optimal solution

INPUT   HIDDEN LAYER   OUTPUT

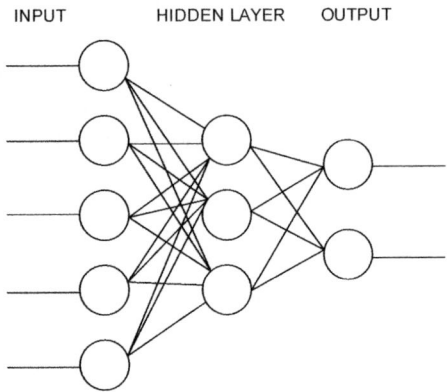

Fig. 3. Neural network [84]

based on experience so far. Artificial neural networks, therefore, learn how to map an input vector to an output. Due to the number of connections between neurons and multiple hidden layers, artificial neural networks can produce complex global behaviours, that emerge from the simple processing capabilities of neurons and the connections between them.

Artificial neural networks can be trained using a range of learning algorithms, including evolutionary algorithms and RL. Therefore, the capabilities of artificial neural networks to learn outputs satisfying multiple objectives is based on the multi-objective capabilities and approach of the underlying learning algorithm.

2.5   Applications of Self-organizing Algorithms

The applicability of the reviewed self-organizing systems to the implementation of self-optimizing behaviour in adaptive systems depends on the type of problem to be addressed.

The most common applications of ant colony optimization algorithms are primarily in systems that need to converge towards the shortest (lowest cost) path and have terminating states (i.e., arriving at the food source). Examples of the use of single-policy ant colony optimization include routing in wired [16] and wireless networks [17], vehicle routing [36], file sharing [2] and load balancing applications [52]. Multi-policy ant colony optimization has been applied in vehicle routing to minimize the number of vehicles involved in delivery, total travel time, and total delivery time [29,4] and in goods transportation to minimize the cost of operation of pick up and delivery fleets [19].

Particle swarm optimization and evolutionary algorithms are suitable for problems that require numerous values of multiple parameters and parameter combinations to be evaluated in order to determine optimal combinations. For example, applications of evolutionary algorithms in multi-agent systems include robot soccer, where the actions of each player are optimized using evolutionary algorithms

[46], large-scale multi-server web services, where evolutionary algorithms are used to optimize session handling strategies [26], and traffic control, where evolutionary algorithms can be used to tune the time sequences for traffic signals [36]. The applications of PSO-based optimization techniques include self-organizing document clustering, where PSO has been used to minimize the intra-cluster distances and maximize the distance between clusters by using swarm particles to represent a possible solution each and evaluating them against a similarity metric [15]. PSO has also been used in self-organizing networked sensor systems to optimize connectivity and minimize sensor overlap [38], as well as in robotic learning for unsupervised learning of obstacle avoidance [59].

Due to capability of artificial neural networks to map vectors of input data to an output, they are suitable for diagnostics problems, for classification of inputs into similar categories, prediction of trends [6], or any other situation where a large number of input parameters and their combinations need to be mapped to a small number of output settings. Examples of artificial neural network application areas include traffic signal control, where traffic conditions are fed as input and traffic signal controller settings were output [71], and in the simulation of autonomous robots, where neural networks were used for trajectory planning by robot manipulators [61].

Self-organizing algorithms are also often used in combination with each other to provide self-adaptivity. For example, particle swarm optimization and evolutionary algorithms can be used to train artificial neural networks (e.g., in [42] and [11], respectively) and artificial neural networks can be used in combination with RL to map states to optimal actions when the state-action space is too large for lookup tables (e.g., in [76]). Evolutionary algorithms can also be used to determine optimal combinations of parameters required by RL-based optimization (e.g., in [37]), while RL can be used to learn and maintain ants' routing tables in routing applications using ant colony optimization (e.g., in [72]).

3   Reinforcement Learning

In this section we focus on the use of RL in the implementation of self-organizing multi-policy optimization. We provide background on multi-policy RL and multi-agent RL, review examples of existing RL applications, and identify open issues for further research in RL-based multi-policy multi-agent optimization.

3.1   Introduction to RL

RL [73] is a learning technique based on trial and error that has been researched and applied in control theory, machine learning and artificial intelligence problems, as well as non-computer science domains such as psychology. RL is a single-agent, unsupervised learning technique, whereby an agent learns how to maximize the long-term sum of the rewards it receives from the environment.

An RL agent's interaction with the environment can be modelled as Markov Decision Process (MDP), as depicted in Figure 4, and consists of the following steps:

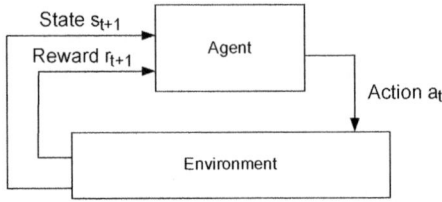

Fig. 4. Reinforcement learning process [73]

- the agent observes the state in which it is currently, $s_t \in$ S (where S is the set of all possible agent's states)
- the agent takes an action $a_t \in$ A (where A is the set of all actions that an agent is capable of taking)
- the agent observes the subsequent state $s_{t+1} \in$ S
- the agent receives a reward r_{t+1} for the action a_t taken in s_t from the environment based on the desirability of the state s_{t+1}. If the reward has a negative value, it can be referred to as a cost, or penalty.

An agent's task is to, based on the rewards received, estimate *value functions* for its state-action pairs. A value function describes how good it is, in terms of future expected return, for an agent to perform a given action in a given state. When estimating a value function, an agent does not only take into account immediate reward, but also takes into account the long-term desirability of the states, taking into account states that are likely to follow, and the rewards available in those states.

Based on experience (i.e., interaction with the environment as described above), at each time step an agent updates its *policy*, denoted Π. Π provides a mapping from each state, $s \in$ S, and action, $a \in$ A, to the probability $\Pi(s, a)$ of taking action a when in state s. The optimal policy is the one that maximizes the total long-term reward of an agent. RL algorithms can be *model-based* or *model-free*, depending on whether they explicitly learn the model of the environment. In RL terms, the environment model consists of a *state transition function* (i.e., the probability of an agent transitioning to state s_{t+1} after taking action a_t in state s_t) and a *reward function* (i.e., the expected numerical value of the next reward, received after taking action a_t in state s_t and transitioning to state s_{t+1}).

Q-learning [83] is a popular model-free RL algorithm in which an agent learns to associate actions with the expected long-term reward of taking that action in a particular state, represented by a so-called Q-value, and denoted $Q(s, a)$. Based on Q-values, an agent selects which action to execute in each given state.

3.2 Multi-policy RL

RL, as described in previous section, is a single-agent, single-policy learning technique. However, an agent might be required to simultaneously optimize towards a number of goals/policies. Therefore, multiple extensions to RL-based techniques have been developed to enable RL to solve single-agent multi-policy optimization problems. Some of the research described in this section refers to this type of problem as a multi-agent problem rather than a multi-policy problem, however, in such work multiple software agents are assumed to control a single set of actuators for which actions must be determined [37].

The techniques described in this section are not only motivated by the need for agents to optimize towards multiple goals, but also by the need to decompose larger complex goals into multiple simpler goals, turning a complex single-goal problem into a multi-goal problem [80].

We identify two major approaches to dealing with multiple policies in RL. Multiple policies can be addressed within a single learning process (by combining their state spaces into a single state space) or using an arbitrated approach where a dedicated learning process is created for each policy. In the latter an arbitrating agent or an action selection method decides, based on some criteria, which of the policies gets control over the actuators at each time step.

Combined State Spaces. Multiple policies on an agent can be combined into a single learning process, where the state space of the joint learning process is a cross product of the state spaces for individual policies. Each resulting state has an associated payoff which is a combination of payoffs for all policies. In order to evaluate each state an agent can have a vector of weights representing the importance of each of its policies [5]. For example, assume that, in a two policy case, a state A has an associated reward of 1 for one of the policies, and a reward of 5 for the other. If the agent's vector of weights assigns a weight of 0.6 to the first policy, and 0.4 to the other, the total payoff an agent will receive for being in state A is $0.6 \times 1 + 0.4 \times 5 = 2.6$.

Assigning the weights is not trivial as there might not be a clear hierarchy of goals, and the weights can change over time (reflecting the changing relative priorities of goals). [27] addresses a problem with fixed goal weights that do not change over time, while [35] and [53] address goals with changing relative weights. [35] considers calculating optimal actions for all combinations of policy weights, however this approach is prone to state explosion. Instead, they use an approximation approach with adaptive error margins, starting off with a large margin to learn in a smaller state set, and reducing the margin at the later stages to increase an accuracy of the results. In [53], a goal's weights, i.e., their relative importance, can change over time. Upon a change of a goal's weight, agents learn the optimal actions for the current weights, by reusing the best policy learnt so far and adjusting it to current weights rather than starting from the initial situation. The approaches above assume that rewards received from the environment for agents' goals are comparable and reflect their relative priority. [69] considers the case where rewards are received from multiple incomparable sources, and provides an algorithm for scaling the rewards so that policies of equal importance get equal weight.

The use of combined-state-space approaches is feasible when the number of objectives and the size of their state spaces are relatively small. However, for situations with a large number of objectives with large state spaces this approach can be computationally expensive. Policies can take a long time to converge on optimal results, and the addition of new policies leads to exponential growth of the state space [14]. As such, for larger and more complex problems, combining state spaces is not a scalable approach and is of limited use in multi-policy optimization. To address this problem, a combined-state-spaces approach can be utilized in combination with various algorithms to reduce the state-space size, or the number of state-action combinations. For example, one way to improve the performance and scalability of the combined-state-space approach is presented in [14], where an algorithm is proposed to reduce the state-action space (of 3.3 million state-action combinations in the authors' example) in order to achieve more feasible convergence times, memory requirements, and improved performance. However, this approach is limited to application domains where eliminating an invalid state-action pair will lead to a significant reduction in search-space size. Work presented in [55] also uses a state-reduction algorithm, in which, instead of eliminating invalid state-action combinations, an agent starts off by perceiving only a single state and learns to split it into multiple states, distinguishing only between the states that need to be distinguished.

Arbitration-based Approaches. In arbitration-based approaches, multiple policies suggest actions to an arbiter, which selects an action for execution based on some criteria. Arbitration-based approaches primarily differ in the way a winning action is decided upon, and in whether a winning action is nominated by one of the policies explicitly, or is a compromise action. Some of the early work in this area includes the subsumption architecture by Brooks [8,9]. Multiple goals in a subsumption architecture are met by separate layers arranged in a hierarchy, where layers higher in the hierarchy always win when they compete with lower-level ones. In [28], the strength of a number of so-called emotions (perceptions of the state of the world) currently "felt" is used to decide on behaviour switching (i.e., to decide which RL process will gain control over the agent). In [60], a gating function is used to assign varying weights to action suggestions by individual policies. Instead of a single learning process being assigned to each goal/sub-goal, [49] uses a family of independent agents for the implementation of each goal, but similarly, results are reported to a negotiation mechanism which selects the action to be executed. It attempts to find an action that satisfies all of the objectives, and failing that, selects a random action from the proposed set.

Some approaches enable a compromise action to be selected, rather than giving full control to only one policy. For example, in [64], policies calculate utilities for each of the outcome states; an arbitrator combines those utilities, and selects the action with the highest combined utility. Similarly, in [66], [41] and [70], policies report Q-values for each of the actions, and the arbitrator selects the action with the highest cumulative Q-value.

The W-learning [37] approach suggests making an action selection based on relative policy weights. In W-learning, weights, instead of being predefined or assigned by an arbitrator, are learnt specific to the particular policy state. Each policy of a W-learning agent is implemented as a separate Q-Learning process with its own state space. Using W-Learning, agents also learn, for each of the states of each of their policies, what happens, in terms of reward received, if the action nominated by that policy is not obeyed. This is expressed as a W-value. High W-values signify the high importance of the next action to a policy given its current state. At each learning step, multiple policies on a single agent nominate their actions, with associated W-values for the states in which they are, and the agent selects an action to execute based on those W-values. The winning policy can be selected in several ways, for example, agents can select the policy with the highest nominated individual W-value or the one with the highest cumulative W-value over all policies.

3.3   Multi-agent RL

In the previous section, we discussed RL-based systems for optimization towards multiple goals (policies) on a single agent. This section covers the optimization of behaviours of multiple agents, which are cooperating in order to meet some global system goal. We do not aim to provide a comprehensive overview of RL in multi-agent systems, but review a few representative examples that allow us to discuss issues related to research into multi-agent multi-policy RL. For a more comprehensive overview of multi-agent systems and multi-agent learning, please refer to [10] and [82].

In single-agent RL, an agent learns the actions with the highest payoffs for each of its states. In the multi-agent case, a global state is a combination of all local states, and a global action is a combination of all local actions, leading to exponential growth of the state-action space [34]. In order to determine jointly optimal actions without explicitly considering all possible combinations of local actions, a number of approaches have been proposed to break down the global problem into a group of local optimization problems. These approaches can be divided into two groups: those where each agent acts independently towards optimizing its local performance and ignores the presence of other agents, so called independent learners (IL), and approaches where agents cooperate with other agents, most commonly their immediate neighbours, in order to ensure that locally good actions are also globally good, so called joint action learners (JAL) [82]. Experiments confirm that the IL approach has practical use [12], however, it achieves poorer performance when compared to communicating agents [75], as communication is often required to ensure locally good actions are also good for the system globally [12]. Below we review some of the JAL-based approaches using different cooperation approaches to achieve self-organizing optimization behaviours.

Collaborative Reinforcement Learning. Collaborative Reinforcement Learning (CRL) [21] enables global system optimization based on cooperation between RL agents. Each agent has its own state space, action space, value function, and only a local view of the environment, while an estimate of the global view of the system is achieved by agents periodically exchanging their value functions. Global optimization of system behaviour arises from optimizing the process of solving smaller tasks introduced locally, represented as Discrete Optimization Problems

(DOPs). DOPs are defined as "selection of minimal cost alternatives from among a finite set of feasible solutions" [22] and are modelled on agents as absorbing MDPs, i.e., MDPs that will enter a terminal state after a finite amount of time. DOPs are introduced into the system on any of the agents, and an agent needs to minimize the cost of solving them by either making an action towards solving it locally, or delegating it to one of its neighbours to solve or delegate further. Delegation incurs the cost of transferring the DOP as well as the cost of estimating whether any of the neighbours can solve it at a lower cost or can find another agent that can do so. Agents periodically advertise their estimated costs of solving DOPs to their neighbours, as this information can change dynamically during system operation (if agents are, for example, solving other DOPs). This process is depicted in Figure 5. Each agent executes its own RL process, i.e., performs actions in the environment and receives rewards for those actions, which it uses to update its value function, i.e., learn the solution to a problem. Periodically, that solution is advertised to its neighbours.

CRL is suited for multi-agent optimization problems where agents delegate actions to each other in order to solve a goal. For example, CRL has been successfully applied in simulations of load balancing [21] and ad-hoc network routing [22], where agents can delegate jobs for processing or packets for delivery to their neighbours.

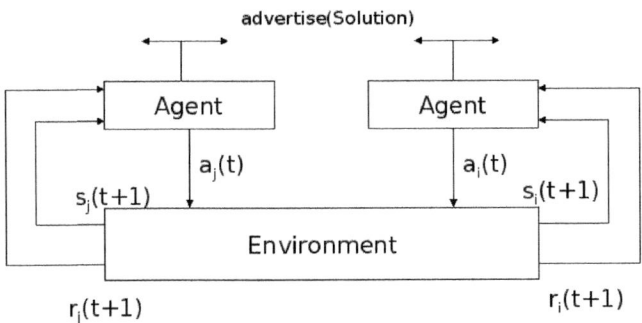

Fig. 5. Collaborative reinforcement learning [21]

Coordination Graphs. Coordination graphs are used to coordinate the actions of multiple agents cooperating to optimize a single large MDP. The action space of the MDP is a joint action space of the entire set of agents, and grows exponentially with an increase in the number of agents in the system. To reduce the number of combinations that need to be taken into account when computing a joint action, [33,34] propose factoring the problem space into smaller more manageable tasks using coordination graphs. They argue that any given agent does not need to coordinate its actions with all of the other agents, but only with a small number of the agents in its proximity. To facilitate this, a global value function is approximated as a linear combination of local value functions, where

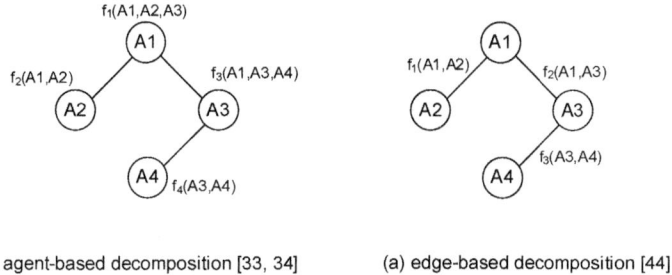

Fig. 6. Coordination graphs

each of the local value functions addresses only a small part of the system controlled by a small number of agents. Using these local value functions, an agent coordinates its action with its neighbours, which in turn coordinate their actions with their neighbours and so on, in order to optimize the behaviour towards a global function. A global MDP is represented as a coordination graph, where agents need to coordinate only with the neighbours to which they are connected on the graph. Local value functions can be modelled on a single agent and a set of its one-hop neighbours (so called agent-based decomposition [33,34]), or on two agents connected by an edge (edge-based decomposition [44]). Two approaches are depicted in Figure 6(a) and Figure 6(b), that show a graph consisting of four agents and three edges connecting them. In agent-based decomposition, the global function is broken down into four local functions, f_1, f_2, f_3, f_4, one per agent, while in edge-based decomposition, it consists of three functions, one per edge, f_1, f_2, and f_3. In agent-based decomposition, as presented in [33,34], it is assumed that dependencies between nodes (i.e., graph connections) are known in advance, while [44] in its edge-based approach provides a means for dependencies to be learnt and reinforced during problem solving.

Coordination graphs with agent-based decomposition have been applied in a simulation of load balancing [34], while coordination graphs with edge-based decomposition have been applied to in a distributed sensor network to coordinate actions of sensors tracking multiple targets [45].

Distributed Value Functions. Work presented in [68] proposes an approach to implementing distributed RL on multiple agents by using shared value functions. Each agent learns its local value function based on its individual actions and the rewards received, but it also incorporates the value functions of its neighbours into the value function updates. In order to do this, each agent needs to have a weight function specifying how much the value functions of other agents contribute to its value function, with the weight of non-neighbours' value functions being zero. In such a way, each agent learns a value function that is an estimate of a weighted sum of its own expected rewards and those of its neighbours. Neighbours' value functions are, in turn, calculated as a weighted sum of their own expected reward and that of their neighbours' expected rewards and

so on. Agents, therefore, can select actions that are good not just locally but
also for the other agents in the system. The distributed value function approach
has been successfully applied in a simulation of power grid management, where
it was used to coordinate the performance of a number of interconnected power
distributors, providers and customers, in order to maximize the rewards received
by providing the desired level of service to their customers.

POMDP-based Approaches. In the work covered in previous sections, agents'
environments are represented as MDPs, however in the case where agents might
only have partial, noisy, or probabilistic observations of the their state space and
limited communication with their neighbours, the environment can be modelled
as partially observable MDP (POMDP) [73]. Solving POMDP-based problems
requires a different set of learning techniques than those used to solve MDPs.
While MDPs are in general solved using value-search RL techniques (i.e., methods based on learning the value function, such as Q-learning), policy-search
techniques, which learn optimal policies without learning a value function [3],
are considered more suitable for solving POMDPs [58]. Overview of these approaches is provided in [39], [58], [31] and [32]. POMDP-based RL approaches
have been applied in, for example, resource allocation for multi-agent wireless
multihop networks [85].

3.4   Learning to Cooperate in Multi-agent RL

In the previous section we have presented a number of cooperative multi-agent
systems, however, in most of those systems agents cooperate with a fixed set of
neighbours (e.g., one-hop neighbours on a graph in 3.3), where those neighbours'
inputs have a fixed weight (e.g., determined by a weight function as discussed
in Section 3.3). A much smaller body of work is concerned with agents in multi-
agent systems learning by themselves with which other agents to cooperate, and
in which situations.

[44] extends the coordination graphs approach discussed in Section 3.3 to
learn the dependencies between the agents, i.e., the strengths of the graph connections, rather than using predefined ones. The dependency graph (i.e., coordination graph) starts off with either no edges, or random edges, and based on
the outcome of the cooperation (joint action decisions made) those edges either
become weaker and disappear or grow stronger.

In [50], agents do not learn with whom to cooperate, but when (i.e., in which
states) to cooperate. This work considers only a two-agent case, but nevertheless proposes an interesting approach to learning when to cooperate with other
agents. For each agent's state, a fictitious action "coordinate" is added. If this
action is selected, an agent senses an other agent's state and bases its local action decision on the combination of the other agent's state and its own state,
rather than only on its own local state. Q-learning proceeds as usual and learns
Q-values for all of the state-action pairs, where one of the actions executed can
be "coordinate" with the other agent. In this way, an agent learns in which states

is it useful to coordinate with the other agent. For example, [50] tests this approach using two robots that need to coordinate their behaviour whilst passing through a narrow doorway. Agents learn that the coordination action is mostly appropriate when they are positioned in the vicinity of a doorway, as that is when they need to make sure not to collide with the other agent.

3.5 RL in Self-organizing Systems

RL is increasingly being used for the implementation of self-adaptive behaviours. MDP-based RL relies on an underlying sequential decision theory that includes the possibility of a current decision having delayed consequences in the future, accounting properly for the dynamic behaviour of large-scale decentralized systems [78]. Additionally, the ability of RL-based methods to learn optimal system behaviours without requiring a model of the system or the environment is removing the need to develop accurate system models, which is often a complex and time consuming task [77]. In this section we describe several applications of RL in self-organizing systems, specifically in resource allocation, load balancing, ad-hoc network routing, and autonomic network repair.

Online Resource Allocation. RL algorithms have been applied for online resource allocation in a prototype distributed data centre in [79]. The system consists of a number of application environments where applications are deployed, and a resource arbiter, whose task is to dynamically assign resources to applications. The goal of the resource arbiter is to maximize the sum of resource utilities, i.e., maximize the long-term expected value of the allocation of a number of servers to given application(s). Each application has its own utility function that expresses the value that a particular application brings to the data center by delivering services to its customers at a particular service level, i.e., the value of being assigned a given number of servers. This work discusses a number of issues common to applications of RL in complex distributed systems, e.g., accurate state-space representation (which is potentially required to incorporate a large number of variables), the duration of training time RL requires in live systems, and a lack of convergence guarantees in distributed RL. However, empirical results in a distributed prototype data centre show feasible training time, and the quality of the solution is comparable to solutions obtained using complex queue-theoretic system performance models that require detailed understanding of system design and user behaviour.

Load Balancing. CRL, a multi-agent collaborative RL technique discussed in 3.3, has been applied in a simulation of a decentralized file storage system to provide load balancing [23]. The system consists of ~50 agents and several server agents, whose storage capacity is ten-fold that of other agents. The goal of the system is to store all inserted loads in as short time as possible. Loads are entered into the system through the agent at position 0. Each agent has actions available to it that allows it to store the load itself, or forward it to one of the 10 neighbours to which each agent has a connection. Unsuccessful store actions result in an agent receiving a high negative reward, and successful ones a reward that is a function of storage space available on an agent. The system was able to successfully store all loads 15 times faster than a random policy, was able to self-adapt to the addition of a new server in the system, and self-heal when connections between agents were broken.

Routing in Ad-Hoc Networks. CRL (as described in 3.3) has also been used to implement self-organizing properties in ad-hoc network routing [22]. The network consists of a number of fixed and mobile agents whose goal is to optimize system routing performance. Each agent can deliver a packet to its destination (and receive a reward), deliver it to an existing neighbour (with an associated cost), or perform a discovery action to find a new neighbour (at an associated cost). The goal of each agent is to minimize the cost, i.e., either deliver a packet or forward it to the lowest cost neighbour. In order to do this, each agent learns and maintains a statistical model of its network links to estimate the cost of a given route, and exchanges the information on route costs with its neighbours. Using this approach agents learn to favour stable routes (consisting of fixed nodes) and re-route the traffic around congested areas of the network.

Autonomic Network Repair. In [47], RL has been used to learn how to efficiently restore network connectivity after a failure. A single agent, called a decision maker, is charged with repairing the network. An agent can perform a number of test actions, used to narrow down the source of a fault, and a number of repair actions, used to repair the fault. Each action is associated with a cost (time required for its execution), and the decision maker's goal is to minimize the cost of restoring the system to proper functioning. It is assumed that the decision maker does not have a complete set of information about faults, and therefore the learning problem is modelled as a POMDP. This approach was implemented in a live network, with a separate program injecting faults, and a decision maker successfully learnt to attempt cheaper repair actions first.

Grid Scheduling. [57] uses a combination of RL and artificial neural network for self-organizing job scheduling on a resource grid. Users submit processing jobs to the grid, and a grid scheduler is charged with selecting jobs for execution. The goal of the scheduler is to maximize user satisfaction (which decreases as a function of the time that it takes to complete the job) and fairness (which is expressed as a difference between actual resources allocated and externally defined resource share that should be given to that user). A number of common issues with RL needed to be addressed in this work. For example, the algorithm was initially trained offline, to overcome bad performance of RL during the training phase, and its convergence was only checked empirically, due to lack of theoretical guarantees. In the authors' simulation, after poor initial performance due to training has been overcome, this approach consistently outperforms the job scheduler currently used in the live system.

Urban Traffic Control. RL has been extensively used in decentralized urban traffic control where each junction uses an RL process to learn the most suitable traffic-light settings for its local traffic conditions and, in collaborative cases, traffic-light settings that are also suitable for the traffic conditions on its neighbouring junctions.

[63] uses RL to optimize average vehicle travel time. Each intersection's environment is modelled as an MDP and each intersection implements an RL process based on that MDP that receives a reward based on the number of vehicles it allows to proceed. Traffic light agents communicate with their immediate neighbours, in order to use neighbours' traffic counts to anticipate their own traffic flows.

[67] uses a technique based on CRL (see Section 3.3) to implement a traffic-control technique called Adaptive Round Robin. On each agent, the proposed algorithm cycles through phases available to a junction, and decides to either skip the phase or sets it with one of a set of predefined durations, based on congestion levels on the approaches served by that phase. An agent is rewarded proportionally to the number of cars that pass through an intersection during the phase, and inversely proportional to the number of cars still waiting at an intersection. Agents periodically exchange information on their performance (their accumulated rewards) with their immediate neighbours and incorporate the information received from neighbours into their own reward. Agents, therefore, receive rewards for the good performance of their neighbours, ensuring cooperation between them.

RL has also been applied in urban traffic control in combination with other self-organizing techniques. For example, in [51] and [86], RL is deployed to learn the settings on individual traffic light agents, while evolutionary algorithms are used to find optimal combinations of parameters for the local RL processes.

3.6 Towards Multi-policy Multi-agent RL

RL-based techniques have been shown to be suitable for use in self-organizing systems due to their ability to learn desired behaviours without requiring extensive domain knowledge and the ability to learn from both the immediate and long-term consequences of executed actions. However, for their wider application in self-organizing systems, we believe RL techniques need to be capable of simultaneous implementation on a number of cooperative agents comprising a multi-agent system, as well as be able to simultaneously address the multiple, potentially conflicting, policies with different characteristics that self-organizing systems might be required to implement.

Existing RL techniques mostly either address multiple policies on a single agent, or are multi-agent techniques but optimize only towards a single explicit system policy. In this section we review a number of multi-policy single-agent approaches as well as a number of single-policy multi-agent approaches that can be used as a starting point for development of a multi-policy multi-agent technique. However, there are a number of issues that arise when developing such a technique.

For example, in order to apply one of the multi-policy techniques in a multi-agent environment, a number of cooperation issues remain to be addressed. For example, policies could be addressed simultaneously on a single agent using W-learning or combined state spaces. However, policies implemented on other agents would also need to be taken into account, as well as how those other agents' policies affect the local agent, or how the local agent's actions affect other agents and their policies. Collaboration between agents might be required in order to address the potential dependencies between the agents in the system and policies that they implement.

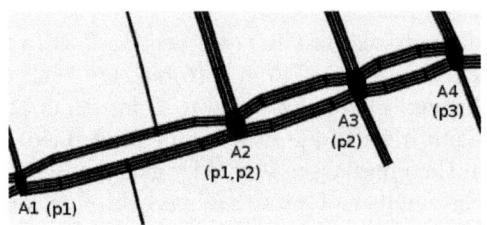

Fig. 7. Agent and policy heterogeneity

To illustrate this point consider Figure 7, representing a part of an urban traffic control system consisting of multiple agents, each controlling a set of traffic lights at a single junction. Due to the shared operating environment, i.e., the road network, the performance of a single agent (junction) can be affected by the performance of other agents. If the junction controlled by agent A2 is oversaturated, traffic can come to a standstill and queues at that junction can spill over upstream to block the junction controlled by A3. No traffic will then be able to go through this junction regardless of the actions of A3, as there will be no space available on the downstream road. Likewise, the performance of A1, A3 and any agents at other junctions that feed traffic to A2 can cause oversaturation at A2 if they are letting through more traffic than A2 is clearing. The dependency can also extend to the agents further upstream from A2, such as A4, as that agent influences the performance of A3, which in turn influences the performance of A2, causing a potential dependency between non-neighbouring agents A2 and A4. Therefore, not only a means of addressing multiple policies on a single agent is required, but also collaboration between these agents needs to be enabled in order to manage the potential dependencies between them.

Similarly, if one of the reviewed multi-agent optimization techniques were to be applied in multi-policy environments, (e.g., CRL or coordination graphs), the means of combining multiple policies on those agents into a single optimization problem would be needed.

If existing multi-policy and multi-agent techniques were combined, i.e., if one of the multi-agent algorithms was implemented globally and one of the multi-policy algorithms locally, heterogeneity of policies and heterogeneity of agents that implement them would open a number of additional issues not present in either multi-agent or multi-policy cases individually. For example, consider again Figure 7. Agent A2 is required to implement regional policies p1 (e.g., prioritize buses) and p2 (e.g., prioritize trams), while one of its neighbours, agent A1, is only implementing p1, and the other neighbouring, agent A3, only p2. The further downstream neighbour A4, implements only its own local policy p3 (e.g., honour pedestrian crossing requests). Such heterogeneity of policies results in heterogeneity of agents, where neighbouring agents might be implementing different policies and policies of different priorities at a given time. Another source of heterogeneity of agents results from the heterogeneity of agents, for example, in Figure 7, from different layouts of junctions controlled by the agents. Heterogeneity of policies and environment layout results in RL agents having different state spaces and action sets, and therefore not having a common understanding of state and action meaning. In such a way heterogeneity is limiting the means by which agents can collaborate and exchange information that is meaningful and useful to receiving agents.

When designing a collaboration mechanism, it also needs to be investigated if an agent should collaborate only with other agents that implement the same policies or with all agents in the system. Should, and how, agents be aware of each other's policies, and how will the priority of multiple policies be maintained if agents implementing lower priority policies are not aware of a higher priority policy being deployed on other agents. Additionally, a way to motivate an agent to collaborate is needed, as, in terms of its local benefit, it might be better for an agent to act selfishly. It is particularly important for an agent to be willing to engage in collaborative behaviour when other agents' policies have a higher priority for the system than its own.

Agents in a MAS also need to know what kind of collaboration and to what degree is useful in a given set of environmental conditions. Agents might need to know if and how their actions influence other agents and their policies (i.e., agents might need to learn the dependencies between them and other agents) and based on that ensure not to negatively affect the performance of other agents and, if possible, to contribute towards other agents' good performance. Agents might need to learn the situations in which collaboration is useful, i.e., learn whether to always collaborate fully, to collaborate only in certain states, or to collaborate only when some conditions have been met or thresholds have been exceeded.

Therefore, an algorithm that simultaneously addresses learning and optimization towards multiple-policies in multi-agent systems needs to draw on techniques from multi-policy optimization (e.g., it needs to balance the action preferences of two policies), techniques from multi-agent optimization (e.g., it needs to ensure that actions executed locally by an agent do not negatively affect another agent, or the system as a whole), as well as address additional issues that arise in multi-policy multi-agent environments as discussed above.

Distributed W-Learning. Some of the issues discussed above have been addressed in Distributed W-Learning (DWL) [24]. DWL is a self-organizing multi-agent multi-policy optimization technique based on Q-learning and W-learning. Each DWL agent learns the suitability of its actions for all of its local policies as well as the relative importance of the suggested actions to each policy

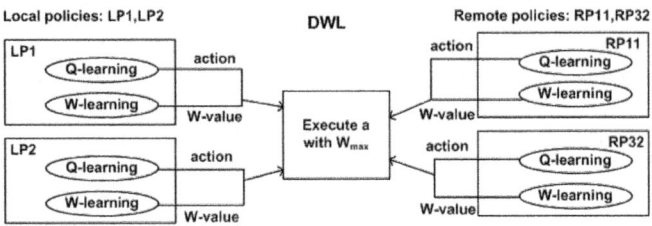

Fig. 8. Distributed W-Learning action selection [24]

based on its current state (i.e., a W-value). As a means of cooperation with its neighbours, each DWL agent also learns the suitability of its actions for all of the policies that its one-hop neighbours are implementing (so-called remote policies) and W-values associated with its neighbours' policies' states. Action selection at each time step is illustrated in Figure 8. Each local and remote policy suggests the action that is the most suitable for its current state, together with the W-value associated with that state. W-values nominated by remote policies can be considered using their full value, or scaled using a cooperation coefficient C (where $0 \leq C \leq 1$), to enable agents to give higher weight to their local preferences. An action with the highest W-value, i.e., an action with highest current importance, is executed. Through use of remote policies, DWL has the ability to address multiple heterogeneous policies on multiple heterogeneous agents and enable collaboration between them, as remote policies do not require the state spaces and action sets of collaborating agents to be the same. DWL also respects relative policy priorities, through use of W-values for action selection. W-values are calculated based on learnt Q-values, which are represented as a function of rewards received by a policy, where higher priority policies receive higher maximum rewards. DWL also enables agents to engage in different degrees of collaboration, through use of the cooperation coefficient, which might be required if there are different degrees of dependencies present between agents. DWL has been successfully applied in a simulation of an urban traffic control system which was required to optimize global traffic flow while prioritizing public transport vehicles. In terms of average vehicle waiting time, DWL outperformed baselines by up to 87%, and non-collaborative multi-policy approaches by up to 90% [24].

Even though DWL addresses some of the issues discussed above, we believe other approaches to RL-based multi-policy multi-agent optimization should be explored. Other collaboration means can be utilized and adapted from single-policy multi-agent optimization, as well as the means of learning how, how much, and with whom to collaborate in a multi-policy multi-agent system.

Open RL Issues. Based on the review of the multi-agent applications of RL in self-organizing systems we also identified a number of issues inherent in RL that any multi-policy multi-agent technique might need to address. RL in complex environments might require long training periods, resulting in poor and unpredictable system performance during this time. To address this issue, use of batch learning algorithms for offline training of multi-agent RL systems could be investigated. One of the main problems in large-scale multi-agent multi-policy RL is a lack of theoretical guarantees of convergence to optimal behaviour. Single-agent single-policy RL has been proved to converge to an optimal policy, however, those guarantees do not hold in multi-agent multi-policy case where multiple agents and multiple policies are affecting each other's environments. Non-stationarity of the agents' environments results in optimal behaviours changing over time, depending on the environment conditions and interactions between agents and between policies. Despite the lack of theoretical convergence guarantees, the RL-based systems we have reviewed in this paper have been empirically shown to be suitable for the applications in the domains discussed. However, some application areas might require stronger behavioural guarantees, for example, providing at least performance boundaries for system behaviour.

4   Summary and Conclusions

In this paper we have reviewed self-organizing approaches to multi-policy optimization in decentralized systems. The most-widely used techniques we have presented are ant colony optimization, evolutionary algorithms, artificial neural networks, particle swarm optimization, and RL. Most of these techniques have a wide range of both single-policy and multi-policy implementations, and therefore we have provided a brief introduction to their general approach to optimization and referred the reader to more comprehensive reviews focusing only on a multi-policy technique in question. Surprisingly, the number of RL-based multi-policy optimization techniques that we identified was very limited, so in this paper we aimed to provide the background relevant for further research into such a technique. We have reviewed techniques that address single policy on multiple agents, as well as multiple policies on a single agents, and identified numerous issues that would need to be addressed if these were to be extended or combined into multi-policy multi-agent RL-based optimization techniques.

Acknowledgments

This work was supported by Science Foundation Ireland grant 03/CE2/I303_1 to Lero - the Irish Software Engineering Research Centre (www.lero.ie).

References

1. Angus, D., Woodward, C.: Multiple objective ant colony optimisation. Swarm Intelligence 3(1), 69–85 (2009)
2. Babaoglu, O., Meling, H., Montresor, A.: Anthill: A framework for the development of agent-based peer-to-peer systems. In: International Conference on Distributed Computing Systems (2002)

3. Baird, L., Moore, A.: Gradient descent for general reinforcement learning. In: Advances in Neural Information Processing Systems, vol. 11, pp. 968–974. MIT Press, Cambridge (1999)

4. Baran, B., Schaerer, M.: A multiobjective ant colony system for vehicle routing problem with time windows. In: Proceedings of IASTED International Conference on Applied Informatics (2003)

5. Barrett, L., Narayanan, S.: Learning all optimal policies with multiple criteria. In: ICML 2008: Proceedings of the 25th International Conference on Machine Learning, pp. 41–47 (2008)

6. Bigus, J.P., Schlosnagle, D.A., Pilgrim, J.R., Nathaniel Mills III, W., Diao, Y.: Able: A toolkit for building multiagent autonomic systems. IBM Systems Journal 41(3), 350–371 (2002)

7. Blum, C., Merkle, D. (eds.): Swarm Intelligence: Introduction and Applications. Natural Computing Series. Springer, Heidelberg (2008)

8. Brooks, R.: Achieving artificial intelligence through building robots. Technical report, Massachusetts Institute of Technology, Cambridge, MA, USA (1986)

9. Brooks, R.A.: How to build complete creatures rather than isolated cognitive simulators. In: Architectures for Intelligence, pp. 225–239. Erlbaum, Mahwah (1991)

10. Busoniu, L., Schutter, B.D., Babuska, R.: Learning and coordination in dynamic multiagent systems. Technical Report 05-019, Delft Center for Systems and Control, Delft University of Technology, Delft, The Netherlands (October 2005)

11. Cantu Paz, E., Kamath, C.: An empirical comparison of combinations of evolutionary algorithms and neural networks for classification problems 35(5), 915–927 (October 2005)

12. Claus, C., Boutilier, C.: The dynamics of reinforcement learning in cooperative multiagent systems. In: Proceedings of the Fifteenth National Conference on Artificial Intelligence, pp. 746–752. AAAI Press, Menlo Park (1998)

13. Coello, C.A.C.: A comprehensive survey of evolutionary-based multiobjective optimization techniques. Knowledge and Information Systems 1, 269–308 (1999)

14. Cuayahuitl, H., Renals, S., Lemon, O., Shimodaira, H.: Learning multi-goal dialogue strategies using reinforcement learning with reduced state-action spaces. International Journal of Game Theory, 547–565 (2006)

15. Cui, X., Potok, T., Palathingal, P.: Document clustering using particle swarm optimization. In: Swarm Intelligence Symposium (2005)

16. Di Caro, G., Dorigo, M.: AntNet: Distributed Stigmergetic Control for Communication Networks. Journal of Artificial Intelligence Research 9, 317–365 (1998)

17. Di Caro, G., Ducatelle, F., Gambardella, L.M.: AntHocNet: An adaptive nature-inspired algorithm for routing in mobile ad hoc networks. European Transactions on Telecommunications, Special Issue on Self-organization in Mobile Networking 16, 443–455 (2005)

18. Di Marzo Serugendo, G., Gleizes, M.-P., Karageorgos, A.: Self-organization in multi-agent systems. Knowl. Eng. Rev. 20(2), 165–189 (2005)

19. Doerner, K., Hartl, R., Reimann, M.: Are COMPETants more competent for problem solving? - the case of full truckload transportation. Central European Journal of Operations Research 11(2), 115–141 (2003)

20. Dorigo, M., Di Caro, G.D.: The Ant Colony Optimization Meta-Heuristic, pp. 11–32. McGraw-Hill, London (1999)

21. Dowling, J.: The Decentralised Coordination of Self-Adaptive Components for Autonomic Distributed Systems. PhD thesis, Trinity College Dublin (2005)

22. Dowling, J., Cunningham, R., Curran, E., Cahill, V.: Building autonomic systems using collaborative reinforcement learning. Knowledge Engineering Review 21(3), 231–238 (2006)

23. Dowling, J., Haridi, S.: Decentralized Reinforcement Learning for the Online Optimization of Distributed Systems. In: Reinforcement Learning. I-Tech Education and Publishing (2008)

24. Dusparic, I., Cahill, V.: Distributed W-Learning: Multi-policy optimization in self-organizing systems. In: Third IEEE International Conference on Self-Adaptive and Self-Organizing Systems (2009)

25. Eiben, A., Smith, J.: Introduction to Evolutionary Computing. Natural Computing Series. Springer, Heidelberg (2003)

26. Eiben, A.E.: Evolutionary computing and autonomic computing: Shared problems, shared solutions? In: Babaoğlu, Ö., Jelasity, M., Montresor, A., Fetzer, C., Leonardi, S., van Moorsel, A., van Steen, M. (eds.) SELF-STAR 2004. LNCS, vol. 3460, pp. 36–48. Springer, Heidelberg (2005)

27. Gábor, Z., Kalmár, Z., Szepesvári, C.: Multi-criteria reinforcement learning. In: ICML 1998: Proceedings of the Fifteenth International Conference on Machine Learning, pp. 197–205. Morgan Kaufmann Publishers Inc., San Francisco (1998)

28. Gadanho, S.C., Hallam, J.: Robot learning driven by emotions. Adaptive Behaviour 9(1), 42–64 (2001)

29. Gambardella, L.M., Taillard, E., Agazzi, G.: MACS-VRPTW: a multiple ant colony system for vehicle routing problems with time windows, pp. 63–76 (1999)

30. Garcia-Martinez, C., Cordon, O., Herrera, F.: A taxonomy and an empirical analysis of multiple objective ant colony optimization algorithms for the bi-criteria tsp. European Journal of Operational Research 180(1), 116–148 (2007)

31. Goldman, C.V., Zilberstein, S.: Optimizing information exchange in cooperative multi-agent systems. In: AAMAS 2003: Proceedings of the Second International Joint Conference on Autonomous Agents and Multiagent Systems, pp. 137–144. ACM, New York (2003)

32. Goldman, C.V., Zilberstein, S.: Decentralized control of cooperative systems: Categorization and complexity analysis. Journal of Artificial Intelligence Research (JAIR) 22, 143–174 (2004)

33. Guestrin, C., Koller, D., Parr, R.: Multiagent planning with factored MDPs. In: 14th Neural Information Processing Systems (NIPS-14), Vancouver, Canada, pp. 1523–1530 (December 2001)

34. Guestrin, C., Lagoudakis, M., Parr, R.: Coordinated reinforcement learning. In: Proceedings of the ICML 2002 The Nineteenth International Conference on Machine Learning, pp. 227–234 (2002)

35. Hiraoka, K., Yoshida, M., Mishima, T.: Parallel reinforcement learning for weighted multi-criteria model with adaptive margin. In: Ishikawa, M., Doya, K., Miyamoto, H., Yamakawa, T. (eds.) ICONIP 2007, Part I. LNCS, vol. 4984, pp. 487–496. Springer, Heidelberg (2008)

36. Hoar, R., Penner, J., Jacob, C.: Evolutionary swarm traffic: if ant roads had traffic lights. In: CEC 2002: Proceedings of the Evolutionary Computation on 2002, CEC 2002. Proceedings of the 2002 Congress, Washington, DC, USA, pp. 1910–1915. IEEE Computer Society, Los Alamitos (2002)

37. Humphrys, M.: Action Selection methods using Reinforcement Learning. PhD thesis, University of Cambridge (1996)

38. Kadrovach, B.A., Lamont, G.B.: A particle swarm model for swarm-based networked sensor systems. In: SAC 2002: Proceedings of the 2002 ACM Symposium on Applied Computing, pp. 918–924. ACM, New York (2002)

39. Kaelbling, L.P., Littman, M.L., Cassandra, A.R.: Planning and acting in partially observable stochastic domains. Artificial Intelligence 101, 99–134 (1995)
40. Kalyanmoy, D.: Multi-Objective Optimization Using Evolutionary Algorithms. Wiley, Chichester (2001)
41. Karlsson, J.: Learning to solve multiple goals. PhD thesis, Rochester, NY, USA (1997)
42. Kennedy, J., Russell, E.C.: Swarm Intelligence, The Morgan Kaufmann Series in Artificial Intelligence. Morgan Kaufmann, San Francisco (March 2001)
43. Kephart, J.O., Walsh, W.E.: An artificial intelligence perspective on autonomic computing policies. In: IEEE International Workshop on Policies for Distributed Systems and Networks (2004)
44. Kok, J.R., 't Hoen, P.J., Bakker, B., Vlassis, N.: Utile coordination: learning interdependencies among cooperative agents. In: Proceedings of the IEEE Symposium on Computational Intelligence and Games (CIG), Colchester, United Kingdom, pp. 29–36 (April 2005)
45. Kok, J.R., Vlassis, N.: Collaborative multiagent reinforcement learning by payoff propagation. Journal of Machine Learning Research 7, 1789–1828 (2006)
46. Lekavy, M.: Optimising Multi-agent Cooperation using Evolutionary Algorithm. In: Bielikova, M. (ed.) Proceedings of IIT.SRC 2005: Student Research Conference in Informatics and Information Technologies, Bratislava, pp. 49–56. Faculty of Informatics and Information Technologies, Slovak University of Technology in Bratislava (April 2005)
47. Littman, M.L., Ravi, N., Fenson, E., Howard, R.: Reinforcement learning for autonomic network repair. In: ICAC 2004: Proceedings of the First International Conference on Autonomic Computing, Washington, DC, USA, pp. 284–285. IEEE Computer Society, Los Alamitos (2004)
48. Maniezzo, V., Gambardella, L.M., Luigi, F.D.: Ant Colony Optimization. In: New Optimization Techniques in Engineering. Springer, Heidelberg (2004)
49. Mariano, C., Morales, E.F.: A new distributed reinforcement learning algorithm for multiple objective optimization problems. In: Monard, M.C., Sichman, J.S. (eds.) SBIA 2000 and IBERAMIA 2000. LNCS (LNAI), vol. 1952, pp. 290–299. Springer, Heidelberg (2000)
50. Melo, F., Veloso, M.: Learning of coordination: Exploiting sparse interactions in multiagent systems. In: Proceedings of the 8th International Conference on Autonomous Agents and Multi-Agent Systems (2009)
51. Mikami, S., Kakazu, Y.: Genetic reinforcement learning for cooperative traffic signal control. In: International Conference on Evolutionary Computation, pp. 223–228 (1994)
52. Montresor, A., Meling, H., Babaoglu, O.: Messor: Load-balancing through a swarm of autonomous agents. Technical Report UBLCS-02-08, Departement of Computer Science, University of Bologna, Bologna, Italy (May 2002)
53. Natarajan, S., Tadepalli, P.: Dynamic preferences in multi-criteria reinforcement learning. In: ICML 2005: Proceedings of the 22nd International Conference on Machine Learning, pp. 601–608. ACM, New York (2005)
54. Oxford. The Oxford English Dictionary. Oxford University Press (2000)

55. Paquet, S., Bernier, N., Chaib-draa, B.: Multi-attribute decision making in a complex multiagent environment using reinforcement learning with selective perception. In: Tawfik, A.Y., Goodwin, S.D. (eds.) Canadian AI 2004. LNCS (LNAI), vol. 3060, pp. 416–421. Springer, Heidelberg (2004)
56. Parsopoulos, K.E., Vrahatis, M.N.: Particle swarm optimization method in multi-objective problems. In: SAC 2002: Proceedings of the 2002 ACM Symposium on Applied Computing, pp. 603–607. ACM, New York (2002)
57. Perez, J., Germain-Renaud, C., Kegl, B., Loomis, C.: Grid differentiated services: A reinforcement learning approach. In: CCGRID 2008: Proceedings of the 2008 Eighth IEEE International Symposium on Cluster Computing and the Grid, Washington, DC, USA, pp. 287–294. IEEE Computer Society, Los Alamitos (2008)
58. Peshkin, L., Eung Kim, K., Meuleau, N., Kaelbling, L.P.: Learning to cooperate via policy search. In: Proceedings of the 16th Annual Conference on Uncertainty in Artificial Intelligence (UAI 2000), pp. 489–496. Morgan Kaufmann, San Francisco (2000)
59. Pugh, J., Zhang, Y., Martinoli, A.: Particle swarm optimization for unsupervised robotic learning. In: Swarm Intelligence Symposium, pp. 92–99 (2005)
60. Raicevic, P.: Parallel reinforcement learning using multiple reward signals. Neurocomputing 69(16-18), 2171–2179 (2006)
61. Ramdane-Cherif, A.: Toward autonomic computing: Adaptive neural network for trajectory planning. International Journal of Cognitive Informatics and Natural Intelligence 1(2), 16–33 (2007)
62. Reyes-Sierra, M., Coello, C.A.C.: Multi-objective particle swarm optimizers: A survey of the state-of-the-art. International Journal of Computational Intelligence Research 2(3), 287–308 (2006)
63. Richter, S.: Learning traffic control - towards practical traffic control using policy gradients. Technical report, Albert-Ludwigs-Universitat Freiburg (2006)
64. Rosenblatt, J.K.: Optimal selection of uncertain actions by maximizing expected utility. Autonomous Robots 9(1), 17–25 (2000)
65. Russell, S., Norvig, P.: Aritifical Intelligence - A Modern Approach. Prentice Hall, Englewood Cliffs (2003)
66. Russell, S.J., Zimdars, A.: Q-decomposition for reinforcement learning agents. In: Fawcett, T., Mishra, N. (eds.) International Conference on Machine Learning, pp. 656–663. AAAI Press, Menlo Park (2003)
67. Salkham, A., Cunningham, R., Garg, A., Cahill, V.: A collaborative reinforcement learning approach to urban traffic control optimization. In: IEEE/WIC/ACM International Conference on Web Intelligence and Intelligent Agent Technology (WI-IAT), vol. 2, pp. 560–566 (2008)
68. Schneider, J., Wong, W.-K., Moore, A., Riedmiller, M.: Distributed value functions. In: Proceedings of the Sixteenth International Conference on Machine Learning, pp. 371–378. Morgan Kaufmann, San Francisco (1999)
69. Shelton, C.R.: Balancing multiple sources of reward in reinforcement learning. In: Neural Information Processing Systems, pp. 1082–1088 (2000)
70. Sprague, N., Ballard, D.: Multiple-goal reinforcement learning with modular Sarsa(0). In: International Joint Conference on Artificial Intelligence (2003)
71. Srinivasan, D., Choy, M.C., Cheu, R.L.: Neural networks for real-time traffic signal control. IEEE Transactions on Intelligent Transportation Systems 7(3), 261–272 (2006)
72. Subramanian, D., Druschel, P., Chen, J.: Ants and reinforcement learning: A case study in routing in dynamic networks. In: IJCAI (2), pp. 832–838. Morgan Kaufmann, San Francisco (1998)

73. Suton, R.S., Barto, A.G.: Reinforcement Learning: An Introduction. A Bradford Book/The MIT Press, Cambridge (1998)
74. Tan, K.C., Lee, E.F.K., Heng, T.: Multiobjective Evolutionary Algorithms and Applications, Advanced Information and Knowledge Processing. Springer, New York (2005)
75. Tan, M.: Multi-agent reinforcement learning: Independent vs. cooperative agents. In: Proceedings of the Tenth International Conference on Machine Learning, pp. 330–337. Morgan Kaufmann, San Francisco (1993)
76. Tesauro, G.: Pricing in agent economies using neural networks and multi-agent Q-learning. In: Proceedings of Workshop ABS-3: Learning About, From and With other Agents (1999)
77. Tesauro, G.: Reinforcement learning in autonomic computing: A manifesto and case studies. IEEE Internet Computing 11(1), 22–30 (2007)
78. Tesauro, G., Chess, D.M., Walsh, W.E., Das, R., Segal, A., Whalley, I., Kephart, J.O., White, S.R.: A multi-agent systems approach to autonomic computing. In: International Joint Conference on Autonomous Agents and Multiagent Systems, pp. 464–471 (2004)
79. Tesauro, G., Das, R., Walsh, W.E., Kephart, J.O.: Utility-function-driven resource allocation in autonomic systems. In: International Conference on Autonomic Computing, pp. 342–343 (2005)
80. Tham, C.K., Prager, R.W.: A modular Q-learning architecture for manipulator task decomposition. In: Proceedings of the Eleventh International Conference on Machine Learning. Morgan Kaufmann, San Francisco (1994)
81. Van Veldhuizen, D.A., Lamont, G.B.: Multiobjective evolutionary algorithms: Analyzing the state-of-the-art. Evolutionary Computation 8(2), 125–147 (2000)
82. Vlassis, N.: A Concise Introduction to Multiagent Systems and Distributed Artificial Intelligence. Morgan and Claypool Publishers (2007)
83. Watkins, C.J.C.H., Dayan, P.: Technical note: Q-learning. Machine Learning 8(3), 279–292 (1992)
84. Weijters, A.J.M.M., Hoppenbrouwers, G.A.J.: Backpropagation networks for grapheme-phoneme conversion: a non-technical introduction. In: Artificial Neural Networks: An Introduction to ANN Theory and Practice, London, UK, pp. 11–36. Springer, Heidelberg (1995)
85. Yagan, D., Tham, C.-K.: Coordinated reinforcement learning for decentralized optimal control. In: IEEE International Symposium on Approximate Dynamic Programming and Reinforcement Learning (2007)
86. Yang, Z., Chen, X., Tang, Y., Sun, J.: Intelligent cooperation control of urban traffic networks. In: Proceedings of 2005 International Conference on Machine Learning and Cybernetics, pp. 1482–1486 (2005)

A Bio-inspired Algorithm for Energy Optimization in a Self-organizing Data Center

Donato Barbagallo, Elisabetta Di Nitto,
Daniel J. Dubois, and Raffaela Mirandola

Politecnico di Milano, Dipartimento di Elettronica e Informazione
Piazza Leonardo da Vinci 32, 20133, Milano, Italy
{barbagallo,dinitto,dubois,mirandola}@elet.polimi.it

Abstract. The dimension of modern distributed systems is growing everyday. This phenomenon has generated several management problems due to the increase in complexity and in the needs of energy. Self-organizing architectures showed to be able to deal with this complexity by making global system features emerge without central control or the need of excessive computational power. Up to now research has been mainly focusing on identifying self-* techniques that operate during the achievement of the regular functional goals of software. Little effort, however, has been put on finding effective methods for energy usage optimization. Our work focuses specifically on this aspect and proposes a bio-inspired self-organization algorithm to redistribute load among servers in data centers. We show, first, how the algorithm redistributes the load, thus allowing a better energy management by turning off servers, and, second, how it may be integrated in a self-organizing architecture. The approach naturally complements existing self-management capabilities of a distributed self-organizing architecture, and provides a solution that is able to work even for very large systems.

Keywords: energy optimization, bio-inspired algorithms, autonomic computing, self-organization, data centers.

1 Introduction

The constant growth of energy usage in industrialized countries is creating problems to the sustainability of the Earth development. The problem of energy use concerns many fields in human activities, for this reason some new disciplines such as green computing are growing up to study how to consume less energy by providing the same quality of service.

In general, research in energy saving focuses towards two different directions:

- developing new technologies that need less energy to work, e.g. energy saving lamps;
- developing new techniques to make a rational and efficient use of existing tools, e.g., car sharing is a way to use cars in a more sustainable way.

D. Weyns et al. (Eds.): SOAR 2009, LNCS 6090, pp. 127–151, 2010.

In the context of software engineering, the first direction is addressed by the development of new software architectures and algorithms that are able to support energy saving, while the second one corresponds to new management processes that may be used to instrument existing architectures and algorithms to achieve a final energy reduction without having to modify or replace existing software components.

In this work we present two distinct contributions: the first is an algorithm for reducing power consumption in a large data center, the second is the integration of this algorithm into a decentralized platform supporting the management of the applications installed in the data center.

The general idea for the algorithm is to move the load from a server to another one to maximize the power efficiency of the whole data center. The power efficiency is defined as the available computational power over the energetic power required to provide such computational power. In other words the goal of the algorithm is to bring the network from a situation of randomly distributed load to a situation in which part of the servers are used with the maximum efficiency, while other servers are turned off to have a maximum reduction of wasted power. The algorithm we have developed takes some ideas from the biological world such as the explorative behavior of swarming insects in their search for a better place [1]: in our context migrating entities are the virtual machines of each server (for us they represent the server load), while the possible migration sites are the servers.

For what concerns the architectural integration problem, the approach we propose starts from the use of an existing autonomic framework (SelfLet architecture [2,3]). The reason for this choice is that such architecture is highly decentralized and therefore scalable to the dimensions of a large data center.

In order to assess our approach, we have set up some preliminary experiments that show the algorithm behavior under different conditions. Based on the results, we can argue that our approach inherits some classical benefits from bio-inspired solutions such as scalability, complete decentralization, and capability to achieve good results even in extreme conditions.

The remainder of this paper is structured as follows: Section 2 shows the motivations and the objectives of this work; Section 3 describes the technologies we are using, and in particular it introduces the architectural model and the algorithm model; Section 4 depicts the integration work that has been done both on the architectural and on the algorithmic level; Section 5 reports and discusses the simulations that have been carried out to test our approach. Section 6 presents the most representative state-of-the-art works aimed at reducing power consumptions in data centers. Finally, Section 7 depicts conclusions and future works.

2   Motivations and Approach

In the past the focus of microelectronics industry has been the miniaturization of components and the design of better performing devices. Such an evolution led to very fast IT systems. However, these systems are often used inefficiently [4].

As shown in [5], the interest towards efficient use of technology is motivated by some alarming trends:

- recent researches have demonstrated how IT is responsible for more than 2% of global CO_2 emissions [6,7,8];
- power is the second-highest operating cost in 70% of all data centers;
- data centers are responsible for the emission of tens of millions of metric tons of carbon dioxide annually, more than 5% of the total global emissions.

Figure 1 shows the partition of global expenditure for servers from 1996 to 2010 as predicted by Josselyin *et al.* [9]. From an economical perspective it can be noticed that whereas the cost of hardware has only slightly grown in the last 12 years, the cost of power and cooling has grown four times.

As described in [10], data centers are found in nearly every sector of the economy, including financial services, media, high-tech, universities, and government institutions. Dramatic servers growth in data centers is indicated by well-known web services such as Google, Amazon, and eBay. Estimates indicate that Google maintains more than 450,000 servers, arranged in several data centers located around the world. According to the research firm IDC [9] by 2012, for every 1$ spent on hardware, 1$ will be spent on power and cooling. For this reason, much of the interest in Green IT comes from the financial return on green data center investments.

In order to follow these new needs of companies, also hardware manufacturers are studying more efficient solutions. In 2007 some big companies like AMD, Dell,

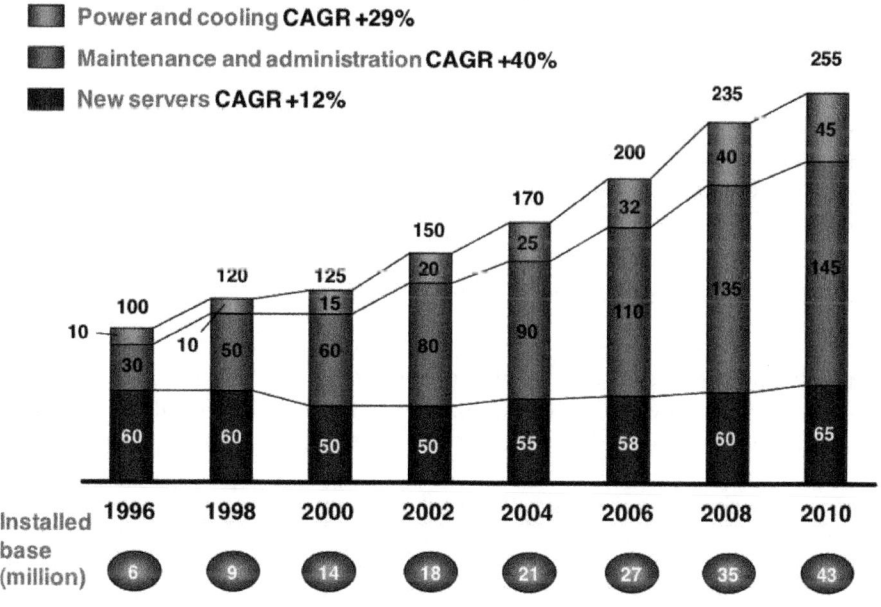

Fig. 1. Global spending for servers, prediction by [9]

IBM, Sun Microsystems, and VMware have founded The Green Grid, a global consortium which has the goal to investigate new green-oriented solutions [11].

The work in [10] indicates several dimensions, which should be exploited when trying to obtain good green results. Such dimensions are divided into organizational tasks (e.g., the creation of a responsible position for the Green IT), technical tasks (e.g., measuring the power-usage effectiveness), and technological tasks (e.g., consolidation and virtualization). The last mentioned one is the one that motivates the approach we are proposing. Virtualization allows a high decoupling between the application layer and the physical layer of any existing system. This allows applications to be moved and executed on different types of hardware, thus enabling the possibility of optimizing power consumption without the need to modify the applications themselves. The next two paragraphs describe more in detail the motivations of the architectural and algorithmic choices we have made in our work.

2.1   The Importance of Self-organizing Architectures

The solution we propose is framed into the Autonomic Computing field. The term Autonomic applied to the IT sector has been first used by IBM [12] to mean systems able to self-manage, self-configure, self-protect, and self-repair without any user intervention. Most architectures that follow this paradigm tend to add a supervision layer to existing software architectures that is able to "reason" using the data provided by the underlying system, and to "react" by ordering the system to perform some sort of actions. This last approach has the problem of keeping all the complexity concentrated in this supervising layer, therefore more recent architectures tend to offer the opportunity to distribute such complexity to all the elements of the system. In such a way the system is able to achieve complex global behaviors using only simple rules at the level of each component. This last approach is the one at the basis of our work since we are dealing with systems that may have thousands of nodes and where system self-reorganization should occur spontaneously. In particular our reference architecture is the SelfLet one that will be described in detail in Section 3.

2.2   The Importance of Self-organizing Algorithms

The architectures described above provide software primitives for interfacing with existing systems, as well for monitoring events, taking actions, and exchanging messages. However they are completely useless without a self-organizational logic that defines the goals to achieve and algorithms to achieve these goals.

For the purpose of this work we have used a bio-inspired algorithm since this class of algorithms is usually intrinsically self-organizing and characterized by scalable and fault-tolerant behavior. This kind of algorithms is particularly important because their final results often emerge from individual behaviors of the "colony" that is running the algorithm (the most common example is the Ant Colony Optimization algorithm [13]). The main drawback of these algorithms is that they are usually expressed in such a way that even if they can be modeled and studied in a theoretical way, there are many assumptions on the running architecture that make difficult to integrate them into existing self-organizing systems. Moreover there are too few attempts of using these techniques in research targeted on Green IT.

The idea behind this work is to model the data center as a peer-to-peer network of nodes that collaborate to each other using a bio-inspired algorithm based on the scout-worker migration method described in [14]. In this algorithm some entities of the system are workers (which correspond to the virtual machines) and other entities are scouts (which are entities allowed to move from one physical node to another to collect information): the information collected by the scouts is then used by workers to take corrective actions (such as the migration of workers, and the switching-on and off of physical machines). This algorithm is particularly suitable in decentralized systems scenarios for the following reason: the absence of global coordinators, the absence of global knowledge, and the fact that choices are emergent perfectly fit the situation in which the data center is characterized by large dimensions (up to thousands servers, like in the case of telecommunication companies or governmental agencies) and high dynamism (e.g., requests peaks or machine failures). A more detailed description of this algorithm will be provided in Section 3.

2.3   Approach and Objectives

Summarizing, the goals of this work are:

- model the data center architecture as a distributed system and assign its management to a decentralized system built on top of the SelLet architecture;
- define a proper bio-inspired algorithm that may be run in such system to obtain actual energy savings;
- define the integration of such algorithm into the self-organizing architecture of the data center pointing out the integration issues that have arisen and that are common when dealing with such type of algorithms;
- run the proposed algorithm in a simulation environment to evaluate its performance, and compare the results with theoretical optimal solutions.

3   The Model

In this section we show the model of the system from an architectural and an algorithmic point of view.

3.1   Architectural Model

From the architectural point of view the system is composed of a set of physical servers interconnected to each other: this is the physical layer of the system. Each server may run virtual machines and is supported by a self-organizing component called SelfLet [2,3], in charge of executing self-organizing behavior.

Server. The server is the entity in charge of computation of the normal load in a data center. In addition to that, each server interacts with the scouts, which are located on it time by time.

In this paper, we assume a simplified but realistic server architecture. Nowadays many data centers have undergone a process of server consolidation in order to reduce the number of physical machines. Such a process involves the introduction of a software layer called Virtual Machine Monitor (VMM), which is a system providing an environment to support an operating system, and its applications and processes. Figure 2 reports the description of a single server machine, which is able to run three different kinds of servers (e.g., they could be a database server, a web server, and an application server), running two different operating systems and their own sets of applications. Such a configuration allows considering servers as applications, which can be moved and run by another machine equipped with an appropriate VMM.

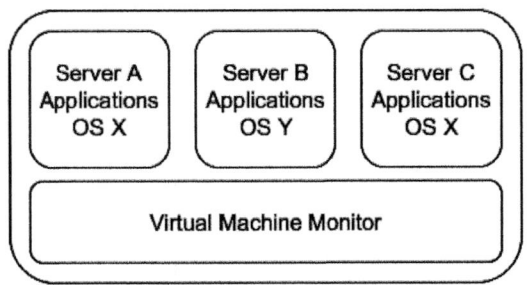

Fig. 2. Schema of a consolidated server

Servers are linked to each other as in a normal data center and each one has the following set of characteristics:

- *Load*: each physical server contains a list of running virtual machines (VMs), which can be migrated from a server to another. In particular, VMs in a single physical server are logically grouped into packages, which are the actual object of migration. The dimension of the package can be customized depending on the characteristics of the physical machines.
- *Capacity*: it is the maximum amount of resources available on the server. In the current model it is considered as a single parameter grouping together different characteristics like local memory, CPU computational power, etc.
- *Power consumption class*: it is a value indicating the technology used to implement the server and is considered as an indicator of power consumption. Power consumption class is an integer value spanning between 1 (most power saving machines) and 4 (most power consuming machines).
- *State*: the status of a server at a certain moment can be *on*, i.e. it is normally computing its load, and *hibernated*, i.e. the machine is put into a particular power saving mode because it has moved all its VMs.

The data center is modeled using a logical classification based upon layers, similarly to what is described in [15]. Servers in the model can be hierarchically organized in: (i) a storage network area, i.e., a set of machines that share storage devices and that have a small cost of VM migration, (ii) cluster, i.e. a set of machines logically grouped together because they are linked to the same switch or are managed by the same load balancer, and (iii) data center level, i.e., all servers that belong to a given data center and are located in the same physical site. As it can be noticed from Figure 5, this hierarchical structure is not mandatory and depends on the particular data center (for instance, a small data center could not need the concept of cluster), nevertheless, it is useful in the attempt to model real world scenarios, to consider heterogeneous configurations (e.g., bandwidth differences among clusters), and to model constraints for VMs migration (e.g., not every server could be able to run all the VMs).

SelfLet. The autonomic architecture of this paper is based on the concept of SelfLet, which can be defined as an autonomic architectural component able to dynamically change and adapt its internal behavior according to modifications in the environment. A SelfLet is also able to interact and cooperate with other SelfLets to achieve high-level goals.

SelfLets have the following characteristics: they are identified univocally by an ID and can be defined as belonging to one or more types, which is an application-dependent property. At any given time a SelfLet can also belong to a Group, constituted by its neighboring SelfLets, to which communication can take place. A SelfLet executes one or more possible Behaviors. A Behavior can be seen as a workflow that involves some actions such as the invocation of Abilities, i.e. services, which can be local or remote. A Behavior usually results in the fulfillment of one or more Goals. Possible changes in the internal state of the SelfLet or in the environment may trigger particular intelligent rules, called Autonomic Rules, which may modify the Behavior, and add/remove/run Abilities. Specific kinds of Abilities are those that perform some autonomic tasks, for example, they can create an aggregation of neighbors respecting certain properties.

Figure 3 represents the internal architecture of a SelfLet. The communication with the other SelfLets takes place thanks to the Message Handler. The Negotiation Manager is in charge of publishing the availability of achieving certain Goals in a SelfLet or of requesting the achievement of some Goal to other Self-Lets if needed. The Behavior Manager is the one that executes Behaviors while all Abilities are managed by the Ability Execution Environment: it offers the primitives to install, store, and uninstall Abilities without restarting the Self-Let. The Internal Knowledge is basically an internal repository, which can be used to store and retrieve any kind of information, needed by any of the SelfLet components. Finally, the Autonomic Manager has the task of monitoring the SelfLet and its neighbors and of dynamically adjusting the SelfLet according to a given policy by firing Autonomic Rules.

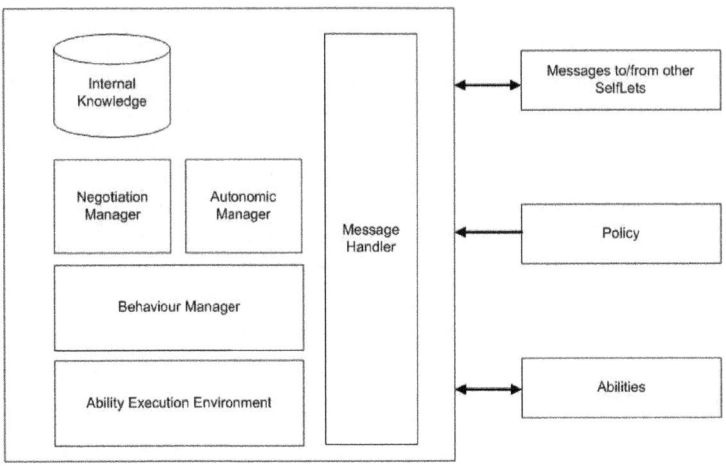

Fig. 3. Architecture of a SelfLet

In our work we plan to instrument the SelfLet by expressing the self-organization algorithm as SelfLet Behaviors and Autonomic Rules, and by implementing the interfaces for monitoring server parameters and for starting VM migrations as SelfLet Abilities.

3.2 Algorithm Model

From the algorithmic point of view the model we are proposing takes some elements from current research in the field of biologically inspired self-organization algorithms, in particular from collective decision making in animals colonies. Nature offers several examples of groups of mammals and birds, which are able to behave collectively and take all together some decisions like whether to migrate or not from a site to another, whether to stay or leave a group, etc. The case of migration in some insects is characterized by the following behaviors:

- Some individuals, called scouts, leave their groups and go looking for different and better sites.
- These individuals evaluate new sites comparing them with their origin sites. When they go back home, they can decide to convince the rest of the group to start a migration.

The algorithmic model presented in this work is based on the same principle. There are three types of entities:

- the *community of VMs* (i.e., the colony in biological terms) residing on a certain server;
- the *scouts*, which explore different physical servers and compare them with the origin one;

– the *server managers*, which permanently reside on each physical machine, and decide to move a community of VMs based on the information brought back by scouts.

When scouts find a better place for the community, they notify the server manager that will then make the decision whether to migrate the VMs to the new site.

Scout. The scout is the entity in charge of investigating the data center characteristics and sharing them among the servers. This information is used to make decisions about migrations of load among the servers. A scout is created in all the physical nodes by server managers during the algorithm initialization and, differently from the original algorithm presented in [14], it does not belong to the server, but it cruises the data center following the links among the physical machines. During the exploration each scout queries the server managers to know the actual configuration of the current location. The following list reports the main characteristics of a scout:

– Current location: a scout can be located only on one server at a given time. The location is identified by the server identification number, which could be its IP address.
– Lifetime: a scout has a parameter indicating the maximum length of its life. At the current state of the model, this value is set to infinite, indicating that scouts never die, but a finite value could be useful to limit the number of homeless scouts in the data center in case of a server failure or crash.
– List of preferred servers: each scout stores the information obtained during the data center investigation into a memory, which classifies the visited servers according to two parameters:
  • *power consumption class*: this is the first characteristic to be evaluated. In our case a server with a lower benchmark data consumption is preferred;
  • *amount of free resources*: this is the second issue to be evaluated. In our case the server with more unused resources is preferred.

The way a scout works is illustrated by Algorithm 1: during each iteration a scout performs the task of monitoring and storing the information of the visited server and then moves to another server. Figure 4 shows a configuration in which servers are represented by large grey circles, scouts and VMs are depicted with triangles and diamonds respectively. Links among servers are the only way scouts can use to move from a server to another for the investigation.

Algorithm 1. Scout Thread

---
parameters: preferenceList
while TRUE **do**
    scout.preferenceList.update(current_node.Id,current_node.Status)
    scout.move()
end while

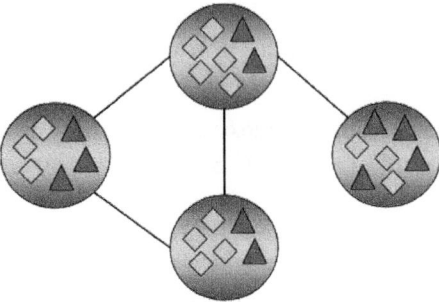

Fig. 4. Example of a small data center configuration on a random network

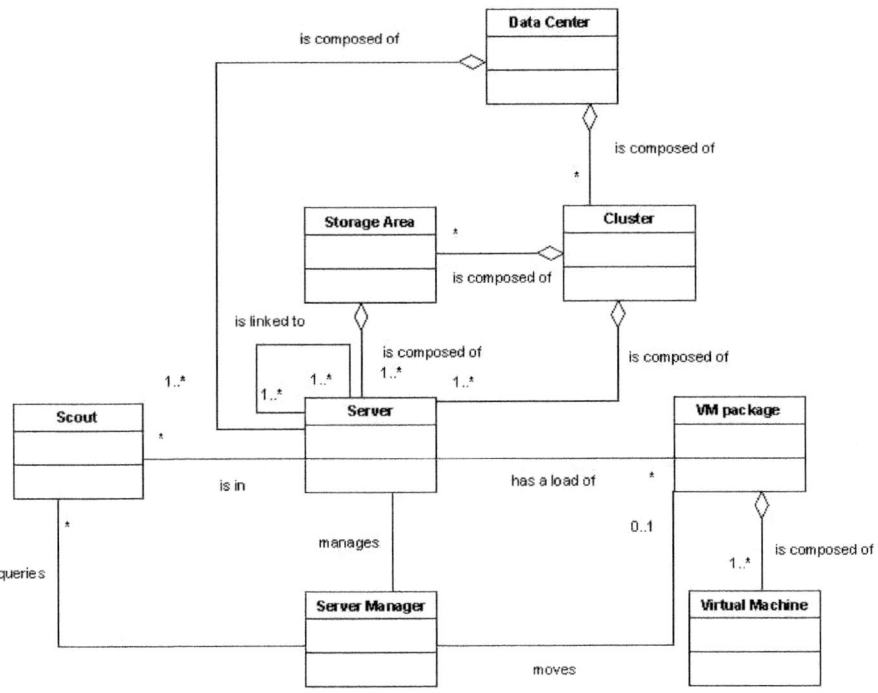

Fig. 5. Conceptual model of the data center

Server Manager. The Server Manager is the entity in charge of communicating with the scouts and of taking decisions about movements of VMs. Decisions on moving or not a VM package is made according to the following parameters:

– Server choice function: server managers could be in the situation of migrating a VM package while two or more scouts are indicating different possible servers. In this case the order of evaluation for the destination machine by the server manager is the following:

1. a server with the lowest power consumption class is preferred;
2. the server that would offer the highest percentage of use of resources is preferred;
3. in case the previous parameters are equal, the destination node will be chosen randomly.

These criteria are fundamental to ensure the system convergence to a stable state, and to avoid continuous switches of servers between states.

– Migration probability: as in many biologically inspired systems, decisions whether to migrate or not a VM package is not deterministic, but follows a probability distribution such as the one showed in Figure 6, which illustrates how the probability of not migrating VMs increases with the load percentage. This non-determinism is needed to avoid situations in which a loaded server goes on migrating all its VMs, with the result of increasing network traffic without improving significantly the global power saving.

– Migration inhibitors: previous choice criteria are not the only reasons to avoid migrations from a given server. In particular there are two other reasons why VMs could be locked in a server: (i) the server manager has a flag variable that indicates that a given set of VMs are not available to migrate, e.g. because they are running important processes and cannot be stopped, or because of some QoS constraints; and (ii) a given VM cannot migrate immediately after its arrival, but the server manager has to wait some time intervals before moving a VM that has just arrived. This

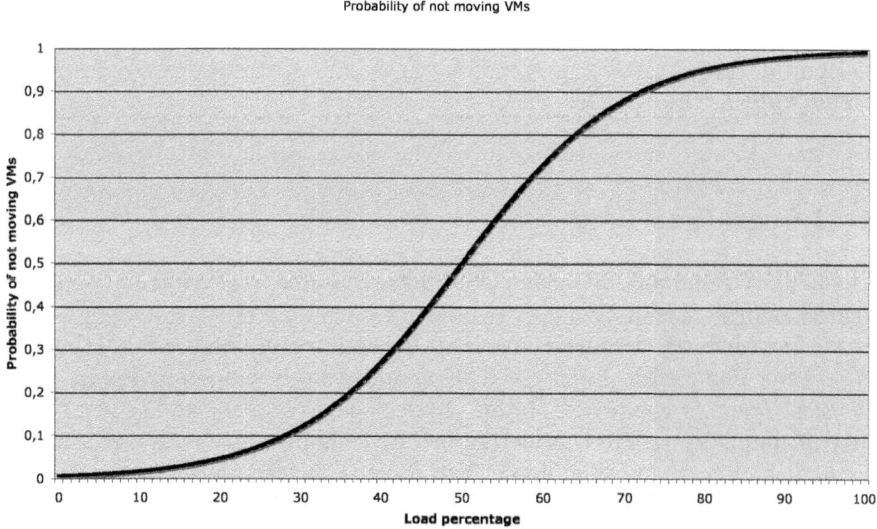

Fig. 6. Probability distribution used to choose if a set of VMs should be moved, depending on the current total load of the machine

behavior has been introduced to avoid bandwidth overload along with its related power consumption, and also vicious circles that would make some VMs move continuously, preventing them from running their jobs.

The way a server manager works is illustrated in Algorithm 2.

Algorithm 2. Server Manager Thread

---

parameters: preferenceList, knownScouts, VMSet, hibernated
scout=new Scout
scout.move()
while TRUE **do**
  if !current_node.isLocked() **then**
    movableVMs=getMovableVMs(VMSet)
    if movableVMs.notEmpty() **then**
      for scout in knownScouts **do**
        preferenceList.update(scout(i).queryPreferenceList())
      end for
      bestLocation=evalLocation(server.preferenceList)
      if bestLocation.notEmpty() **then**
        movableVMs.move(bestLocation)
      end if
    end if
    if VMSet.notEmpty() **then**
      hibernated=false
    else
      hibernated=true
    end if
  end if
end while

---

4   Integration

The SelfLet framework offers an environment suitable to develop and deploy the algorithm of Section 3.2.

We have defined a two layered conceptual architecture that extends the model of a server depicted in Figure 2. This architecture is showed in Figure 7. The *Physical layer* is composed of the physical devices of a data center; for the sake of simplicity we are considering mainly computer machines and a network communication system. The *Application layer* is, in turn, the virtualization level, and it is composed of virtual machines, which can be run by a single physical server. The *SelfLet layer* is composed of SelfLets, which are in charge of managing the autonomic behavior of server machines and their VMs. It should be noticed that the SelfLet and the Application layers are designed at the same logical level as they both need to interact with the Physical Layer.

Communication among the three modules occurs in the following way:

- the communication between the Application layer and the Physical layer is necessary to allow the execution of VMs, which are managed by the VMM;
- the communication between the Application layer and the SelfLet layer allows the latter to know all the properties of each VM, such as its number, and the possibility to move it;
- the communication between the SelfLet layer and the Physical layer is needed to know the global status of the machine, such as power consumption class, status of hibernation, and resources utilization, indeed the modular structure of a consolidated server keeps each VM completely independent from the others, and a single VM is not able to obtain information from the others. Moreover, the SelfLet is in charge of deciding whether to switch the status of the physical machine to hibernation according to the system constraints.

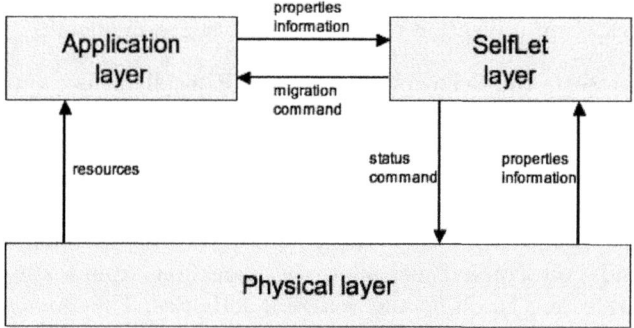

Fig. 7. Layered architecture of the system

To satisfy the needs of the presented algorithm, SelfLet's modules will be used the following way:

- the Behavior Manager is in charge of managing the actual Behavior of each server;
- the Autonomic Manager has the task of monitoring all the other components (and the communications among them) and of firing Autonomic Rules, which may be triggered to dynamically adjust the SelfLet according to a given policy: for example it can take the decision of locking the server status to avoid VMs migrations in particular situations and whether to move a VM package;
- the Internal Knowledge repository can store information on the network learned from moving scouts (forwarded by the SelfLets of other physical servers), such as other servers' status;
- moreover, SelfLets need a set of additional customized Abilities to be able to manage the system:
  • the Scout Management Ability provides the primitives for the SelfLet to create/destroy new scouts, and receive/forward external scouts. This is

done using the communication primitives provided by the physical layer. One of the most important benefits of this Ability is the update of the internal knowledge that may then be used by the monitors described below to compute metrics.

- the Server Interface Ability allows the information exchanges between the Behavior Manager and the Physical layer;
- the Server Monitor Ability has the task of monitoring the status of the server to calculate metrics on server resource utilization and server characteristics such as energy class.
- the Application Interface Ability has the task of managing the information exchanges between the SelfLet and the Application layer;
- the Application Monitor Ability, similarly to the Server Monitor, is in charge of computing metrics on virtual machines or querying for their characteristics, e.g. the kind of server (database server, web server, etc), statistics on utilization rate and resources requirements.

Figure 8 represents the Behavior of a SelfLet. Basically, it is composed of three states: *Active with load*, i.e. the server is working and the SelfLet is communicating with other SelfLets sending and receiving load and scouts; *Active without load*, which is similar to the previous state but the server has not a current set of VMs and is kept on for some constraints or policies set by the datacenter administrator; finally, the SelfLet can turn to *Hibernated* if there is no constraint and it has no load. Autonomic Rules allow the transitions from a state to the next one. These rules are based on the SelfLet's activities. They analyze the status of the server and allow the Autonomic Manager to fire the rules.

Summarizing, the SelfLet model with the proposed extensions has the proper Abilities to communicate and collect data from the other modules of the system (Physical layer and Application layer), to create, send, receive, or destroy scouts that move through the system using diffusive communication algorithms, and to

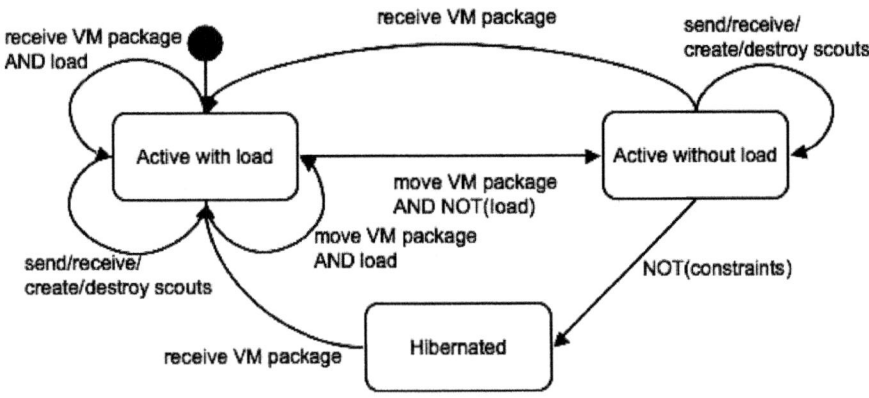

Fig. 8. Behavior of a SelfLet

store and elaborate the data provided by moving scouts. These data are stored in the Internal Knowledge of the system and is used to fire rules that may result in decisions whether to move a VM or not. All the steps of the algorithm are codified as a workflow that is managed by the Behavior manager and that may be sent to other SelfLets (along with the Autonomic Rules) to trigger the same algorithm on neighboring SelfLets in an epidemic way. This epidemic spread is also useful in case of future algorithm improvements, since there is no need to manually change the configuration of each SelfLet. Moreover the Abilities that are required by the algorithm may be automatically migrated from a SelfLet to the other as needed in a seamless way, thus reducing system management costs.

5   Simulations

This section shows some preliminary results that have been obtained from a prototype implementation of the power-saving self-organization algorithm proposed in Section 3.2. The prototype has been implemented using the PeerSim tool [16], which is a peer-to-peer network simulator written in Java, which allows simulations of systems with varying cardinality. The simulation engine that has been used divides the simulation into steps: each step corresponds to atomic operations that involve all the nodes of the systems at the same time. Due to this simplification, these steps cannot be seen as a timing measure, however in this preliminary state, simulation steps are actually the best indicator for studying the algorithm convergence in different situations.

Results obtained by the implemented model have been compared with results obtained on the same network by a static global optimizer based on Java wrapped LpSolve operations research library [17]. This optimizer computes the best possible optimization case by case, spreading the whole network load to the most energy efficient machines, and using the 100% of their resources. Therefore, it is used as an upper limit to understand the performance of the proposed algorithm. The indicators that have been used to evaluate the algorithm's performance are the following:

- *Power consumption*: calculated as the sum of the energy class indicators of each server.
- *Number of hibernated servers*: number of servers that have no VMs at a certain simulation step.
- *Average server usage*: the mean of each server usage percentage.
- *Power saving percentage*: reduction of power consumption with respect to the initial situation.
- *Number of steps to reach a stable state*: the number of steps needed for the data center to reach a stable configuration of hibernated servers.

The remainder of this section shows the results obtained according to the following simulation plan. In the first simulation we report the results of some experiments made to understand the behavior of the algorithm when *varying the number of scouts per each server*. In the following simulations we evaluated the algorithm's performance when *changing the initial load* for each server and the total number of servers. After several runs we have seen that the results obtained in all the experiments stayed coherent with an average discrepancy below 10%.

Varying the number of scouts. The first set of simulations focused on understanding the behavior of the system when varying the initial number of scouts per each node. Simulations confirm our initial hypothesis, since when increasing the number of scouts the final performance is better. Anyway there is a threshold over which system's performance does not increase significantly when such limit is reached, therefore a higher number of scouts does not produce further increase in final savings. For this evaluation the network was configured using 1000 servers with an initial average load of 15%. Results of these simulations are shown in Figure 9: the three subfigures show respectively the total number of hibernated servers, the total power consumption expressed as the sum of the power consumption index of each server, and the average server usage, vs the total number of scouts. From these results we can see that the best performance is obtained by the configuration with 6 initial scouts per server. On the other hand, Table 1 shows how network load, which is strongly dominated by the number of VM migrations, gets higher when increasing the number of scouts. Therefore, it

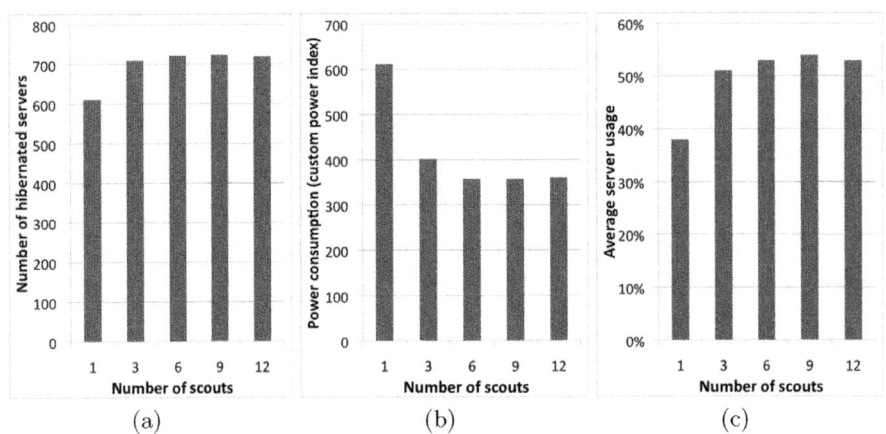

(a)   (b)   (c)

Fig. 9. Comparison of the behavior of the algorithm with a different number of scouts

Table 1. Number of VM migrations after the algorithm has reached a stable state

Number of scouts	Number of migrations
1	8198
3	11510
6	12266
9	13286
12	13458

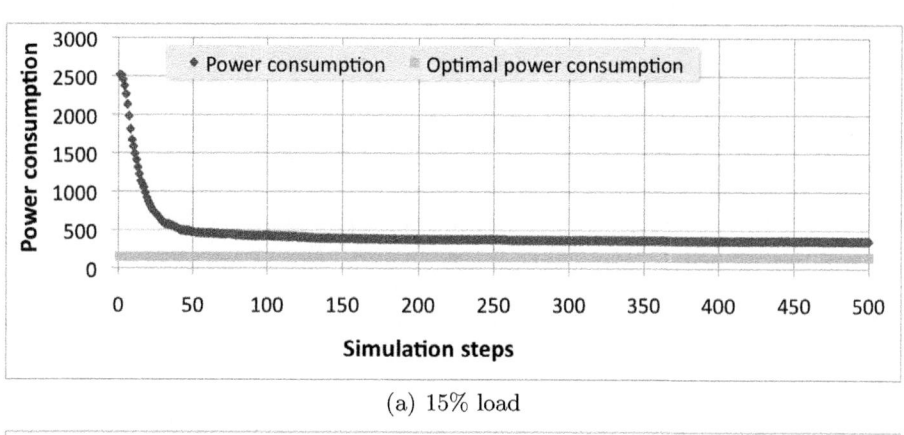

(a) 15% load

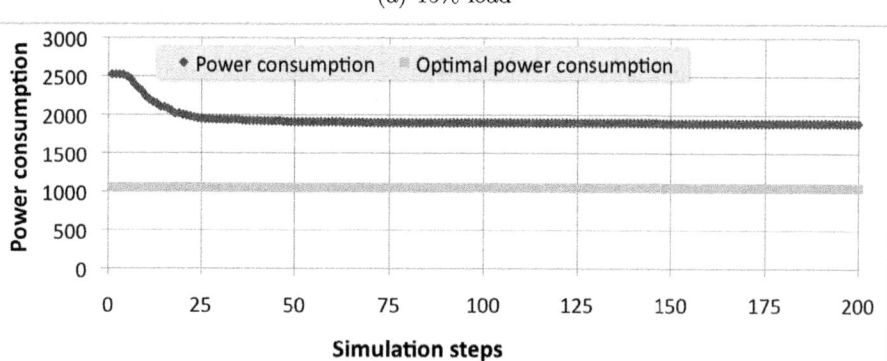

(b) 60% load

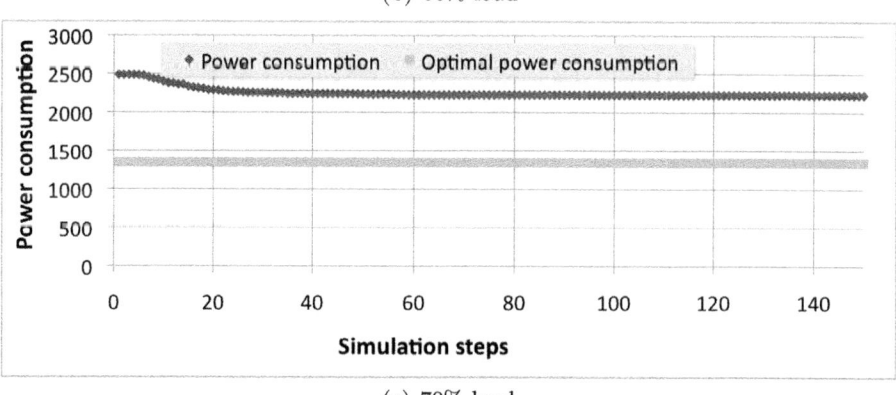

(c) 70% load

Fig. 10. Comparisons of the power consumption between the presented algorithm (dark dots) and the ideal optimal solution (light-gray line) when varying the initial load

is important to establish the number of scouts that plays the role of threshold (6 in the example), so to avoid an increase of the network load without introducing additional energy savings.

Experiments with 15% of initial load. For this simulation we have chosen an initial load for the servers equal to 15% of their capability, to observe how the algorithm is able to perform when conditions do not give narrow margins. Figure 10a shows the behavior of the algorithms in terms of power consumption: the system reaches a stable state after 300 simulation steps and gives an 88% saving. Table 2 shows also the number of hibernated servers and the average server usage (in stable conditions) for different network sizes.

Experiments with 60% of initial load. In this experiment we investigate the system's behavior when servers' initial load is 60%. Figure 10b shows the behavior of the algorithms in terms of power consumption. The system reaches a stable state after 80 simulation steps and gives a 25% saving, but leading servers to a 71% usage of their resources, as shown in Table 2.

Experiments with 70% of initial load. Similarly to previous experiments, in this simulation our goal is to explore the system's behavior in a sort of limit condition. In this case the average initial server load is 70% and there is not a great range of actions that can be carried out to further improve power saving since the algorithm has been tuned not to overload the server and reaching a desired 80% load. Figure 10c shows the power consumption curve: in this case the algorithm obtains a 10% saving in about 50 simulation steps. From Table 2 we can also see that the algorithm manages to hibernate 7% of servers and to increase the average load up to 75%.

Discussion. As previously discussed, the presented approach obtains very good results when the whole load of the data center is low, reaching a percentage of saving up to 88%. In the other situations the results are still good even if the conditions are worse because the servers are saturated. In all the situations

Table 2. Results comparisons

N. servers	Initial avg load	Power saving	Hibernated servers	Final avg load
100	15%	82%	69%	48%
500	15%	85%	72%	52%
1000	15%	88%	72%	53%
100	60%	31%	21%	75%
500	60%	25%	17%	71%
1000	60%	25%	17%	71%
100	70%	5%	3%	72%
500	70%	10%	7%	74%
1000	70%	10%	7%	75%

the algorithm has also shown a great capability to scale with respect to the number of servers, but with a different behavior in presence of different loads: this is probably due to the higher number of alternatives for the load balancing. Another advantage is the fact that the self-organizing logic is scattered across all the nodes, thus increasing the reliability with respect to a centralized solution. Finally the simulation steps needed to reach a stable state can be remarkably low. In less loaded networks stability is reached in about 30-40 steps, but even in larger networks the convergence has a similar rate.

The main drawback of the approach is the increased amount of traffic overhead in the network due to VM migrations. The actual simulations do not show this overhead since each message exchanged is an atomic operation that takes one simulation step, however some parameters that may affect overhead are the size of scout information, the size of each VM, and the number of scouts per each server. Decisions need to be calibrated based on this information to find a trade-off value between power saving efficiency and the amount of network overhead, therefore new and smarter policies to bound the number of migrations should be introduced. Moreover, it is clear that having a high capacity communication network layer among servers become a necessary condition, thus representing the most critical limitation of this work.

6   Related Work

This section contains an overview of the state of the art of existing works that are related to the approaches we have used in this paper. The section is organized as follows: Section 6.1 shows existing approaches for optimizing power consumption in data centers, Section 6.2 shows an overview of the field of bio-inspired self-organization, and finally Section 6.3 presents some alternative architectures that may be integrated into existing systems with the purpose of running bio-inspired self-organization algorithms. Although our results are preliminary, they can be considered indicative of the contribution that our work gives. Indeed our results are comparable to Muse [18], which is able to reduce server energy consumption by 29%-78% for representative Web workloads.

6.1   Approaches for Power Optimization in Data Centers

The typical approaches in literature are divided into two main classes: centralized and distributed.

Centralized Approaches. These approaches are the oldest ones and the ones able to give results that are the closest to the optimum; however having a central element creates coordination and fault-tolerance problems in presence of a very high number of servers. In this context our approach is able to address this limitation, but with the drawback of having a less optimal final result when compared to centralized approaches. Hereafter, we shortly describe some of these works.

In [19,20] the authors have tried to identify workload time series to dynamically regulate CPU power and frequency to optimize power consumption. This attempt led to a reduction of consumption by to up 36%.

Also the work presented by Pinheiro *et al.* [21] addresses the problem of power consumption in server clusters. In this case the objective is turning off the highest number of servers making a redistribution of the load, taking into account a maximum degradation of performance. In this case a machine is in charge to choose which server should be turned on and off per each cluster of the network, being able to measure the overall performance.

Another centralized approach is introduced by Chase *et al.* [18]. In this case an economic perspective is taken to manage shared service resources in hosting centers. In the presented ad-hoc solution, the Muse operating system, a central unit called *Executive* takes requests as inputs and takes resource allocation decisions, while other three modules are in charge of: (*i*) monitoring and estimating the load, (*ii*) switching incoming traffic to the selected server in a transparent way for the services and (*iii*) managing a pool of generic and interchangeable resources. The economic similarity is due to a utility function that maximizes "revenues".

In [22], Elnozahy *et al.* present five different policies to manage energy saving in a server cluster environment. This work (and the one presented in [21]) aims at optimizing voltage scaling by dynamically adjusting the operating voltage and frequency of the CPU to match the intensity of the workload.

The work shown in [23] by White and Abels presents a computing model called VDC in which a centralized dynamic control system is in charge of managing resources for the applications, which are seen as isolated virtual machines. In this approach the resources are virtualized so that they can be provisioned and resized dynamically to applications as required. Moreover applications may frequently migrate from one set of resources to another. VDC implements applications as isolated virtual machines, which are built on top of their virtualized resources (server, storage, and switches), and under their virtualized service layer (user and service levels) across the entire data center.

Distributed Approaches. These approaches aim to solve the power saving problem in a distributed way, so that they may scale well in presence of large-scale system. The problem of these approaches is that they are often based on networks that are usually organized as a hierarchy or as a lattice. This solves scalability and fault-tolerance problems, but has the drawback of being difficult to maintain in presence of very dynamic situations since hierarchies/matrices are difficult to be reconstructed when they are broken. In our approach every node of the network is equal to each other and there are no hierarchies: this means that there are almost no penalties if nodes are added/removed or they change their status (e.g., from hibernated to active) in a fast and unpredictable way. The following are some of the existing distributed approaches.

The work done by Khargharia *et al.* in [24] presents a theoretical framework that optimizes both power and performance for data centers at runtime. In particular the approach is based on a local management of small clusters and on a hierarchy to manage power. The hierarchy is composed of three levels: (i) cluster, (ii) server, and (iii) device, each one with a fine granularity because it is able to address consumption of processor, memory, network and disks.

Bennani and Menasce [25] present a similar hierarchical approach with respect to [24], addressing the problem of dynamically redeploying servers in a continuously varying workload scenario. In this case servers are grouped according to an application environmental logic, and a so-called *local controller* that is in charge of managing a set of servers. In turn, all the local controllers report the workload prediction and the value of a local utility function to the global controller, which takes decisions on the collective resource allocation.

Das *et al.* [5] present a multi-agent system approach to the problem of green performance in data center. As for aforementioned papers, the framework is based on a hierarchy, according to which a resource arbiter assigns resources to the application managers, which, in turn, are in charge of managing physical servers.

6.2   Bio-inspired Self-organization Algorithms

Bio-inspired algorithms are self-organization algorithms that share some principles with phenomena of the natural world [1]. This kind of biological behaviors are often taken as inspiration to develop new algorithms because they are usually built upon very simple elements that are instructed with very simple rules. Moreover each element has very limited memory and computational capabilities. The advantages of these algorithms are the fact that the complexity is spread across all the nodes of the system, and that, if an element is removed from the system or behaves in an unpredictable way, the system keeps working without problems. These algorithms are often characterized by some probabilistic behaviors that are needed to explore the environment (and the possible solutions) and have the chance of finding (or improving) their goal [26,27]. Some studies of these principles have been proposed in [28,29]: these works clearly show these advantages. The main drawbacks of bio-inspired techniques lie in the fact that they are difficult to implement, study and calibrate in real systems. Furthermore they are not able to obtain good results in presence of static environments with a small number of elements. This happens because in this way the chances of finding a good solution by randomly exploring the solution space are much lower.

An example of algorithm that uses the migration principle has been proposed by Saffre in [14]. Our work shows how such algorithm that has been previously described using a toy example (workers, scouts and sites are expressed in generic terms in the previous work) may be modified and tailored to tackle a concrete problem such as the power optimization problem. Other studies that propose some integration principles to use bio-inspired techniques in existing systems have been proposed in [30,31].

6.3   Distributed Self-organization Architectures for Bio-inspired Algorithms

In this work we propose to add to the existing datacenter architecture a self-organizing framework (the SelfLet [2] architecture). We chose to use it because it has been developed with the idea of decentralization and all other bio-inspired principles in mind, so it is the natural complement for this class of bio-inspired algorithms. Beside that, most of existing self-organizing architectures – included the ones based on the IBM Autonomic model [12] (although with some more effort) – may be tailored to execute such algorithms.

Other self-organizing architectures include Autonomia [32], which uses a two layer approach, where the first contains the execution environment and the other manages the resources.

AutoMate [33], which has a multi-layered architecture optimized for scalable environments such as decentralized middleware and peer-to-peer applications.

CASCADAS Toolkit [34], which has been designed to have runtime changing goals, plans, rules, services and a behavior modeled as a finite state machine that may change overtime.

BIONETS [35] is an architecture based on two elements: tiny nodes (T-nodes, characterized by limited computational and communicational capacity like sensors and tags) and user nodes (U-nodes, that are terminal user nodes such as laptops and phones, and have full communication and computational capacity). U-nodes host services and interact with the environment through the T-nodes in their proximity, while U-nodes can also communicate between themselves to exchange information. Anthill [36] is a self-organizing architecture explicitly based on the analogy with ant colonies: it is composed of a network of interconnected nests (peers), which handle requests originated by local users by creating ants, autonomous agents that perform their tasks by traveling across the network. In this context the ants are the autonomic elements of the system and can be modified and evolve to better respond to the requested service.

In conclusion choosing the most suitable self-organization architecture depends on the deployment scenario and on the type of the existing system.

7   Conclusions and Future Work

In this work we have shown a possible way to exploit the self-* properties provided by a bio-inspired algorithm in a self-organizing architecture through their application to a real world use case scenario related to Green IT. In particular, it has been widely discussed how modern data centers are continuously growing up and how their dimensions represent a serious risk for their future development and management. Therefore a new bio-inspired algorithm has been introduced, based on the idea of colonies migrations. The advantages of the presented algorithm are mainly related to its self-management characteristics and to its ability to react autonomously to changes in the environment and in the requirements.

We have also shown how to benefit from these advantages by integrating this algorithm into the architecture of a modern data center composed of physical nodes, which run virtualized machines on top of them with the support of the SelfLet self-organizing framework. The only primitives that the whole system relies on are the possibility to move a virtual machine from a node to another one, and the possibility to hibernate idle machines and resume the operation of hibernated machines if the available ones are not able to deal with the actual workload.

The performance of the presented approach is pretty satisfying. Experiments have shown significant savings in power consumption through a technique based on the hibernation of the highest possible number of servers, preferably among high-consumption ones.

Some limits of the current study that will be further investigated in a future work are the following: (i) introduction of mechanisms to limit the number of VMs migrations, to let tasks run and to decrease the network load; (ii) further investigation on the scouts' life-cycle: in a real world scenario, in which servers can also crash and some scouts can be lost, it could be possible to regenerate them; (iii) further investigation on the scout memory to decide the number of sites to be remembered; (iv) model refinement to introduce QoS constraints and policies to understand if the system is able to respect Service Level Agreements in a changing environment.

In conclusion, from a methodological point of view this paper has shown a clear example of integration of a self-organization algorithmic logic into an architecture, which is supposed to support self-organizing system management. Moreover we have put some evidence and given some clues on the utility of bio-inspired approaches in a raising and important issue like power saving in data centers.

Acknowledgements. This research has been partially funded by the European Commission, Programme IDEAS-ERC, Project 227077-SMScom.

References

1. Camazine, S., et al.: Self-organization in biological systems. Princeton University Press, Princeton (2001)
2. Devescovi, D., Di Nitto, E., Dubois, D.J., Mirandola, R.: Self-Organization Algorithms for Autonomic Systems in the SelfLet Approach. In: Autonomics, ICST (2007)
3. Bindelli, S., Di Nitto, E., Furia, C., Rossi, M.: Using Compositionality to Formally Model and Analyze Systems Built of a High Number of Components. In: 15th IEEE International Conference on Engineering of Complex Computer Systems. IEEE Computer Society, Los Alamitos (2010)
4. Capra, E., Merlo, F.: Green IT: Everything strarts from the software. In: ECIS 2009: Proceedings of the 17th European Conference on Information Systems (2009)
5. Das, R., Kephart, J.O., Lefurgy, C., Tesauro, G., Levine, D.W., Chan, H.: Autonomic multi-agent management of power and performance in data centers. In: AAMAS 2008: Proceedings of the 7th International Joint Conference on Autonomous Agents and Multiagent Systems, Richland, SC, International Foundation for Autonomous Agents and Multiagent Systems, pp. 107–114 (2008)
6. Murugesan, S.: Harnessing green IT: Principles and practices. IT Professional 10(1), 24–33 (2008)
7. Kumar, R.: Important power, cooling and green IT concerns. Technical report, Gartner (January 2007)

8. Brown, E., Lee, C.: Topic overview: Green IT. Technical report, Forrester Research (November 2007)
9. Josselyin, S., Dillon, B., Nakamura, M., Arora, R., Lorenz, S., Meyer, T., Maceska, R., Fernandez, L.: Worldwide and regional server 2006-2010 forecast. Technical report, IDC (November 2006)
10. Lamb, J.: The Greening of IT: How Companies Can Make a Difference for the Environment. IBM Press (2009)
11. Kurp, P.: Green computing. Commun. ACM 51(10), 11–13 (2008)
12. Kephart, J., Chess, D.: The vision of autonomic computing. Computer 36(1), 41–50 (2003)
13. Dorigo, M., Stützle, T.: Ant Colony Optimization. Bradford Book (2004)
14. Saffre, F., Tateson, R., Marrow, P., Halloy, J., Deneurbourg, J.L.: Rule-based modules for collective decision-making using autonomous unit rules and inter-unit communication. Technical report, Deliverable 3.5 - IP CASCADAS (2008)
15. Greenberg, A., Hamilton, J., Maltz, D.A., Patel, P.: The cost of a cloud: research problems in data center networks. SIGCOMM Comput. Commun. Rev. 39(1), 68–73 (2009)
16. Jelasity, M., Montresor, A., Jesi, G.P., Voulgaris, S.: The Peersim simulator, http://peersim.sf.net
17. LpSolve, a Mixed Integer Linear Programming (MILP) solver, http://lpsolve.sourceforge.net
18. Chase, J.S., Anderson, D.C., Thakar, P.N., Vahdat, A.M., Doyle, R.P.: Managing energy and server resources in hosting centers. SIGOPS Oper. Syst. Rev. 35(5), 103–116 (2001)
19. Bohrer, P., Elnozahy, E.N., Keller, T., Kistler, M., Lefurgy, C., McDowell, C., Rajamony, R.: The case for power management in web servers. Kluwer Academic Publishers, Norwell (2002)
20. Lefurgy, C., Rajamani, K., Rawson, F., Felter, W., Kistler, M., Keller, T.W.: Energy management for commercial servers. Computer 36(12), 39–48 (2003)
21. Pinheiro, E., Bianchini, R., Carrera, E.V., Heath, T.: Load balancing and unbalancing for power and performance in cluster–based systems. In: Proceedings of the Workshop on Compilers and Operating Systems for Low Power (2001)
22. Elnozahy, E.N.M., Kistler, M., Rajamony, R.: Energy-efficient server clusters. In: Falsafi, B., VijayKumar, T.N. (eds.) PACS 2002. LNCS, vol. 2325, pp. 179–196. Springer, Heidelberg (2003)
23. White, R., Abels, T.: Energy resource management in the virtual data center. In: ISEE 2004: Proceedings of the International Symposium on Electronics and the Environment, Washington, DC, USA, pp. 112–116. IEEE Computer Society, Los Alamitos (2004)
24. Khargharia, B., Hariri, S., Yousif, M.S.: Autonomic power and performance management for computing systems. Cluster Computing 11(2), 167–181 (2007)
25. Bennani, M., Menasce, D.: Resource allocation for autonomic data centers using analytic performance models. In: Proceedings of the Second International Conference on Autonomic Computing, ICAC 2005, pp. 229–240 (June 2005)
26. Pasteels, J., Deneubourg, J., Goss, S.: Self-organization mechanisms in ant societies (i): trail recruitment to newly discovered food sources. In: Pasteels, J.M., Deneubourg, J.L. (eds.) From Individual to Collective Behavior in Social Insects. Experientia Supplementum, vol. 54, pp. 155–175. Birkhaüser, Basel (1987)
27. Nicolis, S., et al.: Optimality of collective choices: a stochastic approach. Bulletin of Mathematical Biology 65, 795–808 (2003)

28. Babaoglu, O., Canright, G., Deutsch, A., Caro, G.A.D., Ducatelle, F., Gambardella, L.M., Ganguly, N., Jelasity, M., Montemanni, R., Montresor, A., Urnes, T.: Design patterns from biology for distributed computing. ACM Trans. Auton. Adapt. Syst. 1(1), 26–66 (2006)
29. di Nitto, E., Dubois, D.J., Mirandola, R.: On exploiting decentralized bio-inspired self-organization algorithms to develop real systems. In: International Workshop on Software Engineering for Adaptive and Self-Managing Systems, pp. 68–75 (2009)
30. Serugendo, G.D.M., Karageorgos, A., Rana, O.F., Zambonelli, F. (eds.): ESOA 2003. LNCS (LNAI), vol. 2977. Springer, Heidelberg (2004)
31. Nakano, T., Suda, T.: Applying biological principles to designs of network services. Appl. Soft Comput. 7(3), 870–878 (2007)
32. Hariri, X.D., Xue, S.L., Chen, H., Zhang, M., Pavuluri, S., Rao, S.: Autonomia: an autonomic computing environment. In: IEEE International Performance, Computing, and Communications Conference (2003)
33. Parashar, M., Liu, H., Li, Z., Matossian, V., Schmidt, C., Zhang, G., Hariri, S.: Automate: Enabling autonomic applications on the grid. Cluster Computing 9(2), 161–174 (2006)
34. Hoefig, E., Wuest, B., Benko, B.K., Mannella, A., Mamei, M., Di Nitto, E.: On concepts for autonomic communication elements. In: International Workshop on Modelling Autonomic Communications (2006)
35. De Pellegrini, F., Miorandi, D., Linner, D., Bacsardi, L., Moiso, C.: Bionets architecture: from networks to serworks. In: Bio-Inspired Models of Network, Information and Computing Systems, Bionetics 2007, pp. 255–262 (December 2007)
36. Babaoglu, O., Meling, H., Montresor, A.: Anthill: A framework for the development of agent-based peer-to-peer systems. In: International Conference on Distributed Computing Systems, p. 15 (2002)

Towards a Pervasive Infrastructure for Chemical-Inspired Self-organising Services

Mirko Viroli, Matteo Casadei, Elena Nardini, and Andrea Omicini

Alma Mater Studiorum – Università di Bologna
via Venazia 52, 47023 Cesena, FC, Italy
{mirko.viroli,m.casadei,elena.nardini,andrea.omicini}@unibo.it

Abstract. Stimulated by the increasing availability of new mobile computing devices and the corresponding demand of open, long-lasting, and self-organising service applications, recent works proposed the adoption of a nature-inspired approach of chemistry for implementing service architectures [33]. One work in this direction is the chemical tuple-space model [30], by which the existence of data, devices and software agents (in one word, services of the pervasive computing application) gets reified into proper tuples managed by the infrastructure. System behaviour is accordingly expressed by chemical-like reactions that semantically match those tuples and accordingly enact the desired interaction patterns (composition, aggregation, competition, contextualisation, diffusion and decay).

After motivating the proposed approach for situated, adaptive, and diversity-accommodating pervasive computing systems, in this paper we outline an incarnation of this model based on the TuCSoN coordination infrastructure, which can been suitably enhanced with modules supporting semantic coordination and execution engine for chemical-inspired coordination laws.

1 Introduction

The characteristics of the Information and Communication Technology (ICT) landscape – yet notably changed by the advent of ubiquitous wireless connectivity – will further re-shape due to the increasing deployment of computing technologies like pervasive services and social networks: new devices with increasing interaction capabilities will be exploited to create services able to inject and retrieve data from any location of the very dynamic and dense network that will pervade our everyday environments. Addressing this scenario calls for finding infrastructures promoting a concept of pervasive "eternality", namely, changes in topology, device technology, and continuous injection of new services have to be dynamically tolerated as much as possible, and incorporated with no significant re-engineering costs at the middleware level [34,33]. As far as the coordination of such services is concerned, it will increasingly be required to tackle self-organisation (supporting situatedness, adaptivity and long-term accommodation of diversity) as an inherent system property rather than a peculiar aspect

D. Weyns et al. (Eds.): SOAR 2009, LNCS 6090, pp. 152–176, 2010.

of the individual coordinated components. As typical in self-organising computational mechanisms, a promising direction is to take inspiration from natural systems, where self-organisation is intrinsic to the basic "rules of the game". Among the available metaphors (e.g. physical, chemical, biological, social [33]) we focus on chemistry, which we argue that it nicely fits the requirements for self-organisation to emerge.

The concept of chemical tuple spaces is introduced [30], extending the standard tuple space model that is frequently adopted in middleware for situated and adaptive pervasive computing [27,11,18,19]. In chemical tuple spaces, tuples are seen as sort of species in a population, and coordination laws taking the form of chemical reactions semantically apply to such tuples and evolve them over time. Such an evolution behaviour is designed to be *exactly* the same of chemicals in a chemical system [15], hence promoting the exploitation of (natural or idealised) chemical reactions that are known to make interesting self-organisation properties emerge—e.g. auto-catalytic reactions [7]. Additionally, such laws are extended with a mechanism of tuple relocation in a network of tuple spaces, resembling chemical diffusion.

This paper focusses on how a distributed architecture for chemical tuple spaces can be implemented, namely, in terms of a coordination infrastructure providing the *fabric* of chemical tuples along with the stochastic and semantic application of chemical-like laws. As a basis for this implementation we start from the TuCSoN coordination infrastructure [23] supporting the notion of *tuple centre*, i.e. programmable tuple space. In particular, TuCSoN promotes a view of the tuple space as a set of facts in a logic theory, and its program as a set of rules dictating how the tuple set should evolve as time passes and as new interaction events occur. As such, TuCSoN appears a suitable means to enact the two basic ingredients necessary to implement chemical tuples spaces in TuCSoN: *(i)* chemical-inspired stochastic evolution of tuples, which is achieved by implementing the well-known Gillespie's exact simulation algorithm [15] as a tuple space program, and *(ii)* semantic matching (of chemical laws against tuples in the space), obtained by substituting logic unification – as used in TuCSoN – with matching a lá Description Logic [20], namely, seeing a tuple as an individual of an ontology, and a reactant in the chemical reaction as a concept of the ontology [16].

The remainder of this paper is organised as follows. Section 2 motivates the proposed approach by analysing a case study of virtual display infrastructures and outlining its requirements. Section 3 informally introduces the chemical-inspired coordination model (a formal account is already given in [30]), and provides abstract examples of coordination laws inspired by population dynamics [7]. Section 4 grounds the discussed model and examples in a case study for the pervasive display infrastructure, based on advertisements and information services in an airport scenario, showing examples of how we can achieve the desired self-* properties. Section 5 describes how the TuCSoN coordination infrastructure can be tailored to support the proposed model. Finally, Section 6 outlines related works in coordination and middleware for self-adaptive and self-organising systems, and Section 7 concludes discussing a roadmap towards completing the development of the infrastructure.

2 Background and Motivation

2.1 Requirements for Self-organising Pervasive Service Systems

In order to better explain the motivation behind the model presented in this paper, we rely on a case study, which we believe well represents a large class of pervasive computing applications in the near future.

We consider a pervasive display infrastructure, used to surround our environments with digital displays, from those in our wearable devices and domestic hardware, to wide wall-mounted screens that already pervade urban and working environments [12]. However, instead of considering displays as static information servers as usual nowadays (i.e. showing information in a manually configured manner), we envision a truly general, open, and adaptable information service infrastructure. As a reference domain, we consider an airport terminal filled with wide screens mounted in the terminal area, i.e in the shops, in the corridors and in the gates, down to tiny screens installed in each seat of the gate areas or directly on passengers' PDAs.

Situatedness. We first notice that information should be generally displayed based on the current state of the surrounding physical and social environment. For instance, by exploiting information coming from surrounding temperature sensors and passenger profiles/preferences, an advertiser could decide to have ice tea commercials – instead of liquor ones – displayed on a warm day and in a location populated by teenagers.

Thus, a general requirement for pervasive services is *situatedness*. Namely, pervasive services deal with spatially and socially situated users' activities, hence, they should be able to interact with the surrounding physical and social world, accordingly adapting their behaviour. As a consequence, the infrastructure itself should act based on spatial concepts and data.

Adaptivity. Secondly, and complementary to the above, the display infrastructure, and the services within it, should be able to automatically adapt to changes and contingencies in an automatic way. For instance, when a great deal of new information to be possibly displayed emerges, the displayed information should overall spontaneously re-distribute and re-shape across the set of existing local displays, possibly discharging obsolete visualisation services.

Accordingly, another requirement is *adaptivity*. Pervasive services and infrastructures should inherently exhibit self-adaptation and self-management properties, so as to survive contingencies without any human intervention and at limited management costs.

Diversity. Finally, the display infrastructure should be not only intrinsically open to any kind of visualisation services that may be added to the system, but also able to allow users – other than display owners – to upload information to displays so as to enrich the information offer or adapt it to their own needs. For instance, a passenger could watch private content uploaded from her/his PDA to a wider screen close to her/his seat, and may be willing also to share it with people nearby. Put simply, users should act as "prosumers"—i.e. as both consumers and producers of devices, data, and services.

Another general requirement is hence *diversity*. Namely, the infrastructure should tolerate open models of service production and usage without limiting the number and classes of services potentially provided, but rather taking advantage of the injection of new services by exploiting them to improve and integrate existing services whenever possible.

2.2 Chemical-Inspired Tuple Spaces for Pervasive Services

In many proposals for pervasive computing environments and middleware infrastructures, situatedness and adaptiveness are promoted by the adoption of shared virtual spaces for services and component interaction [27,11,18,19,30]. In these approaches, tuple spaces are disseminated in the pervasive environment, one in each network location, and reify the local situation in terms of tuples (structured information items). Depending on the specific proposal, such tuples can represent the occurrence of people nearby, the availability of devices, the state of pervasive services, knowledge, contextual information, signals spread in the network, and so on. Relying on such shared virtual spaces has a main implication: system coordination can be achieved by a rather simple set of rules for managing tuples, acting locally to each space and providing for the "laws" by which such tuples evolve, diffuse, and possibly combine so as to support adaptivity [19].

Among the various tuple space models proposed in literature, we adopt the chemical tuple space model [30], in which tuples – containing semantic information about the "individuals" to be coordinated (services, devices, data) – evolve in a stochastic and spatial way through coordination laws resembling chemical reactions. We observe that this model can properly tackle the requirements sought for adaptive pervasive services.

Concerning *situatedness*, the current situation in a system locality is represented by the tuples existing in the tuple space. Some of them can act as catalysts for specific chemical reactions, thus making system evolution intrinsically context-dependent. Moreover, mechanisms resembling chemical diffusion can be designed to make a node influencing neighbouring ones.

Concerning *adaptivity*, it is known from biology that some complex chemical systems are auto-catalytic (i.e. they produce their own catalyst), providing positive-negative feedbacks that induce self-organisation [10] and lead to the spontaneous toleration of environment perturbations. Such systems can be modelled by simple idealised chemical reactions, e.g. prey-predator systems, Brusselator, and Oregonator [15]. Regarded as a set of coordination laws for pervasive services, this kind of chemical reactions has the potential to intrinsically support adaptivity.

Finally, considering *diversity*, we note that chemical reactions follow a simple pattern: they have some reactants (typically 1 or 2) which combine, resulting in a set of products, through a propensity (or rate) dictating the likelihood for this combination to actually happen. In natural chemistry, this generates a plethora of specific chemical reactions that take into account the diversity of chemical species, and the possibility of creating complex molecular structures. In our framework, general reactions can be designed that can be instantiated for the specific and unforeseen services that will be injected in the system over time—using semantic matching to fill the gap.

Though key requirements seem to be supported in principle, designing the proper set of chemical reactions to regulate system behaviour is crucial. Without excluding the appropriateness of other solutions, in this paper we mostly rely on chemical reactions resembling laws of population dynamics as, e.g. the prey-predator system [15,7]. This kind of idealised chemical reactions has been successfully used to model auto-catalytic systems manifesting self-organisation properties: but moreover, they can also nicely fit the "ecological" metaphor that is often envisioned for pervasive computing systems [5,29,1,33,34]—namely, seeing pervasive services as spatially situated entities living in an ecosystem of other services and devices.

3   The Coordination Model of Chemical Tuple Spaces

In this section, first we informally introduce the coordination model of chemical tuples spaces (Section 3.1), then describe some example applications in the context of competitive pervasive services (Section 3.2).

3.1   Coordination Model

The chemical tuple space model is an extension of standard LINDA settings with multiple tuple spaces [14]. A LINDA tuple space is simply described as a repository of tuples (structured data chunks like records) for the coordination of external "agents", providing primitives used respectively to insert, read, and remove a tuple. Tuples are retrieved by specifying a tuple template—a tuple with wildcards in place of some of its arguments. The proposed model enhances this basic schema with the following ingredients.

Tuple Concentration and Chemical reactions. We attach an integer value called "concentration" to each tuple, measuring the pertinence/activity value of the tuple in the given tuple space: the higher such a concentration, the more likely and frequently the tuple will be retrieved and selected by the coordination laws to influence system behaviour. Tuple concentration is dynamic, as typically the pertinence of system activities is. In particular, tuple concentration "spontaneously" evolves similarly to what happens in chemical behaviour, namely, a tuple with concentration N is handled pretty much in the same way as if it were a chemical substance made of N molecules of the same species. This is achieved by coordination rules in the form of chemical reactions—the only difference with respect to standard chemical reactions is that they now specify tuple templates instead of molecules. For example, a reaction "$X + Y \xrightarrow{0.1} X + X$" would mean that tuples x and y matching X and Y are to be selected, get combined, and as a result one concentration item of y turns into x—concentration of y decreases by one, concentration of x increases by one. According to [15], this transition is modelled as a Poisson event with average rate (i.e. frequency) $0.1 \times \#x \times \#y$ ($\#x$ is the concentration of x). In the general case, such a rate is obtained by multiplying reaction rate by a coefficient depending scontributeolely on the concentration of reactants in the solution: in particular, the contribution of each reactant in the product is $\binom{m}{n}$, where m is the concentration of the reactant in the solution, and n is the number of such reactants existing in the chemical reaction (typically, $n = 1$ or $n = 2$). This model makes a tuple space running as a sort of exact chemical simulator, picking reactions probabilistically: external agents observing the evolution of tuples would perceive something equivalent to the corresponding natural/artificial chemical system described by those reactions.

Semantic Matching. It is easy to observe that standard syntactic matching for tuple spaces can hardly deal with the openness requirement of pervasive services, in the same way as syntactic match-making has been criticised for Web services [24]. This is because we want to express general reactions that apply to specific tuples independently of their syntactic structure, which cannot be clearly foreseen at design time. Accordingly, *semantic matching* can be considered as a proper matching criterion for our coordination infrastructure [24,4,8].

It should be noted that matching details are orthogonal to our model, since the application at hand may require a specific implementation of them—e.g. it strongly depends on the description of application domain. As far as the model is concerned, we only assume that matching is fuzzy [8], i.e. matching a tuple with a template returns a "vagueness" value between 0 and 1, called *match degree*. Vagueness affects the actual application rate of chemical reactions: given a chemical reaction with rate r, and assume reactants match some tuples with degree 0.5, then the reaction can be applied to those tuples with an actual rate of $0.5 * r$, implying a lower match likelyhood—since match is not perfect. Namely, the role of semantic matching in our framework is to allow for coding general chemical laws that can uniformly apply to specific cases—appropriateness influences probability of selection.

Tuple Transfer. We add a mechanism by which a (unit of concentration of a) tuple can be allowed to move towards the tuple space of a neighbouring node, thus generating a *computational field*, namely, a data structure distributed through the whole network of tuple spaces. Accordingly, we introduce the notion of "firing" tuple (denoted $t^{\rightsquigarrow}$), which is a tuple (produced by a reaction) scheduled for being sent to a neighbouring tuple space—any will be selected non-deterministically. For instance, the simple reaction "$X \xrightarrow{0.1} X^{\rightsquigarrow}$" is used to transfer items of concentration of any tuple matching X out from the current tuple space.

3.2   Examples

We now discuss some examples of chemical reactions enacting general coordination patterns of interest for pervasive service systems. We proceed incrementally, first providing basic laws for service matching and competition, which will be then extended towards a distributed setting.

Local competition. We initially consider a scenario in which a single tuple space mediates the interactions between pervasive services and their users in an open and dynamic system. In the pervasive display infrastructure, this example is meant to model the basic case where, given the node where a display is installed, visualisation services are to be selected based on the profile of users nearby the display. We aim at enacting the following behaviour: *(i)* services that do not attract users fade until eventually disappearing from the system, *(ii)* successful services attract new users more and more, and accordingly, *(iii)* overlapping services compete one another for survival, so that some/most of them eventually come to extinction.

   An example protocol for service providers can be as follows. A tuple `service` is first inserted in the space to model publication, specifying service identifier and (semantic) description of the service content. Dually, a client inserts a specific request as a tuple `request`—insertion is the (possibly implicit) act of publishing user preferences. The tuple space is charged with the role of matching a request with a reply, creating a tuple `toserve(service,request)`, combining a request and a reply semantically matching. Such tuples are read by the service provider, which collects information about the request, serves it, and eventually produces a result emitted in the space with a tuple `reply`, which will be retrieved by the client. The abstract rules we use to enact the described behaviour are as follows:

(USE) $\text{SERV} + \text{REQ} \overset{u}{\mapsto} \text{SERV} + \text{SERV} + \texttt{toserve(SERV,REQ)}$

(DECAY) $\text{SERV} \overset{d}{\mapsto} 0$

On the left side (reactants), `SERV` is a template meant to match any service tuple, `REQ` a template matching any request tuple; on the right side (products), `toserve(SERV,REQ)` will create a tuple having in the two arguments the service and request tuples selected, while 0 means there will be no product. Rule (USE) has a twofold role: *(i)* it first selects a service and a request, it semantically matches them and accordingly creates a `toserve` tuple, and dynamically removes the request; and *(ii)* it increases service concentration, so as to provide a positive feedback—resembling the prey-predator system described by Lotka-Volterra equations [7,15]. We refer to *use rate* of a couple service/request as u multiplied by the match degree of those reactants when applying (USE) law, as described in the previous section: as a result, it can be noted that the higher the match degree, the more likely a service and a request are combined. On the other hand, rule (DECAY) makes any concentration item of the service tuple disappear at rate d, contrasting the positive feedback of (USE): here, the overall *decay rate* of a service is d multiplied by the match degree—with no match, we would have no decay at all.

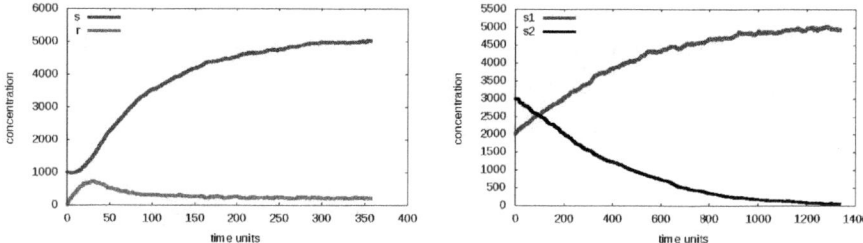

Fig. 1. Service s exploited by matching requests r (left); and competition between services $s1$ and $s2$ (right)

In Figure 1 (left) we consider a scenario in which requests r are injected at average rate 50, and a matching service s exists in the system with initial concentration $1,000$: we additionally have decay rate 0.01 and use rate 0.05. We can observe that after an initial growth, the number of requests which are not served stabilises to few hundreds, while the concentration of s grows to about $5,000$. The behaviour of service concentration can be understood in terms of the positive-negative feedback loop of rules (USE) and (DECAY). It can be shown that service concentration increases while stabilising to about p/d, where p is the rate of injection of requests (pumping rate) and d is the service decay rate (decay rate d).

We now consider a similar scenario but with two services $s1$ and $s2$ initially having concentration $2,000$ and $3,000$ respectively, and matching the same requests, though with different use rate: 0.04 for $s1$, and 0.06 for $s2$. This models the situation in which two different services exist to handle requests, one leading to a better match. The result is that $s1$ and $s2$ engage a competition: this is lost by $s1$ which starts fading until completely vanishing (i.e. being disposed) even though it has an initially higher concentration, as shown in Figure 1 (right). In fact, the sum of the concentration of $s1$ and $s2$ still stabilises to $5,000$, but the contribution of $s1$ and $s2$ changes depending on the number of requests they can serve. Hence, matching degree is key when more services are concerned and the shape and dynamics of user requests is unknown, as it is responsible of the rate at which a service is selected each time, and ultimately, of the evolution of service concentration, i.e. of its competition/survival/extinction dynamics.

Spatial competition. This example can be extended to a network of tuple spaces, so as to emphasise the spatial and context-dependent character of competing services. Suppose each space is programmed with (USE,DECAY) reactions plus a simple diffusion law for service tuples:

$$(\text{DIFFUSE}) \qquad \text{SERV} \overset{m}{\longmapsto} \text{SERV}^{\rightsquigarrow}$$

The resulting system can be used to coordinate a pervasive service scenario in which a service is injected into a node of the network (e.g. the node where service is more urgently needed, or where the producer resides), and accordingly

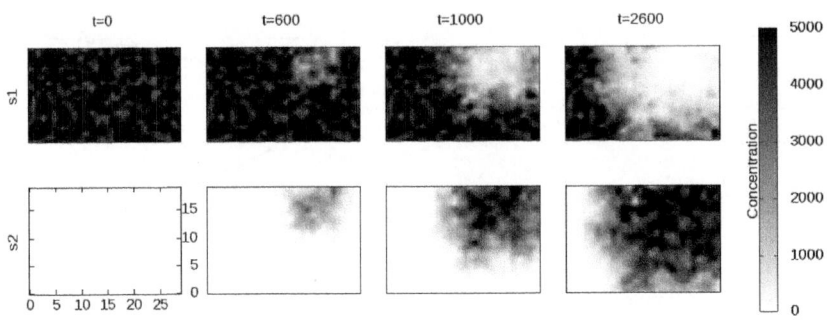

Fig. 2. Spatial competition: after an initial pointwise injection, service s2 (down) globally overcomes s1 (top)

starts diffusing around on a step-by-step basis until possibly covering the whole network—hence becoming a global service. This situation is typical in the pervasive display infrastructure, since a frequent policy for visualisation services would be to show them on any display of the network—although more specific policies might be enacted to make certain services only locally diffuse.

In this system, we can observe the dynamics by which the injection of a new and improved service may eventually result in a complete replacement of previous versions—spatially speaking, the region where the new service is active is expected to enlarge until covering the whole system, while the old service diminishes. In the context of visualisation services, for instance, this would amount to the situation where an existing advertisement service is established, but a new one targeted to the same users is injected that happens to have greater use rate, namely, it is more appropriate for the average profile of users: we might expect this new service to overcome the old one, which accordingly extinguishes.

For the sake of explanation, we start from an abstract case, with a reference "random grid" of 30×20 nodes (i.e., locations). We call a random grid a lattice-like network in which locations are placed as nodes of a square grid (each non-boundary location has in its proximity 8 nodes, 4 in the horizontal/vertical direction and 4 in the diagonal direction), but connection between a location and a neighbouring one is randomly set (with probability 50%). This choice is motivated by the fact that very often computing devices are placed more or less uniformly over the "space" formed by the buildings, corridors, or rooms of the pervasive computing systems of interest, though randomness is useful to tackle heterogeneity of the environment at hand, failures, and so on.

In the case we consider, in every node requests for using a service are supposed to arrive at a fixed rate for simplicity, and a service called s1 is the only available to match the requests (we use the following parameters: use rate $u = 0.01$, decay rate $d = 0.01$, request injection rate $p = 50$, moving rate $m = 0.01$). In particular, in every node, the system stabilises approximately to a concentration of $5,000$ s1 (p/d as usual), in spite of diffusion.

Another service s2 is at some point developed that can serve the same requests, now with use rate 0.1 instead of 0.05, namely, it is a service developed to more effectively serve requests—it matches requests twice as much as s1 does. This service is injected into a randomly chosen node of the network, with an initial very low concentration (10 in our experiment). Figure 2 shows in each column a different snapshot (from left to right), reporting concentration of s1 on top and s2 on bottom: we can observe that s2 starts diffusing where it is injected, until completely overcoming service s1 after about 3,000 time units.

4   Case Study of Long-Term Competition

To better ground the discussion, and emphasise the adaptive and diversity-accommodation character of our model, we consider some application case for the airport display infrastructure, showing how the proposed reactions would work in a more concrete setting.

4.1   Local Competition

We first analyse the behaviour of a single display, located near a gate where passengers wait for the departure of their flight. As soon as a passenger gets nearby, her/his preferences are sensed, and become tuples representing requests for a visualisation service—such a sensing might be due to either the passenger's PDA or the passenger's data which are stored in an RFID (or alike) placed on the boarding pass. Visualisation services are continuously injected in the system (in the long-term, they could be many): they are meant to tackle passengers' preferences (e.g. sport, food, tourism, cars) and accordingly compete with one another, since the display is meant to probabilistically select the best service/passenger match.

Figure 3 shows all the parameters of the considered simulation scenario, in which 100 visualisation services are injected during a year. Parameters r_req and t_stay are inferred from Heathrow statistics in 2008[1]. Match rate is the rate at

Parameter	Value	Description
r_req	$141/137\ min^{-1}$	preference injection rate, as passengers per flight over interval between two flights
t_stay	$50\ min$	passenger time nearby the display
r_ads	$100\ year^{-1}$	advertisement service injection rate
t_show	$30\ sec$	showing time for an advertisement service
c_ad	$1,000$	maximum expected concentration of an advertisement service
r_match	$1,000\ sec^{-1}$	match rate

Fig. 3. Airport scenario: competition of services in one display. Simulation parameters.

---

[1] http://www.caa.co.uk/

which a single match can be performed: note this is negligible with respect to the time between arrival of two passengers' preferences (namely, about 1 minute due to r_req). For the simulation, we used decay rate $d = 1/30,000 \ sec^{-1} = 1/(\text{c\_ad} \times \text{t\_show})$, since the final service concentration c_ad has been shown to be p/d, and the pumping rate for services p is $1/\text{t\_show}$ (service concentration increases by 1 each 30 seconds).

This simulation scenario is relative to a class of advertisement services (e.g., concerning cars), which can match 5 different "marketing targets" (e.g. sport cars, luxury cars, city cars, vans, crossovers). Each passenger is associated to a single marketing target, while each advertisement can cover many marketing targets (e.g. a Ferrari is both a sport and luxury car). Accordingly, each time an advertisement service is created, we randomly draw its match degree with respect to the 5 different marketing targets (a number in between 0 and 1 each), though we keep the sum of such degrees less than the "overlap factor" 1.5 (OF)—to avoid the unrealistic case in which some advertisement perfectly fits all marketing targets (which could happen if $OF = 5$).

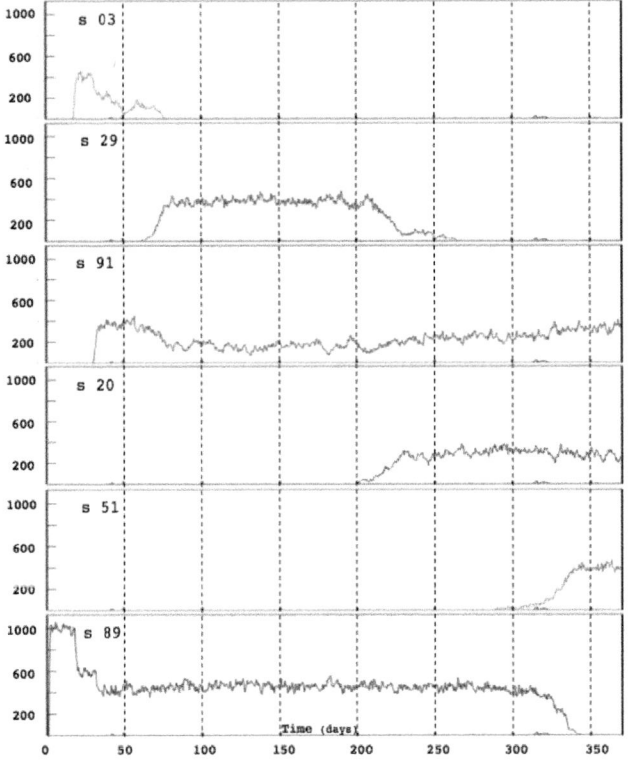

Fig. 4. Airport scenario: competition of services in one display. Charting concentration of the 6 services active throughout the simulation.

Figure 4 shows a simulation over a whole year. We note that: *(i)* only few services are actually active at a given time (i.e., they have non-negligible concentration), for the others get extinguished throughout system evolution, and *(ii)* some new service can overcome an existing and established one, causing its extinction (e.g. $s51$ enters the system at day 290, and makes $s89$ extinguishing at day 350)—results of a larger number of simulations show that the average number of active services in the system is about 3.25. At the end of the year, only the following three services are active (reported with their matching degrees):

s91[0.52,0.11,0.84,0,0], s20[0.62,0.84,0,0,0], s51[0,0,0,0.75,0.70]

Namely, $s91$ is mainly tackling the fourth marketing target (and a good deal of the first), $s20$ is mainly tackling the second marketing target (and a good deal of the first as well), while $s51$ mainly tackles the fourth and fifth targets. Of course, in practice such match factors will be computed from semantic matching module, which ranks the extent to which a newly introduced advertisement deals with the 5 marketing targets. By increasing the number of marketing targets and their overlap factor we can deal with more involved situations; for instance, analogous simulations with 20 marketing targets and $OF = 3$ give an average number of 7 visualisation services for the specific class considered. In general, it is predictable that the services that best tackle one ore more marketing targets will survive, while most of the others will end up extinguishing without unnecessarily overloading the system, and with no human intervention. This sort of "ecological" behaviour is typical of today socially situated domains like social networks, and will be likely to play a key role in future pervasive computing systems [1,29,34].

4.2   Spatial Competition

We now analyse a more concrete example, extending the airport scenario studied in previous section to show the spatial character of our framework. Instead of a single display we now consider an airport terminal with 5 gates in a row, and 25 displays near each gate disseminated in the corridor and gate areas. This is modelled as a 25×5 random grid (gates are at coordinates $(3, 3)$, $(8, 3)$, $(13, 3)$, $(18, 3)$, $(23, 3)$).

Advertisement services are now injected from a random node of the network (taken from 8 nodes in the perimeter of the grid considered as entry point nodes), using same dynamics of previous case. Such services diffuse using (DIFFUSE) reaction.

In a scenario in which passenger preferences are uniformly distributed in space and time, we would expect a behaviour similar to that of Figure 2, where winning services diffuse in the whole network reaching an uniform value. But in a real-life situation preferences are not uniform but context-dependent, and this influences the actual region in which certain services can actually win competition. As an example, we consider a class of services containing news about specific locations in the world, and each passenger's profile – e.g. automatically extracted from

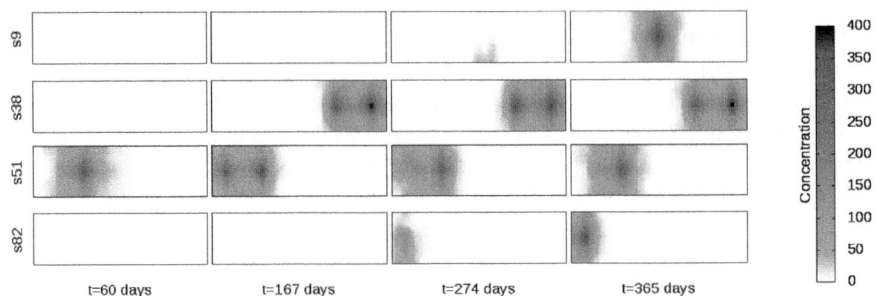

Fig. 5. Airport scenario: competition of services in the terminal. Showing spatial concentration of the 4 surviving services, in 4 snapshots.

the RFID in the boarding pass – is associated to the preference for just one continent, namely, the one where she/he is flying to. As in the previous case, the overlap factor is 1.5 (in that some news service might span more continents). Now assume each gate hosts flights towards a given continent: this means that each gate is a context where passengers will more likely be interested in news on the corresponding continent. This is obtained by making the injection rate of preferences dependent on the distance from the gate: the higher the distance, the smaller the rate (and still r_req in the node of the gate, as in previous section).

A simulation result is shown in Figure 5, which emphasises again the adaptive character of our framework, now also taking into account spatial aspects. Only 4 services are active at the end of the simulation, which are those actually charted. At day 60, $s51$ already established at 2nd gate (from left). At day 167, $s51$ is also establishing on 1st gate, while $s38$ established on gate 4th and 5th. At day 274, $s9$ is appearing on 3rd gate and $s82$ is taking over 1st gate winning competition against $s51$. At the end of simulation, both $s9$ and $s82$ completely established.

Note that in this model, `service` tuples act as a reification of the spatial service state as enacted by the coordination infrastructure: the resulting system features situatedness (success of a service in a location depends on requests and existing services there), adaptivity (the best service actually wins, and unused services fade and get garbage-collected), and accommodation of diversity (the arrival of new services is not foreseen at design time, but automatically managed). In spite the discussed set of chemical reactions appears suitable for the application at hand and its requirements, we believe that different contexts can call for different reactions, without harming the general validity of the proposed model/architecture.

5   An Architecture Based on the TuCSoN Infrastructure

In this section we show that an infrastructure for the chemical tuple space model does not have to be necessarily built from scratch, but can be implemented on top of an existing tuple space middleware, such as TuCSoN. This provides us for free with many model-independent features like tuple spaces life-cycle management, tuple retrieval, agent interactions and so on. In particular, after a general overview of TuCSoN (Section 5.1), we describe how the two basic additional ingredients of the proposed model can be supported on top of TuCSoN: semantic matching (Section 5.2) and chemical engine (Section 5.3).

5.1   Overview of TuCSoN

TuCSoN (Tuple Centres Spread over the Network) [23] is a coordination infrastructure that manages the interaction space of an agent-based system by means of ReSpecT [22] *tuple centres*, which are *Linda* tuple spaces [14] empowered with the ability to define their behaviour in response to communication events. Hence, tuple centres can be thought of as virtual machines for coordination; they enact spaces reifying the existence of agents and the occurrence of their interactions: such tuples are properly created, combined on the fly, evolved and destroyed so as to enable and constrain system coordination.

In the adopted ReSpecT language, a reaction is specified by a *specification tuple* of the kind reaction(E,G,R), which associates (by unification) a reaction body R to an event E if the guard G is satisfied: E expresses the agent interaction to intercept, G the condition that the event must satisfy (concerning status, source, target and time of the event), and R is a list of goals. A goal specifies a basic computation, which could be *(i)* an insertion/retrieval of tuples in the local tuple space (which might itself fire new reactions leading to a Turing-equivalent chain of reaction executions); *(ii)* an insertion/retrieval of tuples in a remote tuple space; and *(iii)* a Prolog algorithmic computation. As an example of a simple ReSpecT program, we consider the following rules, which change the behaviour of the in(X) primitive – removing a single tuple matching X – so that *all* tuples matching the template X are removed:

```
reaction( in(X), (response,from agent), ( out(remove(X))              )).
reaction( out(remove(X)), endo,( in(X), in(remove(X)), out(remove(X)) )).
reaction( out(remove(X)), endo,( no(X), in(remove(X))                 )).
```

Whenever an agent asks to remove a tuple of a generic template X, the first reaction is triggered and executed (the guard predicate response is true when in(X) gets served), causing insertion of tuple remove(X). This triggers the second and third reactions: the former removes a tuple matching X (in(X) succeeds) and re-inserts remove(X), while the latter intercepts the case where no matching tuple is found. Since each reaction is atomically executed, the resulting behaviour is such that all tuples matching X are removed.

Other than supporting the notion of tuple centre, TuCSoN also extends Linda model from the topology viewpoint. While the standard tuple space model was originally conceived for enabling parallel processing in centralised computer network environments [14], TuCSoN has been conceived for providing a suitable infrastructural support for engineering distributed application scenarios [23]. In particular, tuple centres are distributed through the network, hosted in the nodes of the infrastructure, and organised into articulated domains. A domain is characterised by a gateway node and a set of nodes called places: A place is meant

to host tuple centres for the specific applications/systems, while the gateway
node is meant to host tuple centres used for domain administration, keeping
information on the places.

As discussed in [32], TuCSoN supports features that are key for implementing
self-organising systems, some of which are here recapped that are useful for
implementing the chemical tuple space model:

Topology and Locality. Tuple centres can be created locally to a specific
node, and the gateway tuple centre can be programmed to keep track of
which tuple centres reside in the neighbourhood—accessible either by agents
or by tuple centres in current node.

On-line character and Time. TuCSoN supports the so called "on-line coordination services", executed by reactions that are fired in the background of
normal agent interactions, through timed reactions—reactions whose event
E is of kind `time(T)`. When the tuple centre time (expressed as Java milliseconds) reaches T, the corresponding reaction is fired. Moreover, a reaction
goal can be of the kind `out_s(reaction(time(T),G,R))`, which inserts tuple `reaction(time(T),G,R)` in the space, thus triggering a new reaction.
As a simple example, the following reactions are used to insert a tuple `tick`
in the space each second:

```
reaction( time(T), endo,   ( out(tick) )).
reaction( out(tick), endo, ( currentTime(T), NewT is T+1000,
                  out_s(reaction( time(NewT), endo, out(tick))) )).
```

Probability. Probability is a key feature of self-organisation, which is necessary to abstractly deal with the unpredictability of contingencies in pervasive
computing. In TuCSoN this is supported by drawing random numbers and
using them to drive the reaction firing process, that is, making tuple transformation be intrinsically probabilistic. For instance, the following reaction
inserts either tuple `head` or `tail` in the space (with 50% probability):

```
reaction( out(draw), from-agent, (X > 0.5, out(head)) ; out(tail) )).
```

5.2   Semantic Matching

We now describe the extension of the tuple centre model that allows us to perform semantic reasoning over tuples, namely, the ability of matching a tuple with
respect to a template not only syntactically as usual, but also semantically. In
spite some approaches have been proposed to add semantic reasoning to tuple
spaces [21], we here introduce a different model, which is aimed at smoothly
extending the standard settings of tuple spaces [20].

Abstract model. From an abstract viewpoint, a tuple centre can be seen as a
knowledge repository structured as a set of tuples. According to a semantic view,
such knowledge represents a set of objects occurring in the application domain,

whose meaning is described by an ontology, that is, in terms of concepts and relations among them.

In order to formally define the notions of domain ontology and objects, the literature makes available a family of knowledge representation formalisms called *Description Logics* (DL) [2]—we rely on *SHOIN(D)* Description Logic, which represents the theoretical counterpart of W3C's standard *OWL* DL [16]. In DL, an ontology component called *TBox* is first introduced that includes the so-called terminological axioms: *concept* descriptions (denoting meaningful sets of individuals of the domain), and *role* descriptions (denoting relationships among individuals). Concepts can be of the following kinds: $\top$ is the set of all objects, $\bot$ the void set, $C \sqcup D$ is union of concepts, $C \sqcap D$ intersection, $\neg D$ negation, $\{i_1, \ldots, i_n\}$ is a set of individuals, $\forall R.C$ is the set of objects that are in relation (through role R) with only objects belonging to concept C, $\exists R.C$ is the set of objects that are in relation (through role R) with at least one object belonging to concept C, and $\leq nR$ is the set of objects that are in relation (through role R) with no more than n objects (and similarly for concepts $\geq nR$ and $= nR$). Given these constructs, the TBox provides axioms for expressing inclusion of concepts ($C \sqsubseteq D$). For instance, a TBox for a *car domain* [8] can provide the following assertions

$$MaxSpeed \sqsubseteq \{90\text{km/h}, 180\text{km/h}, 220\text{km/h}, 280\text{km/h}\}$$
$$Car \sqsubseteq (= 1 hasMaxSpeed)$$
$$(= 1 hasMaxSpeed) \sqsubseteq Car$$
$$\top \sqsubseteq \exists\, hasMaxSpeed.MaxSpeed$$
$$SlowCar \sqsubseteq Car \sqcap (\exists\, hasMaxSpeed.\{90\text{km/h}\})$$
$$CityCar \sqsubseteq SlowCar$$

which respectly: *(i)* defines the concept *MaxSpeed* as including 4 individuals, *(ii)* defines the concept *Car* and states that all its objects have precisely one maximum speed value, *(iii)* conversely states that any object with one maximum speed value is a car, *(iv)* states that maximum speed value is an object of *MaxSpeed*, *(v)* defines *SlowCar* as a new concept (sub-concept of *Car*) such that its individuals have 90km/h as maximum speed, and finally *(vi)* defines a *CityCar* as a kind of slow car—possibly to be completed with other features, e.g. being slow, compact and possibly featuring electric battery. By these kinds of inclusion, which are typical of the OWL standard approach, one can flexibly provide suitable definitions for concepts of sport cars, city cars, and so on (i.e., filling an ontology for the car advertisement domain of Section 4.1).

Another component of DL is the so-called *ABox*, defining axioms to assert specify domain objects and their properties: they can be of kind $C(a)$, declaring individual a and the concept C it belongs to, and of kind $R(a, b)$, declaring that role R relates individual a with b. Considering the car domain example, the ABox could include axioms $Car(\text{f40})$, $Car(\text{fiat500})$, $hasMaxSpeed(\text{f40}, 280\text{km/h})$, and $hasMaxSpeed(\text{fiat500}, 90\text{km/h})$.

Semantic reasoning of DL basically amounts – among the others – to check whether an individual belongs to a concept, namely, the so-called semantic matching. As a simple example, a DL checker could verify that fiat500 is an instance of *CityCar*. This contrasts syntactic matching, which would have failed since *Car* and *CityCar* are two "types" that do not syntactically match—they rather semantically match due to the definitions in the TBox.

Given the above concepts of DL, we design the semantic extension of tuple centres by the following ingredients, which will be described in more detail in turn: *(ontologies)* an ontology has to be attached to a tuple centre, so as to ground the definition of concepts required to perform semantic reasoning; *(semantic tuples)* a semantic tuple represents an individual, and a language is hence to be introduced to specify individual's name, the concept it belongs to, and the individuals it is related to by roles; *(tuple templates)* templates are to be used to flexibly retrieve tuples, hence we link the semantic tuple template notion with that of concept in the ontology, and accordingly introduce a template language; *(matching mechanism)* matching simply amounts to check whether the tuple is an instance of the concept described by the template (providing a match factor in between 0 and 1), hence we let the notions of *priority* and *optionality* into the language of templates, so as to make it possible to introduce a "fuzzy" notion of matching.

Ontology Language. In our implementation we adopt the OWL [16] ontology language in order to define domain ontologies in TuCSoN. OWL is an XML-based ontology language introduced by W3C for the Semantic Web: relying on a standard language for ontologies is key for the openness aims of the application domains considered in this paper, and moreover, standard automated reasoning techniques can be exploited relying on existing open source tools—like e.g. *Pellet* reasoner [28]. Accordingly, each tuple centre carries an OWL ontology describing the TBox, and that internal machinery can be easily implemented so as to query and semantically reason about it.

Semantic Tuples Language. Instead of completely departing from the syntactic setting of TuCSoN, where tuples are expressed as first-order terms, we design a smooth extension of it, so as to capture a rather large set of situations. The car domain example described above would be expressed by the two semantic tuples f40:'Car'(hasMaxSpeed:'280km/h') and fiat500:'Car'(hasMaxSpeed:'90km/h'), namely, each tuple orderly describes individual name, concept name, and list of role fillers—extending the case of first-order tuples, where each tuple is basically the specification of a "type" (functor name) followed by an ordered list of parameters. Another example is:

```
f550 : 'Car' (hasMaker : ferrari,
              hasMaxSpeed : '285km/h',
              hasEnergyPower in {gasoline, diesel})
```

Note that semantic tuples are basically first-order terms with the introduction of few infix binary functors (":" and "in").

Semantic Templates Language. Another aspect to be faced concerns the representation of semantic tuple templates as specifications of sets of domain individuals (concepts) among which a matching tuple is to be retrieved. The grammar of tuple templates we adopt basically turns DL concepts into a term-like syntax (as in Prolog), as follows:

$$C ::= \text{'\$ALL'} \mid \text{'\$NONE'} \mid \texttt{cname} \mid C,C \mid C;C \mid \texttt{not}(C) \mid \{\texttt{inamelist}\} \mid CR$$
$$CR ::= [\texttt{exists}|\texttt{only}](\texttt{pname in } C) \mid \# \gtrless N : \texttt{pname}$$

Elements `cname`, `iname`, and `pname` are constants terms, expressing names of concepts, individuals and properties. Following the grammar, concepts orderly express *all* individuals, *no* individual, a concept name, intersection, union, negation, an individual list, or a concept specified via role-fillers. Examples of the latter include: "exists P in C" (meaning $\exists P.C$), "only P in C", (meaning $\forall P.C$), "P in C" (meaning $\exists P.C \sqcap \forall P.C$), "$\# \geq 2\text{:P}$" (meaning $\geq 2P$). Additional syntactic sugar is used: "exists $P\text{:}i$" stands for "exists P in $\{i\}$", and "$C(CR_1,..,CR_n)$" stands for "$C,CR_1,..,CR_n$". Examples of semantic templates are as follows:

```
'Car'(exists hasMaxSpeed:{'90km/h','280km/h'})
'Car'(#>1:hasEnergyPower),((hasMaker:ford) ; (hasMaker:ferrari))
```

The former specifies those cars having either `90km/h` or `280km/h` maximum speed, the latter those cars that come with at least two choices of energy power and that have either `ford` or `ferrari` maker.

Semantic matching. In order to enable semantic support in TuCSoN, a tuple centre has to be related to an ontology, to which semantic tuples refer to. In order to encapsulate an ontology, tuple centres exploit the aforementioned *Pellet* reasoner [28]—an open-source DL reasoner based on OWL and written in Java likewise TuCSoN. In particular, *Pellet* can load an OWL TBox and an ABox, and provides the *Jena-API* in order to add and remove individuals by its own ABox.

Hence, each semantic tuple is carried not only in the tuple space likewise syntactic tuples, but also in the ontology ABox, in order to support reasoning. The reasoner is internally called each time we are checking for a semantic match: the semantic template is converted into a SPARQL query (the language used by Jena-API) and the results obtained by the reasoner (the name of individuals) are used to retrieve the actual tuples in the tuple space.

This behaviour is embedded in the tuple centre, such that each time a semantic template is specified into a retrieval operation, *any* semantic tuple can actually be returned—namely, the standard behaviour is still non-deterministic. Additionally, in order to support probabilistic retrieval, such as in the case of our chemical model, Prolog predicates are provided to retrieve the result of fuzzy matching, to be used into ReSpecT specifications to change the default retrieval method. In particular, this is achieved via predicate `retrieve(+SemanticTemplate,-SemanticTuple,-MatchFactor)`, taking the semantic template as an input, providing as an output a matching semantic tuple (as found by the Jena-API) and the corresponding match factor (a number in between 0 and 1). By iterating all solutions, one can retrieve all matching tuples, and use them in ReSpecT e.g. to extract one probabilistically, or the one with the best match.

Our implementation of fuzzy matching is inspired by the work presented in [8,4]. The idea is to annotate the language of semantic templates with numerical values expressing the priority given to certain sub-concepts in it, which are then considered as optional. While computing semantic matching we first obtain those tuples for which the non-optional sub-concepts are all satisfied. For those tuples, we then also check which optional sub-concepts are satisfied, using the result to compute the match factor. As an example, the following semantic template could be used for *CityCars*:

```
'Car',
  'Car'->0.2,
  ((hasMaxSpeed : '90km/h')->0.5;(hasMaxSpeed : '180km/h')->0.1),
  (exists hasEnergyPower : electric)->0.3
```

We first need the tuple to represent a car (the only non-optional sub-concept) and assign a basic match factor 0.2 to it; additionally we add 0.5 if its maximum speed is '90km/h' or 0.1 if its maximum speed is '180km/h'; finally we add 0.3 if it comes with electric energy power. Namely, the more optional concepts are satisfied, the higher the match—and normalisation is performed if needed to finally obtain a real number in $[0, 1]$.

5.3 Chemical Reactions

We now describe how a TuCSoN tuple centre can be specialised to act as a chemical-like system where semantic tuples play the role of reactants, which combine and transform over time as occurring in chemistry.

Coding reactants and laws. Tuples modelling reactant individuals are kept in the tuple space in the form reactant(X,N), where X is a semantic tuple representing the reactant and N is a natural number denoting concentration. Laws modelling chemical reactions are expressed by tuples of the form law(InputList,Rate,OutputList), where InputList denotes the list of the reacting individuals, and Rate is a float value specifying the constant rate of the reaction. As a reference example, consider the chemical laws resembling (USE,DECAY) rules in previous section: $S + R \xrightarrow{10.0} S + S$ and $S \xrightarrow{10} 0$, where S and R represent semantic templates for services and requests. Such laws can be expressed in TuCSoN by tuples law([S,R],10,[S,S]) and law([S],10,[]). On the other hand, tuples reactant(sa,1000), reactant(sb,1000) and reactant(r,1000) represent reactants for two semantic tuples (sa and sb) matching S, and one (ra) matching R. The set of enabled laws at a given time is conceptually obtained by instantiating the semantic templates in reactions with all the available semantic tuples. In the above case they would be law([sa,r],r1a,[sa,sa]), law([sb,r],r1b,[sb,sb]), law([sa],r2a,[]) and law([sb],r2b,[]). The rate of each enabled reaction (usually referred to as *global rate*) is obtained as the product of chemical rate and match degree as described in Section 3. For instance, rate `r1a` can be calculated as $10.0 \times \#sa \times \#r \times \mu(S + R, sa + r)$, where μ is the function returning the match factor between the list of semantic reactants and the list of actual tuples.

As an additional kind of chemical law, it is also possible to specify a transfer of molecules towards other tuple centres by a law of the kind `law([X],10,[firing(X)])`.

ReSpecT engine. The actual chemical engine is defined in terms of ReSpecT reactions, which can be classified according to the provided functionality. As such, there are reactions for *(i)* managing chemical laws and reactants, i.e. ruling the dynamic insertion/removal of reactants and laws, *(ii)* controlling engine start and stop, *(iii)* choosing the next chemical law to be executed, and *(iv)* executing chemical laws. For the sake of conciseness we only describe part (iii), which is the one that focusses on Gillespie's algorithm for chemical simulations [15].

This computes the choice of the next chemical law to be executed, based on the following ReSpecT reaction, triggered by operation `out(engine_trigger)` which starts the engine.

```
reaction( out(engine_trigger), endo, (
          in(engine_trigger),
          chooseLaw(law(IL,_,OL),Rtot),
          rand_float(Tau), Dt is round((log(1.0/Tau)/Rtot)*1000),
          event_time(Time), Time2 is Time + Dt,
          out_s(reaction( time(Time2), endo, out(engine_trigger) )),
          out(execution(law(IL,_,OL),Time))
)).
```

First of all, a new law is chosen by the `chooseLaw` predicate, which returns `Rtot`, the global rate of all the enabled chemical laws, and a term `law(IL,_,OL)`—`IL` and `OL` are bound respectively to the list of reactants and products in the chosen law, after templates are instantiated to tuples as described above. Then, according to Gillespie algorithm, time interval `Dt` – denoting the overall duration of the chemical reaction – is stochastically calculated (in milliseconds) as $log(1/Tau)/Rtot$, where Tau is a value randomly chosen between 0 and 1 [15]. A new timed reaction is accordingly added to the ReSpecT specification and will be scheduled for execution `Dt` milliseconds later with respect to `Time`, which is the time at which `out(engine_trigger)` occurred: the corresponding reaction execution will result in a new `out(engine_trigger)` operation that keeps the chemical engine running. Finally, a new tuple `execution(law(IL,_,OL),Time)` is inserted so that the set of reactions devoted to chemical-law execution can be activated.

The actual implementation of the Gillespie's algorithm regarding the choice of the chemical law to be executed is embedded in the `chooseLaw` predicate, whose implementation is as follows:

```
chooseLaw(Law,Rtot):-
           rd(laws(LL)),
           semanticMatchAll(LL,NL,Rtot), not(Rtot==0),
           sortLaws(NL,SL),
           rand_float(Tau),
           chooseGillespie(SL,Rtot,Tau,Law).
```

After retrieving the list LL of the chemical laws defined for the tuple centre, semanticMatchAll returns the list NL of enabled chemical laws and the corresponding overall rate Rtot, computed as the sum of the global rate of every enabled law. To this end, predicate semanticMatchAll relies on predicate retrieve(+SemanticTemplateList,-SemanticTupleList,-MatchFactor) already described (properly extended to deal with lists of semantic tuples and templates).

The chemical law to be executed is actually chosen via the chooseGillespie predicate if Rtot > 0, i.e. if there are enabled chemical laws. This choice is driven by a probabilistic process: given n chemical laws and their global rates $r_1, ..., r_n$, the probability for law i to be chosen is defined as r_i/R, where $R = \sum_i r_i$. Consequently, law selection is simply driven by drawing a random number between 0 and 1 and choosing a law according to the probability distribution of the enabled laws.

6   Related Work

Coordination Models. The issue we face in this article can be framed as the problem of finding the proper coordination model for enabling and ruling interactions of pervasive services. Coordination models generated by the archetypal LINDA model [14], which simply provides for a blackboard with associative matching for mediating component interactions through insertion/retrieval of tuples. A radical change is instead the idea of engineering the coordination space of a distributed system by some policy "inside" the tuple spaces as proposed here, following the pioneer works of e.g. TuCSoN [23]—in fact, as shown in this paper, TuCSoN can be used as a low-level virtual platform for enacting the chemical tuple-space model. As already mentioned, the work presented in this article is based on [30], which was extended in [31] to deal with self-composition of services, a concept that we can support in our framework though it is not addressed in this paper.

Chemistry has been proposed as an inspiration for several works in distributed computing and coordination over many years, like in the Gamma language [9] and the chemical abstract machine [6], which lead to the definition of some general-purpose architectures [17]. Although these models already show the potential of structuring coordination policies in terms of chemical-like rewriting rules, we observe that they do not bring the chemical metaphor to its full realisation as we do here, as they do not exploit chemical stochastic rates.

Situatedness. In many proposals for pervasive computing environments and middleware infrastructures, the idea of "situatedness" has been promoted by the adoption of shared virtual spaces for services and components interactions. Gaia [27] introduces the concept of active spaces, a middleware infrastructure enacting distributed active blackboard spaces for service interactions. Later on, a number of proposals have extended upon Gaia, to enforce dynamic semantic pattern-matching for service composition and discovery [13] or access to contextual information [11]. Other related approaches include: Egospaces [18], exploiting a network of tuple spaces to enable location-dependent interactions across components; LIME [26], proposing tuples spaces that temporarily merge based on network proximity, to facilitate dynamic interactions and exchange of information across mobile devices; and TOTA [19], enacting computational gradients for self-awareness in mobile networks. Our model shares the idea of conceiving components as "living" and interacting in a shared spatial substrate (of tuple spaces) where they can automatically discover and interact with one another. Yet, our aim is broader, namely, to dynamically and systemically enforce situatedness, service interaction, and data management with a simple language of chemical reactions, and most importantly, enacting an ecological behaviour thanks to the support of diversity in the long term.

Self-organisation. Several recent works exploit the lessons of adaptive self-organising natural and social systems to enforce self-awareness, self-adaptivity, and self-management features in distributed and pervasive computing systems. At the level of interaction models, these proposals typically take the form of specific nature- and socially inspired interaction mechanisms [3] (e.g. pheromones [25] or virtual fields [19]), enforced either at the level of component modelling or via specific middleware-level mechanisms. We believe our framework integrates and improves these works in two main directions: *(i)* it tries to identify an interaction model that is able to represent and subsume the diverse nature-inspired mechanisms via a unifying self-adaptive abstraction (i.e. the semantics chemical reactions); *(ii)* the "ecological" approach we undertake goes beyond most of the current studies that limit to ensembles of homogeneous components, supporting the vision of novel pervasive and Internet scenarios as a sort of cyber-organisms [1].

7   Roadmap and Conclusion

In this paper we described research and development challenges in the implementation of the chemical tuple space model in TuCSoN. These are routed in two basic dimensions, which are mostly – but not entirely – orthogonal.

On the one hand, the basic tuple centre model is to be extended to handle semantic matching, which we support by the following ingredients: *(i)* an OWL ontology (a set of definitions of concepts) stored into a tuple centre which grounds semantic matching; *(ii)* tuples (other than syntactic as usual) can be semantic, describing an individual of the application domain (along with the concept it belongs to and the individuals it is linked to through roles); and *(iii)* a matching function implemented so as to check whether a tuple is the instance of a concept, returning the corresponding match factor.

On the other hand, the coordination specification for the tuple centre should act as a sort of "online chemical simulator", evolving the concentration of tuples over time using the same stochastic model of chemistry [15], so as to reuse existing natural and artificial chemical systems (like prey-predator equations); at each step of the process: *(i)* the reaction rates of all the chemical laws are computed, *(ii)* one is probabilistically selected and then executed, *(iii)* the next step of the process is triggered after an exponentially distributed time interval, according to the Markov property.

The path towards a fully featured and working infrastructure has been paved, but further research and development is required to tune several aspects:

Match factor. Studying suitable fuzzy matching techniques is currently a rather hot research topic, e.g. in the Semantic Web context. Our current support trades off simplicity for expressive power, but we plan to extend it using some more complete approach and in light of the application to selected cases—e.g. fully relying on [8].

Performance. The problem of performance was not considered yet, but will be subject of our future investigation. Possible bottlenecks include the chemical model and its implementation as a ReSpecT program, but also semantic retrieval, which is seemingly slower than standard syntactic one. We still observe that in many scenarios of pervasive computing – like those considered in Section 4 – this is not a key issue.

Chemical language. Developing a suitable language for semantic chemical laws is a rather challenging issue. The design described in this paper supports limited forms of service interactions that will likely be extended in the future. For instance, a general law $X + Y \rightarrow Z$ is meant to combine two individuals into a new one, hence the chemical language should be able to express into Z *how* the semantic templates X and Y should combine—aggregation, contextualisation, and other related patterns of self-organising pervasive systems are to be handled at this level.

Application cases. The model, and correspondingly the implementation of the infrastructure, are necessarily to be tuned after evaluation of selected use cases can be performed. Accordingly, the current version of the infrastructure is meant to be a prototype over which initial sperimentation can be performed. A main application scenario we will considered for actual implementation is a general purpose pervasive display infrastructure.

References

1. Agha, G.: Computing in pervasive cyberspace. Commun. ACM 51(1), 68–70 (2008)
2. Baader, F., Calvanese, D., McGuinness, D.L., Nardi, D., Patel-Schneider, P.F. (eds.): The Description Logic Handbook: Theory, Implementation, and Applications. Cambridge University Press, Cambridge (2003)
3. Babaoglu, O., Canright, G., Deutsch, A., Caro, G.A.D., Ducatelle, F., Gambardella, L.M., Ganguly, N., Jelasity, M., Montemanni, R., Montresor, A., Urnes, T.: Design patterns from biology for distributed computing. ACM Trans. Auton. Adapt. Syst. 1(1), 26–66 (2006)

4. Bandara, A., Payne, T.R., Roure, D.D., Gibbins, N., Lewis, T.: A pragmatic approach for the semantic description and matching of pervasive resources. In: Wu, S., Yang, L.T., Xu, T.L. (eds.) GPC 2008. LNCS, vol. 5036, pp. 434–446. Springer, Heidelberg (2008)
5. Barros, A.P., Dumas, M.: The rise of web service ecosystems. IT Professional 8(5), 31–37 (2006)
6. Berry, G., Boudol, G.: The chemical abstract machine. Theoretical Computer Science 96(1), 217–248 (1992)
7. Berryman, A.A.: The origins and evolution of predator-prey theory. Ecology 73(5), 1530–1535 (1992)
8. Bobillo, F., Straccia, U.: fuzzyDL: An expressive fuzzy description logic reasoner. In: 2008 International Conference on Fuzzy Systems (FUZZ 2008), pp. 923–930. IEEE Computer Society, Los Alamitos (2008)
9. Bonâtre, J.-P., Le Métayer, D.: Gamma and the chemical reaction model: Ten years after. In: Coordination Programming, pp. 3–41. Imperial College Press, London (1996)
10. Camazine, S., Deneubourg, J.-L., Franks, N.R., Sneyd, J., Theraulaz, G., Bonabeau, E.: Self-Organization in Biological Systems. Princeton Studies in Complexity. Princeton University Press, Princeton (2001)
11. Costa, P.D., Guizzardi, G., Almeida, J.P.A., Pires, L.F., van Sinderen, M.: Situations in conceptual modeling of context. In: Tenth IEEE International Enterprise Distributed Object Computing Conference (EDOC 2006), Workshops, Hong Kong, China, October 16-20, p. 6. IEEE Computer Society, Los Alamitos (2006)
12. Ferscha, A., Riener, A., Hechinger, M., Schmitzberger, H.: Building pervasive display landscapes with stick-on interfaces. In: CHI Workshop on Information Visualization and Interaction Techniques (April 2006)
13. Fok, C.-L., Roman, G.-C., Lu, C.: Enhanced coordination in sensor networks through flexible service provisioning. In: Field, J., Vasconcelos, V.T. (eds.) COORDINATION 2009. LNCS, vol. 5521, pp. 66–85. Springer, Heidelberg (2009)
14. Gelernter, D.: Generative communication in linda. ACM Trans. Program. Lang. Syst. 7(1), 80–112 (1985)
15. Gillespie, D.T.: Exact stochastic simulation of coupled chemical reactions. The Journal of Physical Chemistry 81(25), 2340–2361 (1977)
16. Horrocks, I., Patel-Schneider, P.F., Harmelen, F.V.: From shiq and rdf to owl: The making of a web ontology language. Journal of Web Semantics 1 (2003)
17. Inverardi, P., Wolf, A.L.: Formal specification and analysis of software architectures using the chemical abstract machine model. IEEE Trans. Software Eng. 21(4), 373–386 (1995)
18. Julien, C., Roman, G.-C.: Egospaces: Facilitating rapid development of context-aware mobile applications. IEEE Trans. Software Eng. 32(5), 281–298 (2006)
19. Mamei, M., Zambonelli, F.: Programming pervasive and mobile computing applications: The TOTA approach. ACM Trans. Software Engineering and Methodology 18(4) (2009)
20. Nardini, E., Viroli, M., Panzavolta, E.: Coordination in open and dynamic environments with TuCSoN semantic tuple centres. In: Shin, S.Y., Ossowski, S., Schumacher, M., Palakal, M., Hung, C.-C., Shin, D. (eds.) 25th Annual ACM Symposium on Applied Computing (SAC 2010), Sierre, Switzerland, March 22–26, vol. III, pp. 2037–2044. ACM, New York (2010); Awarded as Best Paper
21. Nixon, L.j.b., Simperl, E., Krummenacher, R., Martin-recuerda, F.: Tuplespace-based computing for the semantic web: A survey of the state-of-the-art. Knowl. Eng. Rev. 23(2), 181–212 (2008)

22. Omicini, A.: Formal ReSpecT in the A&A perspective. Electronic Notes in Theoretical Computer Sciences 175(2), 97–117 (2007)
23. Omicini, A., Zambonelli, F.: Coordination for Internet application development. Autonomous Agents and Multi-Agent Systems 2(3), 251–269 (1999)
24. Paolucci, M., Kawamura, T., Payne, T.R., Sycara, K.P.: Semantic matching of web services capabilities. In: Horrocks, I., Hendler, J. (eds.) ISWC 2002. LNCS, vol. 2342, pp. 333–347. Springer, Heidelberg (2002)
25. Parunak, H.V.D., Brueckner, S., Sauter, J.: Digital pheromone mechanisms for coordination of unmanned vehicles. In: Autonomous Agents and Multiagent Systems (AAMAS 2002), vol. 1, pp. 449–450. ACM, New York (2002)
26. Picco, G.P., Murphy, A.L., Roman, G.-C.: LIME: Linda meets mobility. In: The 1999 International Conference on Software Engineering (ICSE'99), Los Angeles (CA), USA, May 16–22, pp. 368–377. ACM, New York (1999)
27. Román, M., Hess, C.K., Cerqueira, R., Ranganathan, A., Campbell, R.H., Nahrstedt, K.: Gaia: a middleware platform for active spaces. Mobile Computing and Communications Review 6(4), 65–67 (2002)
28. Sirin, E., Parsia, B., Grau, B.C., Kalyanpur, A., Katz, Y.: Pellet: A practical OWL-DL reasoner. J. Web Sem. 5(2), 51–53 (2007)
29. Ulieru, M., Grobbelaar, S.: Engineering industrial ecosystems in a networked world. In: 5th IEEE International Conference on Industrial Informatics, pp. 1–7. IEEE Press, Los Alamitos (2007)
30. Viroli, M., Casadei, M.: Biochemical tuple spaces for self-organising coordination. In: Field, J., Vasconcelos, V.T. (eds.) COORDINATION 2009. LNCS, vol. 5521, pp. 143–162. Springer, Heidelberg (2009)
31. Viroli, M., Casadei, M.: Chemical-inspired self-composition of competing services. In: Shin, S.Y., Ossowski, S., Schumacher, M., Palakal, M., Hung, C.-C., Shin, D. (eds.) 25th Annual ACM Symposium on Applied Computing (SAC 2010), Sierre, Switzerland, March 22–26, vol. III, pp. 2029–2036. ACM, New York (2010)
32. Viroli, M., Casadei, M., Omicini, A.: A framework for modelling and implementing self-organising coordination. In: 24th Annual ACM Symposium on Applied Computing (SAC 2009), vol. III, pp. 1353–1360. ACM, New York (2009)
33. Viroli, M., Zambonelli, F.: A biochemical approach to adaptive service ecosystems. Information Sciences 180(10), 1876–1892 (2010)
34. Zambonelli, F., Viroli, M.: Architecture and metaphors for eternally adaptive service ecosystems. In: IDC 2008. Studies in Computational Intelligence, vol. 162, pp. 23–32. Springer, Heidelberg (2008)

Self-adaptive Architectures for Autonomic Computational Science

Shantenu Jha[1], Manish Parashar[2], and Omer Rana[3]

[1] Center for Computation & Technology and Department of Computer Science,
Louisiana State University, USA, and e-Science Institute,
University of Edinburgh, UK
sjha@cct.lsu.edu
[2] Department of Electrical & Computer Engineering, Rutgers University, USA
parashar@rutgers.edu
[3] School of Computer Science, Cardiff University, UK
o.f.rana@cs.cf.ac.uk

Abstract. Self-adaptation enables a system to modify it's behaviour based on changes in its operating environment. Such a system must utilize monitoring information to determine how to respond either through a systems administrator or automatically (based on policies pre-defined by an administrator) to such changes. In computational science applications that utilize distributed infrastructure (such as Computational Grids and Clouds), dealing with heterogeneity and scale of the underlying infrastructure remains a challenge. Many applications that do adapt to changes in underlying operating environments often utilize ad hoc, application-specific approaches. The aim of this work is to generalize from existing examples, and thereby lay the foundation for a framework for Autonomic Computational Science (ACS). We use two existing applications – Ensemble Kalman Filtering and Coupled Fusion Simulation – to describe a conceptual framework for ACS, consisting of *mechanisms, strategies and objectives*, and demonstrate how these concepts can be used to more effectively realize pre-defined application objectives.

1 Introduction

Developing and deploying self-adaptive applications over distributed infrastructure provides an important research challenge for computational science. Significant recent investments in national and global cyberinfrastructure, such as the European EGEE/EGI, the US TeraGrid, the Open Science Grid and the UK National Grid Service, have the potential for *enabling* significant scientific insights and progress. The use of such infrastructure in *novel* ways has still not been achieved however, primarily because of the inability of applications that are deployed over such infrastructure to adapt to the underlying heterogeneity, fault management mechanisms, operation policies and configuration parameters associated with particular software tools and libraries. This problem is only compounded by new and more complex application formulations such as those

D. Weyns et al. (Eds.): SOAR 2009, LNCS 6090, pp. 177–197, 2010.

based on dynamic data. Tuning and adapting an application is often left to the skills of specialist developers, with limited time and motivation to learn the behaviour of yet another deployment environment. In applications where automation has been achieved, this generally involves understanding specific application behaviours, and in some instances, specialised capabilities offered by the underlying resources over which the application is to be executed. Generalising from such applications and developing a more generic *framework* has not been considered. We take an application-centric approach to better understand what such a framework should provide, focusing on how: (i) applications can be characterized, to enable comparison across different application classes – the basis of our previous work [17]; (ii) understanding tuning mechanisms and associated strategies that can be applied to particular application classes. The longer term objective of this work is to derive adaptation *patterns* that could be made available in a software library, and directly made use of when constructing distributed scientific applications.

It is useful to note that the development of self-adaptive systems has generally followed either: (i) a top-down approach, where overall system goals need to be achieved through the modification of interconnectivity or behaviour of system components – realized through a system manager; (ii) a bottom-up approach, where local behaviour of system components needs to be aggregated (without a centralized system manager) to generate some overall system behaviour. In our approach we are primarily focused on (i), as this relates closely with existing approaches within computational science. However, we also believe that approach (ii) could be used as an initial phase, whereby resource ensembles could be dynamically formed within Grid and Web-based communities, using self-organization approaches, and as discussed in Serugendo et al. [24]. Such an approach would enable application characteristics or resource characteristics to be used as an initial phase to cluster resources/applications prior to utilizing an autonomic deployment strategy.

2   A Conceptual Framework for Autonomic Computational Science

A conceptual framework to support autonomic computational science applications is presented in this section. We identify possible architectures, and relate the approaches discussed here to reflective middleware and control loop models. In the context of Grid computing environments considered in this work, the use of a shared, multi-tasking environment is assumed. An application (or user) in such an environment requests access to a pre-determined number of resources (CPUs, memory, etc), and it is the responsibility of the resource management system (generally a batch queuing system) to ensure that access to these resources is granted over the requested time interval. The resource manager does not (in most cases) provide any quality of service guarantees.

2.1 The Autonomic Computing Paradigm

The autonomic computing paradigm is modelled after the autonomic nervous system and enables changes in its essential variables (e.g., performance, fault, security, etc.) to trigger changes to the behavior of the computing system such that the system is brought back into equilibrium with respect to the environment [16]. Conceptually, an autonomic system requires: (a) sensor channels to sense the changes in the internal state of the system and the external environment in which the system is situated, and (b) motor channels to react to and counter the effects of the changes in the environment by changing the system and maintaining equilibrium. The changes sensed by the sensor channels have to be analyzed to determine if any of the essential variables have gone out of their viability limits. If so, it has to trigger some kind of planning to determine what changes to inject into the current behavior of the system such that it returns to the equilibrium state within the new environment. This planning requires knowledge to select the right behavior from a large set of possible behaviors to counter the change. Finally, the motor neurons execute the selected change. Sensing, Analyzing, Planning, Knowledge and Execution are thus the keywords used to identify an autonomic computing system. A common model based on these ideas was identified by IBM Research and defined as *MAPE* (Monitor-Analyze-Plan-Execute) [19]. There are, however, a number of other models for autonomic computing [23], [11] – in addition to work in the agent-based systems community that share commonalities with the ideas presented above. In what follows, we explore the applications of this paradigm to support computational science.

2.2 Conceptual Architectures for ACS

Looking at existing practices in computational science, two corresponding conceptual architectures can be observed, which are described below. These architectures are composed of the application, a resource manager that allocates, configures and tunes resources for the application, and an autonomic manager that performs the autonomic tuning of application and/or system parameters. Figure 1 illustrates the first conceptual architecture, where the application and resources are characterized using a number of *dynamically modifiable* parameters/variables that have an impact on the overall *observed behaviour* of the application. Each of these parameters has an associated range over which it can be modified, and these constraints are known *a priori*. The autonomic tuning engine alters these parameters based on some overall *required behaviour* (hereby referred to as the application objective) that has been defined by the user. Tuning in this case is achieved by taking into account, for example, (i) historical data about previous runs of the application on known resources, obtained using monitoring probes on resources; (ii) historical data about previous selected values of the tunable parameters; (iii) empirically derived models of application behavior; (iv) the specified tuning mechanism and strategy; etc.

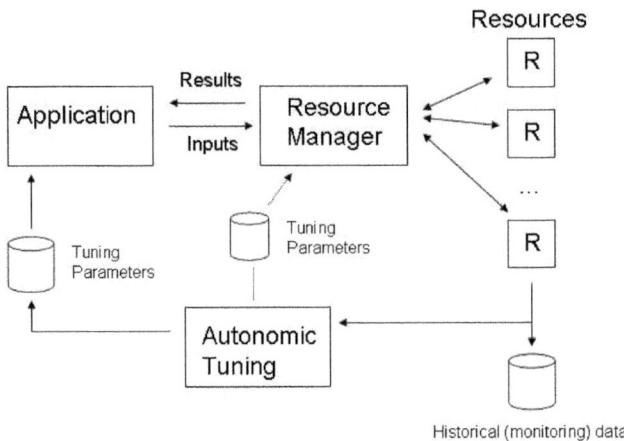

Fig. 1. Autonomic tuning *of* application and resource manager. *R* refers to a computational resource. parameters.

For example, an autonomic tuning mechanism in this architecture may involve changing the *size* of the application (for instance, the number of data partitions generated from a large data set, the number of tiles from an image, etc.), or the set of parameters over which execution is being requested. This tuning is used to make desired tradeoffs between quality of solution, resource requirements and execution time or to ensure that a particular Quality of Service (QoS) constraint, such as execution time, is satisfied.

A variant, also illustrated in Figure 1, involves updating the resource manager based on information about the current state of the application. As an example of such an approach, consider an application using dynamic structured adaptive mesh refinement (SAMR) [9] techniques on structured meshes/grids. Compared to numerical techniques based on static uniform discretization, SAMR methods employ locally optimal approximations and can yield highly advantageous ratios for cost/accuracy by adaptively concentrating computational effort and resources to regions with large local solution error at runtime. The adaptive nature and inherent space-time heterogeneity of these SAMR implementations lead to significant challenges in dynamic resource allocation, data-distribution, load balancing, and runtime management. Identifying how the underlying resource management infrastructure should adapt to changing SAMR requirements (and possibly vice versa) provides one example of the architecture in Figure 1.

Figure 2 illustrates another conceptual architecture, where the application is responsible for driving the tuning of parameters, and choosing a tuning strategy. The autonomic manager is now responsible for obtaining monitoring data from resource probes and the strategy specification (for one or more objectives to be realized) from the application. Tuning now involves choosing a resource management strategy that can satisfy the objectives identified by the application. This approach primarily relates to the *system-level* self-management described

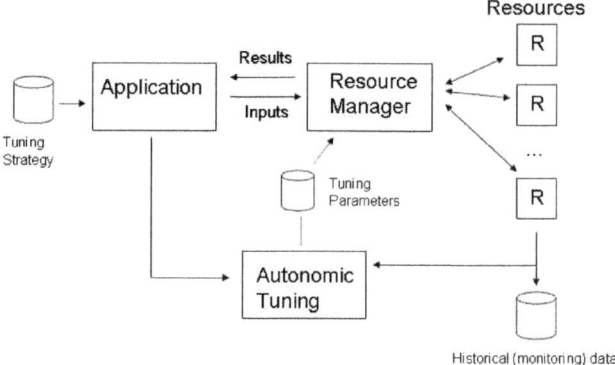

Fig. 2. Autonomic tuning *by* application. R refers to a computational resource.

in Section 1. An example of such an architectural approach is the use of resource reservation to achieve a particular QoS requirement. The G-QoSM framework [1] demonstrates the use of such an architecture, involving the use of a soft real-time scheduler (DSRT) along with a bandwidth broker to make resource reservation over local compute, disk and network capacity, in order to achieve particular application QoS constraints.

We reiterate that the conceptual architectures – tuning by and of applications, defined above are not exhaustive, but provide an initial formulation with a view towards understanding a set of applications that we discuss.

2.3   A Conceptual Framework

A conceptual framework for ACS can be developed based on the conceptual architectures discussed above (and illustrated in Figures 1 and 2), comprised of the following elements:

Application-level Objective (AO): An AO refers to an application requirement that has been identified by a user – somewhat similar to the idea of a "goal" in Andersson et al. [2]. Examples of AO include increase throughput, reduce task failure, balance load, etc. An AO needs to be communicated to the autonomic tuning component illustrated in Figures 1 and 2. The autonomic tuning component must then interact with a resource manager to achieve these objectives (where possible). There may be multiple AOs that need to be satisfied, and there may be relationships between them.

Mechanism: a mechanism refers to a particular action that can be used by the application or the resource manager (from Figures 1 and 2) to achieve an AO. A mechanism is triggered as a consequence of some detected event(s) in the application or it's environment over some pre-defined duration.

Hence, a mechanism m may be characterized in terms of: $\{m_i^e\}$ – the set of events that lead to the triggering (activation) of the mechanism; $\{m_i\}$ – the

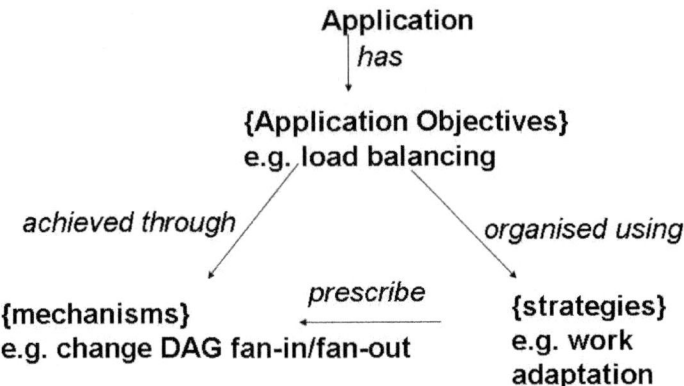

Fig. 3. Relationship between AO, mechanism and strategy. An Application can have multiple application objectives (indicated by {...} notation). Each objective can be achieved through a set of mechanisms based on strategies. The set and organization of mechanisms used to achieve an AO is prescribed by the specific strategy used.

set of data inputs to the mechanism; $\{m_o\}$ – the set of output data generated; and $\{m_o^e\}$ – the set of output events that are generated during the execution of the mechanism or after its completion. An example of a mechanism includes *file staging*. In this mechanism, $\{m_i\}$ corresponds to one or more file and resource references prior to the staging process has started, $\{m_o\}$ corresponds to the file references after staging has completed, $\{m_i^e\}$ refers to the input events that trigger the file staging to begin, and $\{m_o^e\}$ corresponds to output events that are generated once file staging is completed.

Comparing with the view of a mechanism in [2], where the action undertaken is classified as being: (i) structural or parametric (primarily parametric in this framework); (ii) system or human assisted; (iii) centralized or decentralized; (iv) local or global; (v) short, medium or long term; (vi) best effort or guaranteed; and (vii) event or time triggered. All of these classifications also apply in the context of the ACS presented here.

Strategy: One or more strategies may be used to accomplish a particular AO. A strategy is defined in terms of one or more partially-ordered mechanisms. A strategy may be specified manually (by a systems administrator, for instance) or constructed using an autonomic approach. The focus of this particular work is on the latter. A strategy is managed by the autonomic tuning component illustrated in Figures 1 and 2, and may be maintained as a collection of templates that are adapted depending on the application or the resource manager properties.

Note that given AO can be fulfilled by multiple strategies, and the same strategy can be used for different AO. For example, self-configuration can be used for both load-balancing as well as higher-throughput. Figure 3 illustrates the relationship between these concepts.

2.4   Relationship to Reflective Middleware

Reflective middleware [21] attempts to adapt the behaviour of an application depending on the deployment platform. One primary motivation for this work stems from the need to expose the dynamic state of the underlying middleware (and provide tuning capability) to the application, so that it can utilize more effective control strategies to adapt it's behaviour as the underlying deployment infrastructure changes (assuming that an instance of the same reflective middleware is available on these different infrastructures). Two key features enable this capability within the middleware: (i) "reflection" to enable the system to reason about and act upon itself, through a representation of its own behaviour, amenable to examination and change; (ii) "causal connectivity" to enable any changes made to the system's self-representation to impact it's actual state and behavior, and vice-versa. A component-based approach is generally adopted to enable different component instances to be chosen depending on the underlying deployment platform.

Reflective middleware relies on the existence of multiple component implementations that can be deployment over multiple platforms. In addition, such middleware relies on the ability of the underlying platform to expose it's state to the middleware, so that it can be managed externally. This is not always possible, as many scheduling engines, for instance, would not allow an application to alter job execution priorities on a given platform. From Section 2.2, in the conceptual architecture in figure 1, the autonomic tuning engine is considered to be external to the application. In figure 2, the application interacts with the tuning engine and not directly with the resource manager. Hence, behaviour or structural reflection capabilities could be used by the tuning engine, but not directly by the application. In many practical application deployments, it is unrealistic to assume the availability of specialist middleware on externally hosted platforms.

There are three key differences between the approach advocated here and concepts identified in reflective middleware efforts: (i) tuning of the application is *logically* external to the application and undertaken through a tuning engine; (ii) the tuning engine does not rely on the availability of specialist modules to be hosted/executed by the resource manager, instead relying on existing tuning parameters that are already made available by the resource manager; (iii) there is also generally a separation between adaptation policy and mechanisms – both of which can be dynamic, and combined in different ways to get different autonomic behaviors. In the reflective middleware this is not the case - policy and mechanisms are generally integrated in middleware design.

Andersson et al. [3] take a wider view and suggest how self-adaptation could be supported in software systems through reflection. They describe a self-adaptive software system as one that is able to "change it's behaviour by reflecting on itself". In this work, reflection is primarily associated with developing a meta-model of computation undertaken by a system. The meta-model is proposed as the basis for allowing a computational system to reason about and act upon itself. Hence, two types of activities are supported at such a meta-level, "introspection" and "intercession". Introspection is the process of inspection and reasoning about the system, whereas intercession is the subsequent modification of the system's meta-model. Self-Representation, Reflective Computation and Separation of Concerns are used to characterize the properties of a reflective system. The overall framework provided in this work provides a useful basis for developing adaptive systems that can be driven through the development of a meta-model and a suitable representation of a domain-model. However, developing such a meta-model (or domain-model) is difficult in complex applications. In our work, we therefore do not require a meta-model to be created, instead relying on a tuner component that is able to observe the outputs only, and not the internal working of a system. In the same context, when considering the tuning *of* an application, the granularity of what can be modified is also limited in many realistic applications, and access to granularity at the level of classes, objects, methods and method calls (as advocated in [3]) is often not possible.

2.5   Relationship to Control Loop Models

It is useful to note that the MAPE architecture identified in section 2.1 shares similarities with architectures adopted in feedback control, such as Model Reference Adaptive Control (MRAC) [25] and Model Identification Adaptive Control (MIAC) [26]. MRAC relies on comparing existing system state with that of a known model, and using this response to drive a controller that manipulates the system, whereas MIAC relies on deriving known properties of the model from system function (observations and measurements). Both of these approaches have been found to be of most benefit when the system being controlled has limited capability, and where the control loop is explicit. In many computational science applications, control loops are often hidden / abstracted, or hard to identify in the same way as in the types of applications utilizing a *traditional* MRAC architecture. Where control loops do exist, either the construction of a model (to steer the controller) or the tunable parameters that can be externally modified are limited. It is also possible in distributed computational science application for multiple control loops to exist, one associated with each resource (or ensemble).

As outlined in [10] adaptation mechanisms may be external to an application and not hard-wired. Developing an autonomic tuner that is external to the application provides a separation between system capability and tuning strategies. In the conceptual architectures presented in section 2.2, we have identified an explicit control loop external to the application. It is also possible to use the catalogue of self-adaptive mechanisms derived from natural systems (as outlined in [10]), by combining a top down approach (as being advocated in this work) and top-down approaches based on 'emergent' coordination mechanisms.

Model-driven approaches in software engineering have also been proposed to maintain a link between high and low-level views of software. Such approaches advocate the inclusion of particular run-time configurable parameters that are added at design time to a model of the software being adapted. However, as Nierstrasz et al. [22] point out, certain types of anomalies arise only after the software has been deployed (and cannot be known at design time) – thereby making it difficult to anticipate what and where to trace to observe the problematic behaviour. Nierstrasz et al. [22] propose the idea of a *model-centric* view that can take the context of deployment and execution into account, in order to control the scope of adaptations to be made to the software system. Our approach is closely aligned with this thinking, as we believe that the context is important in the types of tuning that can be supported. In the case of the application scenarios we describe, such context is based on the particular computing environment over which the application is deployed. A key difference from the work of Nierstrasz et al. [22] is that our adaptation is not at the same level of granularity – as [22] propose software modification at the level of source code and the dynamic addition of instrumentation code (through the use of an aspect-oriented approach). For many scientific applications, having access to source code is often not possible. Similarly, such applications may use external libraries (such as numeric libraries) that would be difficult to instrument directly. In our approach, therefore, identifying what needs to be monitored has to be specified beforehand – and we cannot overcome the constraints of anticipating where monitoring should take place.

3   Application Case Study

In Jha et al. [18] we provided a discussion of application vectors that may be used to characterize scientific applications. Every distributed application must, at a minimum, have mechanisms to address requirements for communication, coordination and execution, which form the vectors we use: Execution Unit, Communication (Data Exchange), Coordination, and Execution Environment. *Execution unit* refers to the set of pieces/components of the application that are distributed. *Communication (data exchange)* defines the data flow between the executions units. Data can be exchanged by messages (point-to-point, all-to-all, one-to-all, all-to-one, or group-to-group), files, stream (unicast or multicast), publish/subscribe, data reduction (a subset of messaging), or through shared data. *Coordination* describes how the interaction between execution units is managed (e.g. dataflow, control flow, SPMD (where the control flow in implicit in all copies of the single program), master-worker (tasks executed by workers are controlled by the master), or events (runtime events cause different execution units to become active.)) An *execution environment* captures what is needed for the application to run. It often includes the requirements for instantiating and starting the execution units (which is also referred to as *deployment*) and may include the requirements for transferring data between execution units as well as other runtime issues (e.g. dynamic process/task creation, workflow execution, file transfer, messaging (MPI), co-scheduling, data streaming, asynchronous data I/O, decoupled coordination support, dynamic resource and service discovery, decoupled coordination and messaging, decoupled data sharing, preemption, etc.).

Two applications are described in sections 3.1 and 3.2 which demonstrate how the conceptual architectures described in section 2.2 are utilized, and which make use of the application vectors described above.

3.1   Ensemble Kalman Filters

Ensemble Kalman filters (EnKF) are widely used in science and engineering [14]. EnKF are recursive filters that can be used to handle large, noisy data; the data can be the results and parameters of ensembles of models that are sent through the Kalman filter to obtain the true state of the data. EnKF-based History Matching for Reservoir simulations [13] is an interesting case of an application with irregular, hard-to-predict run time characteristics. The variation in model parameters often has a direct and sizable influence on the complexity of solving the underlying equations, thus varying the required runtime of different models (and consequently the availability of the results). Varying parameters sometimes also leads to varying systems of equations and entirely new scenarios. This increases the computational size requirements as well as memory requirements. For example as a consequence of the variation in size, the underlying matrix might become too large or even effectively lead to doubling the number of the system of equations, which could more than double the memory required to solve these system of equations. The forecast model needs to run to completion – which is defined as convergence to within a certain value. The run time of each model is unpredictable and uncorrelated with the run-time of models running on the same number of processors. At every stage, each model must converge, before the next stage can begin. Hence dynamically load-balancing to ensure that all models complete as close to each other as possible is the desired aim. The number of stages that will be required is not determined a priori. In the general case the number of jobs required varies between stages. Table 1 provides the tuning mechanisms used in the EnKF application.

We define T_c as the time in seconds it takes to complete a defined application workload – comprising of a fixed ensemble size (one hundred members) and number of stages (five iterations of ensemble runs followed by KF). Any consistent reduction in the total time to completion will eventually have a greater impact on larger runs with more stages. There are three main components that are necessary to consider in order to understand T_c . The first is the time that it takes to submit to the queuing system and time for file-transfers (in and out) – labelled as $t_{overhead}$, and which is typically small for these large, long-running

Table 1. Tuning Mechanisms in EnKF

Vectors	Mechanisms
Coordination	Centralized Data-store (SAGA)
	Pilot-Job (BigJob) abstraction
Communication	File staging, File indexing
Execution	Centralized Scheduler
Environment	Resource Selection/Management,
	Task re-execution,
	Task migration, Storage management,
	File caching, File distribution,
	Checkpointing

simulations. The second component is the time that the submitted jobs wait in queue for resources requested to become available – labelled as t_{wait}; the final component is the run-time that simulations actually take to complete – labelled as t_{run}. Thus, $T_c = t_{overhead} + t_{wait} + t_{run}$.

Experiments to determine T_c using different number of machines working concurrently towards a solution of the same application instance were performed; an increase in performance was measured by a reduced T_c for up to three machines. Although there were fluctuations in both the wait-time in queue and the time to complete the work-load, the fluctuations were dominated by the former. Therefore in an attempt to minimise wait-times in queue and thus to lower the overall T_c , this application attempts to launch jobs on multiple TeraGrid (TG) resources using a Batch Queue Predictor (BQP) [4] [8]– a tool available on a number TG resources that allows users to make bounded predictions about the time a job of a given size and duration will spend in the queue. The objective is to run a number of jobs corresponding to a different stage of EnKF execution. BQP-based prediction is given with a degree of confidence (probability) that the job will start before a certain deadline (i.e. the time in the queue) and quantile. Quantile value is a measure of repeatability; more precisely it is an indication of the probability that jobs of similar sizes and durations will have the same wait time. This information is vital when submitting jobs to various machines as the longer a job sits in the queue, the longer the delay for the entire stage. BQP provides the ability to predict, with given probability, which resource and when a job (with a specified number of processors and estimated run-time) is most likely to finish.

Figure 4 shows the results when using three TG machines: Ranger, Queen-Bee and Abe. Ranger is the flagship machine of the Texas Advanced Computing Centre; QueenBee (QB) the flagship machine of LONI and Abe a large machine at NCSA. This figure demonstrates how machine combinations can be used, with and without BQP. To ensure that each machine is used efficiently, it is necessary to undertake load balancing that takes account of the properties of each machine (such as queue times and predictions derived from BQP). This therefore becomes an application objective that could be achieved manually by an application scientist or supported through an autonomic strategy, as identified in Table 2. Hence, either the size of a job/task could be adapted based on what machines are available to run the application, or suitable resources could be discovered using queue prediction information derived from BQP. It is also possible to make use of BQP to autonomically chose resources for a particular job (based on resource properties), and to guide the selection of resource configuration on a predetermined machine. When using more than one machine, e.g., RQA-BQP, both the selection of the resources and the selection of resource configuration are variables. For RQA-BQP, it is possible that even though three resources are available, all jobs will be submitted to a single resource with much higher capacity or temporary lower load-factors (e.g, after a power-down/start-up). These results are further explained in [13].

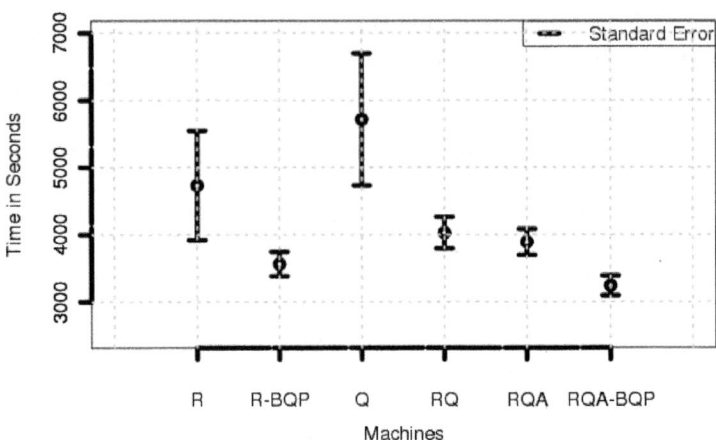

Fig. 4. Time to completion for different configurations. Left to Right: (i) Ranger (ii) Ranger when using BQP, (iii) QueenBee, (iii) Ranger and QueenBee, (iv) Ranger, QueenBee and Abe, (v) Ranger, QueenBee and Abe when using BQP. [12]

Table 2. EnKF application management through autonomic strategies

Application Objective	Autonomic Strategy
Load Balancing	**1. Adapt task mapping granularity based on system capabilities/state** File staging, File splitting/merging Task rescheduling, Task migration File distribution and caching, Storage Management **2. Resource Selection** Resource selection (using BQP), resource configuration update, Task rescheduling, Task migration File distribution and caching Storage Management
Scientific Fidelity	**Algorithmic Adaptivity** Change solvers -

This application therefore demonstrates that the use of queue information to guide the selection of resources and to configure resources, provides a better overall job execution performance. Rather than postponing such a decision until run-time, utilizing prior information to enable an autonomic manager to support resource selection can lead to better overall quality of service. This application also demonstrates the use of the conceptual architecture in Figure 1, where each `R` utilizes a queuing system with BQP, and collects historical data. The `Autonomic Tuning` manager can then communicate with a `Resource Manager` to determine which resource to select. The EnKF application utilizing the resource manager does not need to be modified, as the tuning is undertaken external to the application.

3.2   Coupled Fusion Simulation

The DoE SciDAC CPES fusion simulation project [20] is developing an integrated, Grid-based, predictive plasma edge simulation capability to support next-generation burning plasma experiments, such as the International Thermonuclear Experimental Reactor (ITER). The typical application workflow for the project consists of coupled simulation codes, i.e., the edge turbulence particle-in-cell (PIC) code (GTC) and the microscopic MHD code (M3D), which run simultaneously on thousands of processors on separate HPC resources at supercomputing centers, requiring data to be streamed between these codes to achieve coupling. Furthermore, the data has to be processed en-route to the destination. For example, the data from the PIC codes has to be filtered through "noise detection" processes before it can be coupled with the MHD code. As a result, effective online management and transfer of the simulation data is a critical part for this project and is essential to the scientific discovery process.

As a result, a core requirement of these coupled fusion simulations is that the data produced by one simulation must be streamed live to the other for coupling, as well as, possibly to remote sites for online simulation monitoring and control, data analysis and visualization, online validation, archiving, etc. The fundamental objective being to efficiently and robustly stream data between live simulations or to remote applications so that it arrives at the destination just-in-time – if it arrives too early, times and resources will have to be wasted to buffer the data, and if it arrives too late, the application would waste resources waiting for the data to come in. A further objective is to opportunistically use in-transit resources to transform the data so that it is more suitable for consumption by the destination application, i.e., improve the quality of the data from the destination applications point of view. Key objectives/constraints for this application can be summarized as: (1) Enable high-throughput, low-latency data transfer to support near real-time access to the data; (2) Minimize overheads on the executing simulation; (3) Adapt to network conditions to maintain desired QoS – the network is a shared resource and the usage patterns typically vary constantly. (4) Handle network failures while eliminating loss of data – network failures usually lead to buffer overflows, and data has to be written to local disks to avoid loss, increasing the overhead on the simulation. (5) Effectively schedule and manage in-transit processing while satisfying the above requirements – this is particularly challenging due to the limited capabilities and resources and the dynamic capacities of the typically shared processing nodes.

These objectives can be effectively achieved using autonomic behaviors [5,6] based on a range of strategies and mechanisms. Autonomic behaviors in this case

Table 3. Tuning mechanisms in the Coupled Fusion Simulation application

Vectors	Mechanisms
Coordination	Peer-2-Peer interaction
Communication	Data Streaming, Events
Execution Environment	Storage Selection (local/remote), Resource Selection/Management, Task migration, Checkpointing Task execution (local/remote) Dynamic provisioning (provisioning of in-transit storage/processing nodes)

span (1) the application level, e.g., adapting solver behavior, adapting iteration count or using speculative computing by estimating the delayed data and rolling back if the estimation error exceeds a threshold; (2) the coordination level, e.g., adapting end-to-end and in-transit workflows as well as resources allocated to them; and (3) the data communication level, e.g., adaptive buffer management and adaptive data routing. Furthermore, this application also involves the use of a hybrid autonomic approach that combines policy-based autonomic management with model-based online control [7].

A conceptual overview of the overall architecture is presented in Figure 5. It consists of two key components. The first is an application-level autonomic data streaming service, which provides adaptive buffer management mechanisms and proactive QoS management strategies, based on online control and governed by user-defined polices, at application end-points. The second component operates at the in-transit level, and provides scheduling mechanisms and adaptive

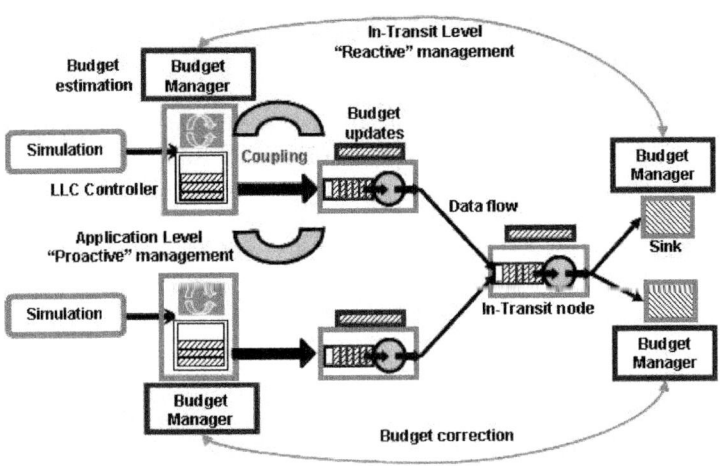

Fig. 5. Conceptual overview of the self-managing data streaming and in-transit processing service

runtime management strategies for in-transit data manipulation and transformation. These two components cooperate to address overall application constraints and QoS requirements.

Application level autonomic data streaming: The application level autonomic data streaming service combines model-based limited look-ahead controllers (LLC) and rule-based autonomic managers, with adaptive multi-threaded buffer management and data transport mechanisms at the application endpoints. The service manager monitors the state of the service and its execution context, collects and reports runtime information, and enforces the adaptation actions determined by its controller. Augmenting the manager with an LLC controller allows human defined adaptation polices, which may be error-prone and incomplete, to be combined with mathematically sound models and optimization techniques for more robust self-management. Specifically, the controller decides when and how to adapt the application behavior, and the service managers focus on enforcing these adaptations in a consistent and efficient manner.

In-transit data processing: The in-transit data manipulation framework consists of a dynamic overlay of in-transit nodes, which is on the path between the source and the destination. The in-transit node services run on commodity clusters (or possibly public clouds) with heterogeneous capabilities, interconnects and loads, and are shared between multiple scientific workflows. They perform simple operations such as processing, buffering, and forwarding. The processing performed by a node on a data item depends on the node's capacity and capability, and the amount of processing outstanding for the data item. The latter information is contained in the data item itself as metadata. Any processing that is outstanding when the data item reaches the sink will have to be performed at the sink. The processing capabilities themselves can be pre-staged at the in-transit nodes or can be dynamically deployed. Our initial explorations have assumed the former.

Cooperative self-management: Coupling application level and in-transit management: The application level and in-transit management can be coupled to achieve cooperative end-to-end self-management. Such a coupling has many benefits, especially in cases of congestion, which often occurs at a shared links in the data path between the sources and sink nodes. Without such a coupling, the application level controller would detect congestion by observing a decrease of the effective bandwidth, and in response, it would advise the service manager to reduce the amount of data sent on the network and increase the amount of data written to the local storage, to avoid data loss. While this would eventually reduce the congestion in the data path, it would require that the data blocks written to the local storage be separately transferred to and processed at the sink. By contrast, using coupling between the application and in-transit management levels, the in-transit node signals the application level controller at the source in response to the local congestion that it has detected by observing its buffer occupancy, and sends the information about its current buffer occupancy.

This allows the application level controller to detect congestion much earlier, rather than having to wait until the congestion propagates back to the source, and in response, it can increase its buffer size and buffer data items until congestion at the in-transit nodes is relieved. This, in turn, reduces the amount of data that has to be written to the local disk at the source and improves QoS at the sink.

Summary of results: An experimental evaluation demonstrating the feasibility and benefits of the concepts and the framework described above, as well as investigating issues such as robustness, stability and overheads are presented in [5,7,6]. These experiments are conducted using a data streaming service that is part for the CPES FSP workflow and streams data between applications running between Rutgers and PPPL in NJ, ORNL in TN, and NERSC in CA. The data transport service managed the transfer of blocks of data from applications buffers at NERSC to PPPL/ORNL or to local storage. Service adaptations included the creation of new instances of the streaming service when the network throughput dipped below a certain threshold. Furthermore, during network congestion, adaptations included changing the buffer management scheme used to ensure better network throughput and lower average buffer occupancy. In cases of extreme congestion and buffer overflows, the data was moved to local storage rather than over the network to prevent further congestions and buffer overflows and thus maximizes the amount of data reaching the sink. These adaptation resulted in an average buffer occupancy of around 25% when using a combination of model and rule based self-management. This leads to lower application overheads and this avoids writing data to shared storage. The percentage overhead due to the service at the application level was less than 5% of the computational time. We also performed experiments with in-transit data processing, which demonstrated processing at the in-transit nodes reduced buffering time from 40% to 2% (per data item) during network congestion. Using cooperative management, the & buffer occupancy further reduced by 20% and 25% fewer data items had to be diverted to local storage at end-points in spite of network congestions and high loads. Furthermore, the "quality" of data reaching "in-time" at the sink increased, effectively reducing the processing time at the sink by an average of 6 minutes per run.

This application demonstrates the use of the conceptual architecture in Figure 2, where the application can interact with the Autonomic Tuning engine to undertake model correction, modify solver behaviour or iteration count, for instance. In this architecture, the tuning engine also utilizes performance data to undertake buffer management and data routing.

4   Discussion and Analysis

In Sections 3.1 and 3.2, we identify how two scientific applications make use of the conceptual architectures illustrated in figures 1 and 2. In the first application scenario, a tuning engine is responsible for interacting with a resource manager

Table 4. Coupled fusion simulation application management using autonomic strategies

Application Objective	Autonomic Strategy
Maintain latency-sensitive data delivery	**Resource Management** adaptive data buffering (time, size), adaptive buffering strategy, adaptive data transmission & destination selection
Maximize data quality	**Resource Management** opportunistic in-transit processing adaptive in-transit buffering
Scientific Fidelity	**Algorithmic Adaptivity** in-time data coupling model correction using dynamic data solver adaptations -

to support resource selection, whereas in the second, the tuning engine can modify the application behaviour directly based on the observed outcome. In both instances, the tuning engine is external to the application and resource management components, thereby making it more general purpose and re-usable. In the first application scenario, we make use of a centralized tuning engine, while in the second application scenario, we have multiple coordinated tuning engines.

It is also important to emphasize that as an application can have multiple objectives, an application can operate in multiple *usage* modes. Hence, the examples of application scenario mappings to particular conceptual architectures presented above were selected to provide illustrative examples of *particular* usage. For example, an alternative version of the EnKF application (presented in Section 3.1) could utilize the conceptual architecture in figure 2. This involves combining individual jobs into *job aggregates* (batches) based on their particular properties. Hence, if the same application was to be deployed over a combination of the TeraGrid and commercial Cloud computing infrastructure, such as Amazon EC2 and S3 for instance, it would be possible to match jobs to resource characteristics. In this approach, computation intensive jobs could be mapped to the TeraGrid, whilst those that require a quick turn around may be mapped to EC2. As discussed in Ref. [15], in this scenario the autonomic tuning engine is responsible for modifying the behaviour of the application to aggregate computations based on an estimate of which resource type is: (i) currently available; (ii) most likely to respond within some time threshold; and/or (iii) within cost/allocation budgets. Such characterisation may also be used for capacity planning, for instance to determine the number of virtual machine/EC2 instances required for a particular group of jobs. This would be supported through an abstraction such as customized pilot-jobs, allowing a sufficiently large number of resources to be instantly available for executing jobs in the future, thereby minimising data transfer and queuing overheads. Utilizing job and resource properties in this way would also lead to a complementary self-organizing architecture – i.e. one that is able to adapt application properties and job scheduling based on the characteristics of resources on offer. The two proposed architectures could also be integrated, so that the tuning *by* and *of* approaches could be applied during different periods of execution for the same application. Integrating these architectures also provided mulitple redundant pathways for adaptation. For example, in case of the coupled fusion simulations, tunning *by* the application is used to adapt buffer management strategies or in-transit processing based on data productions rates and tunning *of* the application is used to adjust data production rates to react to congestion or in-network loads. Also, increased buffer occupances due to link congestion can be tolerated by adapting in-transit paths, buffering strategies, and/or data production rates.

As illustrated in figure 3, an application objective is *organized using* an application tuning strategy. However, it is important to emphasize that when supporting autonomic tuning, it is necessary to chose tuning strategies that are able to satisfy multiple application objectives concurrently. For instance, in the case of the EnKF application, it is necessary to support *both* load balancing and scientific fidelity, within some pre-defined bounds. In future work we will discuss how the same application instance can use multiple pathways towards achieving multiple objectives.

5   Conclusion

The need for self-adaptation within computational science applications is outlined, along with two conceptual architectures that can be used to support such adaptation. The approach utilizes ideas from control theory and reflective middleware, although it differs from these approaches by considering practical concerns in realistic application deployments – where the application tuning mechanism needs to be separated from the application itself. Such an architectural approach also renders it more general purpose, and therefore re-usable. Two specific real world applications are used to demonstrate the use of the two conceptual architectures, outlining the basis for a framework for autonomic computational science – consisting of mechanisms, strategies and objectives. It is also important to emphasize that reflective middleware as well as the control models may be used as specific mechanisms to achieve autonomic behaviors.

In section 2.5 we identify how our approach aligns with that of Nierstrasz et al. [22], primarily considering a model-centric view that takes account of deployment and execution. The primary reason is the particular focus we adopt in section 2.2, which involves managing the execution of such an application on computational resources. However, the general autonomic computing concepts identified in section 2.1 are much broader in scope and a model-driven approach could also considered more generally. We believe such an approach would be useful to better understand how an application could be re-formulated, for instance, based on different scientific objectives (e.g. time to solution, error tolerance, accuracy etc); such criteria being mostly application domain specific.

Our motivation has come from existing work in executing computational science applications over distributed infrastructure. Using autonomic computing approaches to improve this execution has been the primary focus of the conceptual architectures outlined in section 2.2 and the conceptual framework in section 2.3. Developing an adaptive design methodology that extends these would be the next step in our work. Such a methodology would provide a set of stages that a designer of an application would need to follow to map application-level objectives to autonomic strategies and mechanisms. Furthermore, the conceptual architecture presented in figures 2 and 1 represents a centralized autonomic tuner – which may be co-located with the application. However, an implementation of such an architecture may have a tuning process that is distributed – for instance, each resource may itself use a tuning strategy for improving queuing times for particular types of jobs.

Acknowledgment

This paper is the outcome of the e-Science Institute sponsored Research Theme on Distributed Programming Abstractions. We would like to thank Murray Cole, Daniel Katz and Jon Weissman for being partners in the DPA expedition and having contributed immensely to our insight and understanding of distributed applications and systems, which form the basis of their application specifically to Autonomic Computational Science. We would also like to thank Hyunjoo Kim (Rutgers University), Viraj Bhat (Yahoo! Research) and Yaakoub El Khamra (TACS, Univ. of Texas Austin) for the two applications described in this paper.

References

1. Al-Ali, R.J., Amin, K., von Laszewski, G., Rana, O.F., Walker, D.W., Hategan, M., Zaluzec, N.J.: Analysis and Provision of QoS for Distributed Grid Applications. Journal of Grid Computing 2(2), 163–182 (2004)
2. Andersson, J., de Lemos, R., Malek, S., Weyns, D.: Modeling dimensions of self-adaptive software systems. In: Cheng, B.H.C., de Lemos, R., Giese, H., Inverardi, P., Magee, J. (eds.) Software Engineering for Self-Adaptive Systems. LNCS, vol. 5525, pp. 27–47. Springer, Heidelberg (2009)
3. Andersson, J., de Lemos, R., Malek, S., Weyns, D.: Reflecting on self-adaptive software systems. In: Proceedings of Workshop on Software Engineering for Adaptive and Self-Managing Systems (SEAMS), Vancouver, BC, Canada. IEEE, Los Alamitos (2009)
4. Batch Queue Predictor, http://nws.cs.ucsb.edu/ewiki/nws.php?id=Batch+Queue+Prediction (last accessed: May 2010)
5. Bhat, V., Parashar, M., Khandekar, M., Kandasamy, N., Klasky, S.: A Self-Managing Wide-Area Data Streaming Service using Model-based Online Control. In: 7th IEEE International Conference on Grid Computing (Grid 2006), Barcelona, Spain, pp. 176–183. IEEE Computer Society, Los Alamitos (2006)

6. Bhat, V., Parashar, M., Klasky, S.: Experiments with In-Transit Processing for Data Intensive Grid workflows. In: 8th IEEE International Conference on Grid Computing (Grid 2007), Austin, TX, USA, pp. 193–200. IEEE Computer Society, Los Alamitos (2007)
7. Bhat, V., Parashar, M., Liu, H., Khandekar, M., Kandasamy, N., Abdelwahed, S.: Enabling Self-Managing Applications using Model-based Online Control Strategies. In: 3rd IEEE International Conference on Autonomic Computing, Dublin, Ireland, pp. 15–24 (2006)
8. Brevik, J., Nurmi, D., Wolski, R.: Predicting bounds on queuing delay for batch-scheduled parallel machines. In: Proc. ACM Principles and Practices of Parallel Programming (PPoPP), New York, NY (March 2006)
9. Chandra, S., Parashar, M.: Addressing Spatiotemporal and Computational Heterogeneity in Structured Adaptive Mesh Refinement. Journal of Computing and Visualization in Science 9(3), 145–163 (2006)
10. Cheng, B.H.C., de Lemos, R., Giese, H., Inverardi, P., Magee, J.: Software Engineering for Self-Adaptive Systems: A Research Roadmap. In: Cheng, B.H.C., de Lemos, R., Giese, H., Inverardi, P., Magee, J. (eds.) Software Engineering for Self-Adaptive Systems. LNCS, vol. 5525, pp. 1–26. Springer, Heidelberg (2009)
11. Dobson, S., Denazis, S.G., Fernández, A., Gaïti, D., Gelenbe, E., Massacci, F., Nixon, P., Saffre, F., Schmidt, N., Zambonelli, F.: A survey of autonomic communications. ACM TAAS 1(2), 223–259 (2006)
12. El-Khamra, Y., Jha, S.: Developing autonomic distributed scientific applications: A case study from history matching using ensemble kalman-filters. In: GMAC 2009: Proceedings of the 6th International Conference on Grids Meets Autonomic Computing. ACM Press, New York (2009)
13. El-Khamra, Y., Jha, S.: Developing autonomic distributed scientific applications: a case study from history matching using ensemblekalman-filters. In: Proceedings of the 6th International Conference on Autonomic Computing (ICAC); Industry session on Grids meets Autonomic Computing, pp. 19–28. ACM, New York (2009)
14. Evensen, G.: Data Assimilation: The Ensemble Kalman Filter. Springer, New York (2006)
15. Kim, S.J.H., Khamra, Y., Parashar, M.: Autonomic approach to integrated hpc grid and cloud usage. Accepted for IEEE Conference on eScience 2009, Oxford (2009)
16. Hariri, S., Khargharia, B., Chen, H., Yang, J., Zhang, Y., Parashar, M., Liu, H.: The autonomic computing paradigm. Cluster Computing 9(1), 5–17 (2006)
17. Jha, S., Cole, M., Katz, D., Parashar, M., Rana, O., Weissman, J.: Abstractions for large-scale distributed applications and systems. ACM Computing Surveys (2009) (under review)
18. Jha, S., Parashar, M., Rana, O.: Investigating autonomic behaviours in grid-based computational science applications. In: GMAC 2009: Proceedings of the 6th International Conference on Grids Meets Autonomic Computing, pp. 29–38. ACM Press, New York (2009)
19. Kephart, J.O., Chess, D.M.: The vision of autonomic computing. Computer 36(1), 41–50 (2003)
20. Klasky, S., Beck, M., Bhat, V., Feibush, E., Ludäscher, B., Parashar, M., Shoshani, A., Silver, D., Vouk, M.: Data management on the fusion computational pipeline. Journal of Physics: Conference Series 16, 510–520 (2005)
21. Kon, F., Costa, F., Campbell, R., Blair, G.: A Case for Reflective Middleware. Communications of the ACM 45(6), 33–38 (2002)

22. Nierstrasz, O., Denker, M., Renggli, L.: Model-centric, context-aware software adaptation. In: Cheng, B.H.C., de Lemos, R., Giese, H., Inverardi, P., Magee, J. (eds.) Software Engineering for Self-Adaptive Systems. LNCS, vol. 5525, pp. 128–145. Springer, Heidelberg (2009)
23. Parashar, M.: Autonomic grid computing. In: Parashar, M., Hariri, S. (eds.) Autonomic Computing – Concepts, Requirements, Infrastructures. CRC Press, Boca Raton (2006)
24. Serugendo, G.D.M., Foukia, N., Hassas, S., Karageorgos, A., Mostefaoui, S.K., Rana, O.F., Ulieru, M., Valckenaers, P., Aart, C.: Self-organising applications: Paradigms and applications. In: Di Marzo Serugendo, G., Karageorgos, A., Rana, O.F., Zambonelli, F. (eds.) ESOA 2003. LNCS (LNAI), vol. 2977, Springer, Heidelberg (2004)
25. Sevcik, K.: Model reference adaptive control (mrac), http://www.pages.drexel.edu/~kws23/tutorials/MRAC/MRAC.html (last accessed: August 12, 2009)
26. Söderström, S.: Discrete-Time Stochastic Systems - Estimation and Control, 2nd edn. Springer, London (2002)

Modelling the Asynchronous Dynamic Evolution of Architectural Types[*]

Cristóbal Costa-Soria[1] and Reiko Heckel[2]

[1] Dept. of Information Systems and Computation, Universidad Politécnica de Valencia, Spain
[2] Department of Computer Science, University of Leicester, UK
ccosta@dsic.upv.es, reiko@mcs.le.ac.uk

Abstract. Self-adaptability is a feature that has been proposed to deal with the increasing management and maintenance efforts required by large software systems. However this feature is not enough to deal with the longevity usually these systems exhibit. Although self-adaptive systems allow the adaptation or reorganization of the system structure, they generally do not allow introducing unforeseen changes at runtime. This issue is tackled by dynamic evolution. However, its support in distributed contexts, like self-organizing systems, is challenging: these systems have a degree of autonomy which requires asynchronous management. This paper proposes the use of asynchronous dynamic evolution, where both types and instances evolve dynamically at different rates, while preserving: (i) type-conformance of instances, and (ii) the order of type evolutions. This paper describes the semantics for supporting the asynchronous evolution of architectural types (ie. types that define a software architecture). The semantics is illustrated with PRISMA architecture specifications and is formalized by using typed graph transformations.

Keywords: dynamic evolution, asynchronous updates, software architecture, typed graph transformations.

1 Introduction

A well-known property of current and future software systems is their increasing scale and complexity [42]. Larger size and complexity requires more management and maintenance efforts, which increase the costs of such systems [39]. In order to alleviate the management issues, self-adaptability is being proposed as a feasible solution [23,24].

Self-adaptability is the ability of a system to manage itself at runtime, that is, to adapt its structure or organization in response to changing operating conditions or user requirements. There are two main engineering approaches realizing self-adaptability, which differ on how the interaction patterns of the system are managed. On the one

---

[*] This work has been partially supported by the Spanish Department of Science and Technology under the National Program for Research, Development and Innovation project MULTIPLE (TIN2009-13838), and by the Conselleria d'Educació i Ciència (Generalitat Valenciana) under the contract BFPI06/227.

D. Weyns et al. (Eds.): SOAR 2009, LNCS 6090, pp. 198–229, 2010.

hand, top-down self-adaptable approaches rely on a centralized representation/model, which describes the relations among the different elements of the system. Self-adaptability is guided by this model: (global) decisions are taken upon this model and changes are applied on the system globally. An example of top-down approaches are self-adaptive systems [12,32], which are based on an architecture model and a set of (high-level) goals to guide the adaptation process. On the other hand, bottom-up approaches are fully decentralized: interactions are managed locally by the elements of the system. Thus, self-adaptability emerges from the local adaptation decisions taken by each component. An example of bottom-up approaches are self-organizing systems [38,41], which are based on algorithmic functions to guide the (local) adaptation process. These systems, usually of a distributed nature, are composed of several autonomous instances, which run concurrently and organize themselves according to different criteria. However, although self-adaptability can support the management of large software systems, it is not enough to deal with another property: their longevity. Due to their high development costs and mission critical nature, many systems are being used for a long period and with very little time for maintenance. Unforeseen maintenance operations may be required by technology changes, new requirements or necessary corrective measures. Although self-adaptive and self-organizing systems allow the adaptation or reorganization of the system structure, they will not generally allow to introduce new functionality at runtime.

What happens if the behaviour of the components, or the reorganization algorithms, need to be changed without the system being stopped? Then, dynamic evolution is required. Dynamic evolution enables the modification of running software artifacts, thus allowing the introduction of new, non-predicted changes. However, the support for dynamic evolution in distributed contexts, like self-organizing systems, is challenging. Due to the autonomous nature of self-organizing instances, their dynamic evolution cannot be performed synchronously with respect to other instances. That is, while one instance may be able to accommodate new changes, another one may not (due to various constraints, e.g. it might be disconnected). In this context, it is difficult and time-costly to stop (and synchronize) all the distributed entities to introduce a new update.

The solution we propose is the use of *asynchronous dynamic evolution*. Dynamic asynchronous evolution is a feature that allows introducing changes in a running system *concurrently*, but maintaining the order of changes. It allows any type of the system and its instances to evolve at different times. Then, a type may be dynamically updated several times, while each one of its instances: (i) remains out of date, (ii) continues evolving to a previous version, or (iii) starts evolving to the last version. Asynchronous evolution is useful for those situations where dynamic evolutions are required, but synchronisation is not possible and evolutions should be postponed. For instance, this is the case for distributed and/or autonomous systems, because they are composed of entities that sometimes may be inaccessible (i.e. distributed systems) or busy (autonomous systems). Furthermore, asynchronous evolution is also useful for building systems with a dynamic nature, that is, with a high rate of dynamic changes. For instance, systems that are incrementally or collaboratively built at runtime, like dynamic web systems, require to concurrently introduce new changes on the system specification while its instances are still running or evolving to previous versions.

In this paper we introduce a semantic model for asynchronous dynamic evolution of software architectures. The contribution of this paper is twofold: (i) a general and abstract model of asynchronous evolution; and (ii) its application to the evolution of types in PRISMA architectural specifications. Specifically, asynchronous evolution is modelled by the definition of patterns of changes in terms of typed graph transformations [17,21]. This paper describes the approach, but due to lack of space does not contain the complete model. The approach is illustrated by its application to the dynamic evolution of PRISMA architecture specifications (i.e. the type) and its configurations (i.e. its instances).

This paper is organized as follows. Next section introduces the asynchronous evolution of types, the differences with synchronous evolution, and the challenges it poses. Section 3 describes PRISMA architectural types, which are used to illustrate our approach. Section 4 introduces the formalization of asynchronous evolution of architectural types by using graph transformations. Then, sections 5 and 6 discuss our approach and related work. Finally, section 7 presents the conclusion and further works.

2 Asynchronous Evolution of Types

In order to describe what dynamic asynchronous evolution is, first the concept of types and instances should be introduced. A type is an abstract concept which defines the structure (i.e. the state) and behaviour (i.e. how this state is modified) of a software artifact. A type is comprised of two elements: a specification, which is the high-level description of a software artifact, and an executable code, which is the realization of this software artifact (and the implementation of the specification). In a Model-Driven-Development approach [6], the executable code is automatically generated from a (possibly partial formal) specification, which reflects how both elements are closely related. The executable code allows the creation and execution of different instances of the software artifact. An instance is the execution of a type on a concrete platform: it behaves as defined by the type specification, and it is characterized by an internal state (i.e. the data stored in the instance), which is different from other instances.

When a type, that is deployed and active in a software system (i.e. it has been instantiated), needs to be changed or updated, *dynamic software evolution* [11,40] (or online software evolution [48]) is used. Dynamic software evolution is a feature that allows changing a type without the need to shut down the system. This is performed by the following evolution process: (1) the modification of the current type specification; (2) the update or regeneration of the executable code of the type, so that instances could be created according to the updated type; and (3) the evolution or migration of the running, stateful instances to integrate the changes. The last step is the longest, because it entails the quiescence (i.e. the safe stopping) of all the instances that are going to evolve. This quiescent [25] or tranquil [47] status[1] of an

---

[1] As introduced by Vandewoude et al. in [47], we will use the distinction between the internal *state* of an element (which is migrated or transformed in the evolution process), and the *status* that describes its condition with respect to the evolution process (idle, active, quiescing, passive, ...).

instance guarantees that there are no pending or running transactions which could be affected by the evolution process.

The dynamic evolution process can be synchronous or asynchronous, depending on how a type and its instances evolve with respect to each other. In *dynamic synchronous evolution*, a type and its instances are evolved sequentially, i.e., the evolution of a type is followed immediately by the update of its instances before the type can evolve any further. In *dynamic asynchronous evolution*, a type and its instances are evolved independently. The type may evolve again before all of its instances have applied the previous changes. Asynchronous evolution advantages synchronous evolution when evolution requests are frequent and/or when instances are highly distributed or partly reachable. On the one hand, in asynchronous evolution changes can be performed earlier: a type can be evolved as soon as required, without waiting to update all of its instances. This is useful for developing highly flexible systems, i.e. those systems with a high probability of changes. On the other hand, changes can be propagated and applied asynchronously, without requiring the instances to be permanently reachable. This is particularly useful for supporting dynamic changes in distributed systems and/or mobile systems.

In order to better understand the implications of synchronous and asynchronous evolution approaches, we need to detail how they manage the execution of multiple evolution processes. But first let us introduce the concepts of type version, activeness, and evolution process.

The different evolutions of a type over time are reflected by means of *type versions*. Each time a type is evolved, a new version is created/generated. This version contains the new (evolved) type specification and the executable code from which new instances will be created. A type version is kept in memory while it is *active*. A version is *active* while: (i) it is the most recent type version, or (ii) it has running instances. The most recent type version is active by default (i.e. loaded in the system) to allow its evolution: only the latest version can be evolved. The previous versions are outdated and are only kept in the system to allow the execution of out-of-date instances. As soon as the out-of-date instances are updated or deleted, then the old type version is not active anymore and can be safely unloaded from the system. The evolution of a type and its instances is performed by an *evolution process*, which: (i) evolves the latest version of the type (generating a new version), and (ii) evolves the instances from the previous version to the new version, preserving their states. In general, the end of an evolution process implies the inactivity of the type version that has been updated, due to the evolution of all of its instances to the new type version.

Fig. 1 describes visually the execution of several evolution processes and how they are managed in synchronous (Fig. 1(a)) and asynchronous (Fig. 1(b)) approaches. This figure shows the evolution of a type, called T, and the evolution of its instances, identified by numbers *1, 2, 3*. In Fig. 1 the different versions of the type T (i.e. its evolutions) are depicted as squares (T_0, T_1, T_2 and T_3). The position of each square represents the time when the corresponding version was introduced in the system. An adjacent, diamond-ended, dotted line represents the lifetime of a version (i.e. the time that a version is active in the system). Circles with identical number represent instantiations or evolutions of an instance. The vertical position of each circle depicts the version that the instance is conformant to, whereas the horizontal position depicts time: (i) the moment when the instance was created; or (ii) the moment when the instance evolved from a previous version (the instance had a state which has been migrated to the new version). However, the figure does not describe the reason why an instance delays its evolution: this may be due to that: (i) the instance has not yet received the evolution request (i.e. in distributed systems), or (ii) the instance is waiting to reach a quiescent status. Finally, solid round-ended lines (see Fig. 1) depict evolution processes, which start with the generation of a new version and end when all the instances of the old version have been evolved to the new version.

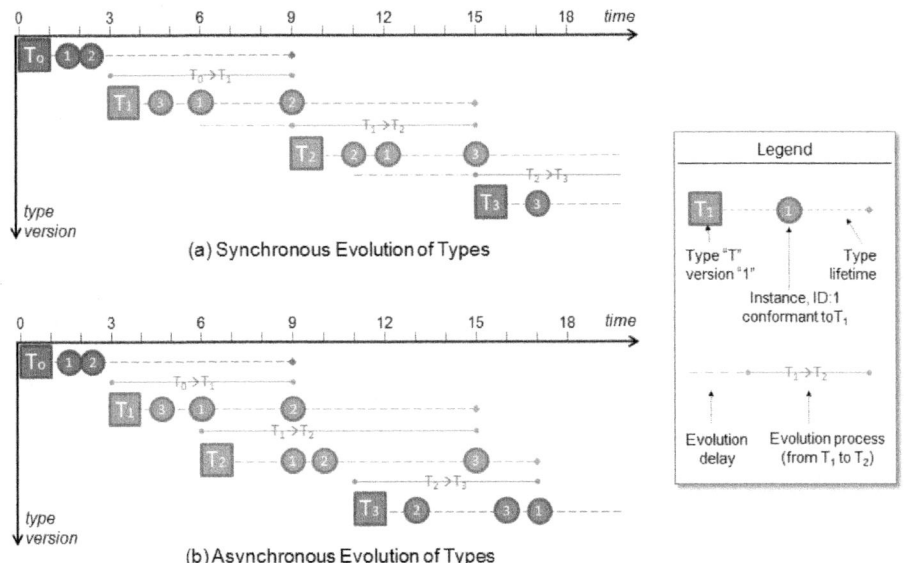

Fig. 1. Synchronous vs Asyncronous Evolution

Fig. 1(a) shows the dynamic evolution of the type T following a synchronous approach, as implemented in most of current approaches (e.g. [31,48]). In this approach, a new evolution process cannot be started until the completion of the previous evolution process. This delay is depicted in the figure as a dotted line at the beginning of each evolution process. For instance, although the evolution request "$T_1 \rightarrow T_2$" have been received at time instant 6 (see Fig. 1(a), 2nd evolution process), the evolution process cannot start (i.e. it cannot generate the new version, T_2) until time instant 9, when the previous evolution process "$T_0 \rightarrow T_1$" finishes the evolution of the last instance of T_0, identified as 2. As it can be seen from the figure, the main disadvantage of synchronous evolution is the time that is needed until a new change can be introduced in the system. This may be an important drawback when evolving either distributed or mobile systems, where network fluctuations and/or the reachability of the instances (e.g. some may be disconnected) may increase the time needed to propagate new changes, and thus, the time needed to perform several dynamic changes.

On the other hand, Fig. 1(b) shows the dynamic evolution of the type T following an asynchronous approach (e.g. [37]). In an asynchronous evolution approach, type and instances evolve at different times. This allows evolution processes to overlap their execution: as soon as a type has been evolved, it can be evolved again, although some of its instances may still be applying the previous changes. Several examples of overlapping evolution processes are shown in Fig. 1(b). For instance, at time instant 6 the evolution process "$T_0 \rightarrow T_1$" is still active (see the 1st evolution process, started at instant 3), because the instance 2 is still pending to evolve T_0 to T_1. However, as soon as the change request "$T_1 \rightarrow T_2$" is received (see the 2nd evolution process), it is immediately executed: the version T_1 is evolved despite there are instances pending to evolve to this version (i.e. the instance 2). Another example of overlapping evolution processes (and type versions) is shown at time instant 13, where there are three active type versions: T_1, T_2 and T_3. These versions provide the behaviour of the instances 3, 1, and 2 respectively. Two pending evolution processes, "$T_1 \rightarrow T_2$" and "$T_2 \rightarrow T_3$", are still active: the former to evolve the instance 3 to version T_2, and the latter to evolve the instance 1 to version T_3. Each instance can evolve independently of the other instances, without waiting for the other instances to finish the previous evolution process(es). As soon as an instance is ready to evolve to the next type version, it does, although other instances remain in k-previous versions.

That is, while in synchronous approaches only one version and one evolution process is active, in asynchronous approaches several versions and evolution processes can coexist at the same time. However, this does not necessarily mean that several evolution paths (i.e. version branching) are allowed. Our approach is focused on an incremental evolution setting: the set of changes performed at time t (both at type-level and instance-level) is performed on the result of changes performed at time t-1. At the end, all the instances must integrate all the sequence of changes introduced over time. In order to do this, only a single evolution path must be allowed. Only in this case we are able to determine the correct version an instance must be evolved to. A single evolution path can be guaranteed by constraining the evolution to only the latest type version, protecting it from concurrent evolution requests. Thus, evolution processes are sequenced: a new evolution process cannot be started until the previous one has generated at least the new type version (it makes no sense to evolve something that has not been generated yet). And an evolution process cannot finish until the end of the previous evolution process. This is because the previous evolution process may still be evolving instances, which must be also evolved by the newer evolution process in order to integrate the newer changes.

Asynchronous evolution takes a step forward from synchronous evolution: it allows us to introduce multiple (but ordered) change requests, deferring them until they can be effectively applied. For instance, this will allow different stakeholders to update different parts (i.e. types) of a system at different times, without taking into account the update of the running (and perhaps distributed) instances. This update will be carried out by the evolution infrastructure.

However, the consequence of this is that asynchronous evolution is also the most challenging. Most of related works have been focused on the challenges posed by synchronous evolution: the adaptation of runtime data structures and code [27,31,40], the state consistency before and after a dynamic change [25,47], or the migration of the state [37,46]. Asynchronous evolution builds on the features provided by synchronous evolution, but adds new features (asynchrony) which in turn poses new challenges:

- *Type conformance.* Since types evolve at different rates with respect to their instances, it is then difficult to check if the instance-level is conformant to the type-level.
- *Version management.* Different evolutions of a single type entail the existence and management of different versions of this type at runtime (at least where there are running instances of such type), which adds more complexity to the process.
- *Order of evolution processes.* In such a case where a type could require a high rate of evolutions, it could happen that an instance might have two or more pending evolutions. In this case, the order of pending evolutions must be preserved. This is important in distributed systems, where instances may receive newest evolution changes before the older ones.
- *Coherence of interactions.* Interactions among instances that are in different versions can produce incorrect behaviours. The context where instances evolve is also important.

These challenges are addressed in Section 4. Next we introduce the context of this work, which will allow us to illustrate how we have modelled asynchronous evolution.

3 PRISMA Architectural Types

Software Architecture [36,44] provides techniques for (i) describing the structure (or architecture) of complex software systems (i.e. the key system elements and their organization), and (ii) reflecting the rationale behind the system design. The structure of a software system is mainly described in terms of architectural elements (i.e. components and connectors) and their interactions with each other. This structure is formally specified using an Architecture Description Language (ADL), which is used later to build the executable code of the software system.

Our work is focused on supporting the dynamic evolution of this structure: evolving both the formal system description (i.e. the ADL specification) and its executable code. Among the different formal ADLs from the literature [28], we selected the PRISMA ADL [33,34] because of the benefits it provides for supporting dynamic evolution. First, the PRISMA language allows modelling the functional decomposition of a system (by using architectural elements), and the system's crosscutting concerns (by using aspects), which results in more simple, clear, and concise system specifications. This allows us to separate parts of the software that exhibit different rates of change, and evolve only the interesting parts [29]. Second, PRISMA does not only allow modelling the structure (i.e. the architecture) of a system, but also allows describing precisely the internal behaviour of each architectural element. The behaviour is specified by using a modal logic of actions [43] (for describing services) and π-calculus [30] (for describing interactions among services); see [35] for more details. Thus, since the internal behaviour is formally described, this allows us to automatically interleave the actions required to perform the runtime evolution of its instances: (i) actions to achieve quiescence, and (ii) actions to perform the state migration. Lastly, the PRISMA ADL is supported by a Model-Driven Development framework [6], which allows the automatic generation of executable code from PRISMA models/specifications. This also benefits the support for dynamic evolution. The code generation templates can include not only the code for supporting the runtime evolution of the system, but also the code to reflect the changes on the formal system specification, keeping both in sync. In particular, in this paper we cover the semantics for supporting the asynchronous dynamic evolution of PRISMA architectures. For this reason, next we introduce the main concepts of the PRISMA ADL and a small example, which is used later to illustrate our approach.

The PRISMA ADL defines the architectural elements of a software system at different levels of abstraction: the type definition level and the configuration level. The type definition level defines architectural types, which are instantiated in specific architectures or are reused by other architectural types. The configuration level defines the architecture of a concrete software system, by creating and connecting instances of the architectural types defined at the type definition level. In other words, the configuration level specifies the topology of a specific architectural instance. This separation among the type level and the configuration level allows to easily differentiate-and implement-changes in a type and changes in a configuration (i.e. an *architectural* instance).

An architecture is defined at the type-level as a pattern, so that it can be reused in any other system or architectural type. The architectural type that defines this pattern is called *System*[2]. A System can be used in other architectural types as a single unit, and be treated like other simple architectural types (i.e. components and connectors). This allows PRISMA to support the compositionality, or hierarchical composition, of its architectural elements: the architecture of a complex software system can be described as a composition of several architectural elements which, in turn, can be described as the composition of other architectural elements. Thus, a complex system can be recursively defined as an architecture of architectures, because each composition describes an architecture.

The pattern of a System defines: (i) a set of ports for communicating with its environment (or with other architectural elements); (ii) the set of architectural types it is composed of and the number of instances that can be created of each type; and (iii) the set of valid connections among the architectural types and the number of connections allowed. A port defines a point of interaction among architectural elements; it publishes an interface with a set of provided and/or required services. A connection can be of two kinds: an Attachment, if it links two architectural elements; or a Binding, if it links an (internal) architectural element with one of the ports of the system (i.e. allowing the communication with external architectural elements). For instance, Fig. 2-top shows a System called *Sys*. It consists of two architectural element types, *A* and *B*, with a cardinality of *1..1* and *1..n*, respectively. These types are connected to each other by an Attachment called *Att_AB* with a cardinality *1..1* and *1..n*. These cardinalities mean that only one instance of A is allowed, which can be connected to several instances of B. Finally, *Sys* interacts with its environment by means of the port *p1*. The behaviour of this port is provided by the architectural type *A*, which is connected by means of the Binding called *Bin_p1A*.

The instantiation of a System is defined at the configuration-level, and is called *Configuration*. A Configuration instantiates a concrete architecture from the different combinations allowed by the pattern: it instantiates each of the architectural types

---

[2] In order to avoid confusions, we will use capitalized letters when referring to concepts of the PRISMA metamodel (e.g. System and Configuration).

defined in the pattern and connects them appropriately. For instance, Fig. 2-bottom
shows two Configurations, *C1* and *C2*, of the System *Sys*. The PRISMA ADL specification of this example is shown in Fig. 3.

In order to later illustrate the asynchronous evolution process, we will use this example, assuming that these two Configurations, *C1* and *C2*, are instantiated (and running) in different parts of a large software system.

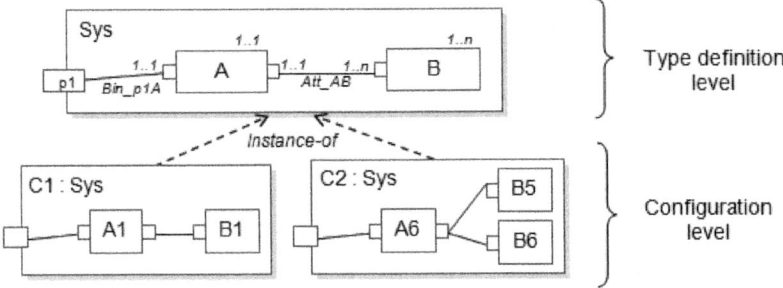

Fig. 2. Example of a PRISMA System and two Configurations

(a) Specification of the System Sys	(b) Specification of the Configurations C1 and C2
| ```
System Sys

Ports
  P1 : interface1;
End_Ports;

Import Architectural Elements
  A:TA(1,1), B:TB(1,n);

Attachments
  Att_AB: A.PServ(1,1) <--> B.PServ(1,n);
End_Attachments;

Bindings
  Bin_p1A: P1(1,1) <--> A.PClient(1,1);
End_Bindings;

New() { /* Constructor definition */ }
Destroy() { /* Destructor definition */ }

End_System Sys;
``` | ```
Architectural_Model_Configuration C1 =
  new Sys {
    A1 = new A();
    B1 = new B();

    att_A1B1 = new Att_AB(A1, B1);
    bin_A1 = new Bin_p1A(A1);
  }

Architectural_Model_Configuration C2 =
  new Sys {
    A6 = new A();
    B5 = new B();
    B6 = new B();

    att_A6-B5 = new Att_AB(A6, B5);
    att_A6-B6 = new Att_AB(A6, B6);
    bin_A6 = new Bin_p1A(A6);
  }
``` |

Fig. 3. Example of PRISMA ADL specifications

4 Formalization of the Evolution Process

Dynamic evolution is not a feature of a software system, but a feature of *its* architec-
tural types. The reason is that a software system is composed of several elements,
probably heterogeneous, and not all of them require dynamic change support. Some
of them may be evolvable whereas others may not. If a type is evolvable, it requires a
specific infrastructure to support its dynamic evolution, which is independent of the
other types of the system. The structure of a type (i.e. its state space) is different from
other types, so different (state-migration) functions will be required to evolve the

instances of each type. In our approach, evolvable types are generated with the required evolution infrastructure. We are not going to describe the details of such infrastructure, since it has been covered in other works [13,14]. Here we describe the semantics of the asynchronous dynamic evolution process performed by this evolution infrastructure and how to cope with the problems it poses.

An (asynchronous) dynamic evolution process is triggered when a meta-service called *Reflection,* provided by each evolvable architectural type, is invoked. This service requires a new specification for the type to evolve, which is described in terms of modifications on the current type specification[3]: additions, deletions or updates. An optional parameter allows specifying a set of instances to exclude from the evolution process. This is useful in situations where it is preferable to not evolve an instance but destroy it when it is not longer needed.

The set of modifications that can be performed on an architectural type is defined by its metamodel (recall that a metamodel describes how an architectural type is defined). According to the PRISMA metamodel [35] and the definition of the System element, we have defined 13 evolution operations for System architectural types (see Fig. 4). Each operation involves runtime changes both at type-level and instance-level.

```
AddArchitecturalElement(aeName, type, minCard, maxCard)
AddAttachmentType(attName, sourceAE, sourcePort, srcMinCard, srcMaxCard, targetAE, targetPort, trgMinCard,
trgMaxCard)
AddBindingType(bindName, srcPort, targetAE, trgMinCard, trgMaxCard)
AddPort(portName, interface, [playedRole])
RemoveArchitecturalElement(aeName)
RemoveAttachmentType(attName)
RemoveBindingType(bindName)
RemovePort(portName)
UpdateArchitecturalElement(aeName, newType, [newMinCard, newMaxCard])
UpdateAttachmentType(attName, srcMinCard, srcMaxCard, trgMinCard, trgMaxCard)
UpdateBindingType(bindName, trgMinCard, trgMaxCard)
```

Fig. 4. PRISMA evolution operations for System types

For instance, Fig. 5 shows two evolution processes which modify the System *Sys* described previously in Fig. 2. The first evolution process, started in time *t1*, introduces a new architectural type *C,* with a minimum and maximum cardinality of 1 and n, respectively. This type is attached (attachment type called *'A-C'*) to the port *'pA'* of type *A*, with these cardinalities: C$\rightarrow$A[1..1], A$\rightarrow$C[1..n]. This evolution process also removes the type *B* and the attachment from *A* to *B*. The second evolution process, started in *t5*, introduces a new type *D* that is attached to the type *C*, added in the previous evolution process.

The successful execution of each evolution process creates a new version (i.e. a new specification) of the architectural type being evolved. Type versions are identified from each other by means of a version number, which is increased each time an evolution process is applied on the type successfully. Each instance also contains an attribute which keeps track of the type version that it is currently an instance of. This

---

[3] There is another meta-service, called *Reify*, which provides –at runtime- the current type specification, so it can be evolved.

is used to control that only the evolution operations leading to the next type version (from the instance perspective) are taken into account, thus preserving the order of evolutions.

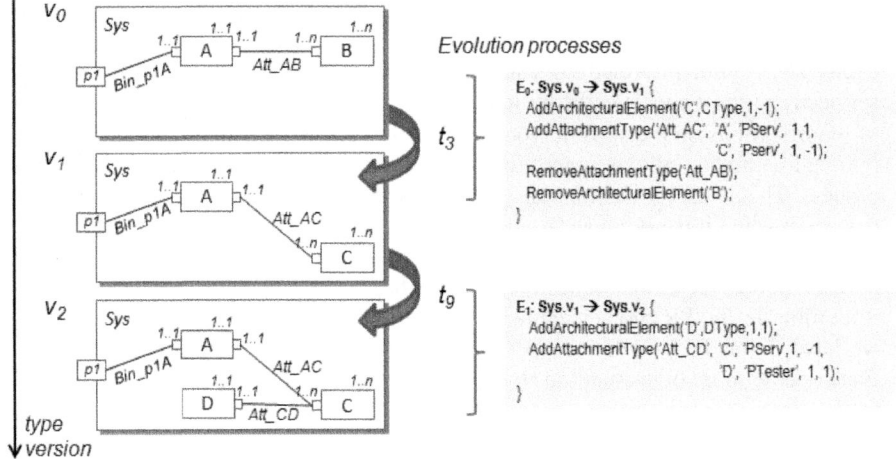

Fig. 5. Example of two evolution processes

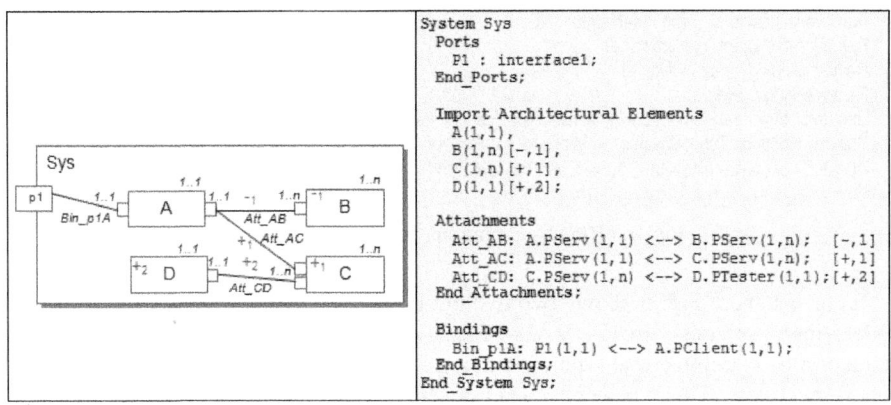

Fig. 6. Example of an architectural type with evolution tags

In our approach, only one single type specification is needed to describe both the current and previous type versions. This is possible because each change that is performed on an architectural type is also reflected on its specification, by means of evolution tags. An evolution tag is linked to one of the elements a type specification consists of (e.g. in the case a System: architectural elements, connections and ports), and describes: (i) the kind of evolution operation performed (i.e. addition, removal or update), and (ii) the type version where the evolution operation took place. Fig. 6 shows an example of such a tagged specification, which describes the different type versions produced by the evolution processes presented in Fig. 5. For instance, the tag *[+,1]* near the declaration of the type *C* in the specification of System *Sys* (see Fig. 6, right) tells us that the type *C* has been imported in version 1 of *Sys*.

With such a tagged type specification, the specification of a type *t* at version *v* can be obtained applying this function:

$$Spec_t(v) = Upd_t\big(\{(Spec_t(v_0) \cup Add_t(v_0, v)) \setminus Rem_t(v_0, v)\},\ v_0,\ v\big) \qquad (1)$$

This function builds a set with: (i) the elements of the first version of the type *t* (i.e. before the execution of any evolution operation): $Spec_t(v_0)$; and (ii) all the elements that have been added to the type *t* since the first version, v_0, until version *v*: $Add_t(v_0, v)$. Then, from this set are excluded all the elements that have been removed from the type *t* since the first version, v_0, until version *v*: $Rem_t(v_0, v)$. Finally, from the last resulting set are replaced the elements that have been updated since the first version, v_0, until version *v*: $Upd_t(Spec_t, v_0, v)$. Then, an instance *i*, whose version is v_i, will be *conformant with a type t* if its structure satisfies the specification returned by the function $Spec_t(v_i)$ defined above.

4.1 Describing the Semantics of Evolution

As described above, an evolution process comprises the execution of several evolution operations; each evolution operation impacts one of the elements an architectural type consists of. The complete set of evolution operations of System types has been formalised by means of typed graph transformations [21]. Graph transformations combine the idea of graphs as a modelling paradigm with a rule-based approach to specify the evolution of graphs. They are supported by an established mathematical theory and a variety of tools for its execution and analysis [17]. The main reasons to choose typed graph transformations as the basis for the formalisation are the following: (i) software architectures can be easily formalised as graphs, as shown in other works [22,49]; (ii) graph transformations are asynchronous, i.e., each rule can be applied once its preconditions are satisfied, which benefits the formalisation of asynchronous evolution; and (iii) typed graphs capture the relation among types and instances, required to model the evolution both at the type-level and instance-level.

In this work, typed graph transformations have been used to describe the *observed* behaviour of evolution operations: i.e. how a System type and its instances change as the execution of evolution operations. These graph transformations are presented using an architecture-based concrete syntax, which is more concise and easier to understand than the graph-based abstract syntax. Instead of using only vertices and nodes (as provided by the graph-based abstract syntax), a concrete syntax allows describing the same behaviour using concepts closer to the area of software architectures (i.e. components, ports, attachments, etc.), but extended with concepts required to deal with asynchronous evolution. The mapping of this architecture-based concrete syntax to the graph-based abstract syntax will be described later in section 4.2.

The semantics description of the entire set of evolution operations (11 in total, see Fig. 4) has produced 43 transformation rules, concerned with the evolution management of the type-level, the instance-level or both. Due to space reasons, we will only describe the formalization of the operations involved with the addition of a new architectural type to the composition of a System, and with the update of an architectural type. We have chosen these operations because they are enough to illustrate how the different challenges described in section 2 are addressed by their respective transformation rules. Each rule is self-contained and can be understood without requiring the description of the complete set of rules.

4.1.1 Addition of a New Architectural Type

The addition of a new architectural type to a System is started by the execution of the *AddArchitecturalElement* operation. This operation modifies the composition of a System type to add a new, unforeseen, architectural type at runtime. This operation requires four parameters: *AEName*, the alias of the type to add; *type*, the executable code of the type to add; *minCard*, the minimum number of instances of this type that must exist in each system instance; and *maxCard*, the maximum number of instances that can exist in each system instance (i.e. in each Configuration). Its semantics is described by the rule R1 (see Fig. 7), which acts at the type-level. The left hand side of the rule (see Fig. 7, left side) specifies the condition(s) that must be satisfied to execute such rule: the type to add (i.e. *AEName*) must not already be present in the pattern defined by the System type. The System type to modify is represented as the top-left component named 'S', and the set of its instances as a multiobject (i.e. a collection of elements) named 'S1' at the bottom-left. The right hand side of the rule (see Fig. 7, right side) specifies the consequence of applying the rule: the pattern of the System 'S' has been modified to include a new architectural type, identified by *AEName*. This new type can be instantiated in each one of the System instances, but only while satisfying the *minCard* and *maxCard* constraints.

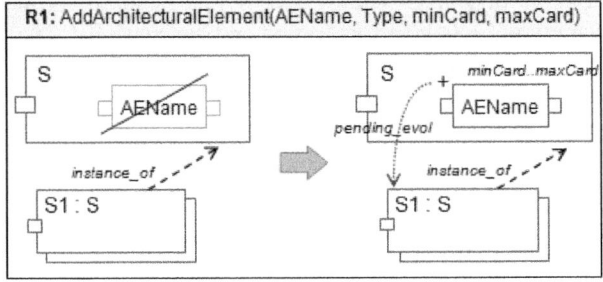

Fig. 7. Rule R1 - AddArchitecturalElement

As described before, a System type keeps track of the changes performed on it, in order to describe its evolution over time and to control the asynchronous evolution of its instances. This is also described in the transformation rules. When the rule R1 is executed, a new *add* tag (represented with the symbol '+') is created and attached to the new architectural element (see Fig. 7, right side). This tag is initialized with the current version number of the System type (whenever a new evolution process is started, the version number of the type is increased; the new evolved type is identified by this version number). This tag is also provided with a link to every instance of the type that must be evolved to the new type specification (remember that we can exclude instances from new evolutions). This relationship is reflected in R1 with the 'pending_evol' link to the set of instances. The use of this link will be shown later.

Changes at the instance-level are performed by reconfiguration operations. A reconfiguration operation can change the configuration (i.e. the topology) of a System instance (those where the service has been invoked), but only while conforming to the pattern defined in its System type. In addition, if a System instance has pending evolutions to apply, a reconfiguration operation also constraints the set of allowed changes in order to converge to the new type. For instance, if a type has been removed from the System pattern, the reconfiguration operation *CreateArchitectural Element* will not allow creating instances of such type. In order to carry out these checkings, a reconfiguration operation takes into account the system instance current version and the set of evolution tags corresponding to the instance version + 1 (i.e. to check valid type conformance and to lead the instance convergence to the next type version).

Instances of architectural elements are created by invoking the reconfiguration operation *Create ArchitecturalElement*. This service requires three parameters: *AEType*, the name of the architectural type which to create an instance; *id*, a unique identifier of the instance to create; and *params*, the parameters required to create an instance of the type *AEType*. The semantics of this reconfiguration service is described by the rule R2 (see Fig. 8). In this rule, *S1* is a Configuration (i.e. an instance of a System), and its type is the System *S*.

The left-hand side of the rule describes which conditions must be satisfied to allow the instantiation of an architectural type, *AEType*, in the Configuration *S1* (the context where the reconfiguration service has been invoked). These conditions are the following:

(i) The architectural type *AEType* must be defined in the type of *S1*: *AEType* exists in the pattern defined by the System *S* (see Fig. 8, top, in the box called *S*)

(ii) The instance *S1* cannot have another instance of *AEType* with the same identifier *id*.

(iii) The number of instances of *AEType* created in the Configuration *S1* is lower than the maximum cardinality *maxCard* defined in the System *S* (see Fig. 8, condition 1).

(iv) The type *AEType* has not been removed from the System *S* (i.e. it has not been tagged for deletion, see Fig. 8, cond. 2, first part). Or, if it has been removed, it has not been done in the subsequent evolution of the System type (i.e. it has been removed in a version later than the Configuration version + 1, see Fig. 8, condition 2, second part). Thus, the rule takes into account not only the conformance of a Configuration to its System type (i.e. same version), but also its convergence to the subsequent evolution of its System type.

(v) If the type *AEType* has been added at runtime (i.e. it is tagged with an addition tag), it has been done in previous versions of the System type (i.e. the version of this addition tag is less than or equal to the Configuration version, see Fig. 8, condition 3) or in the subsequent evolution of the System type (i.e. the version of the *AEType* addition tag is equal to the Configuration version + 1). In other words, the type *AEType* cannot be instantiated if it has been added in future versions of the System type (with respect to the Configuration perspective). Thus, the order of runtime type evolutions is preserved.

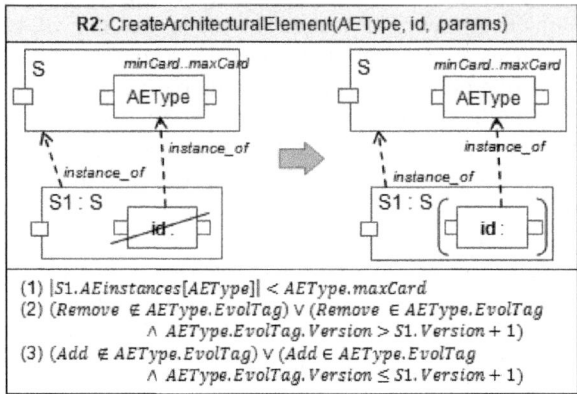

Fig. 8. Rule R2 - CreateArchitecturalElement

The right-hand side of the rule describes how the *AEType* type is instantiated in a Configuration (a System instance). Each new instance created is left in a quiescent state [25] by default, i.e. it cannot start or process any transaction. This is represented by using the symbols '[' and ']' around the software artifact being quiescent (see Fig. 8, right-hand side). The creation of instances stopped by default is to guarantee that the minimum cardinality constraint defined in the System type is satisfied first. Thus, several instances can be created, but they will not be started until their number would not be enough.

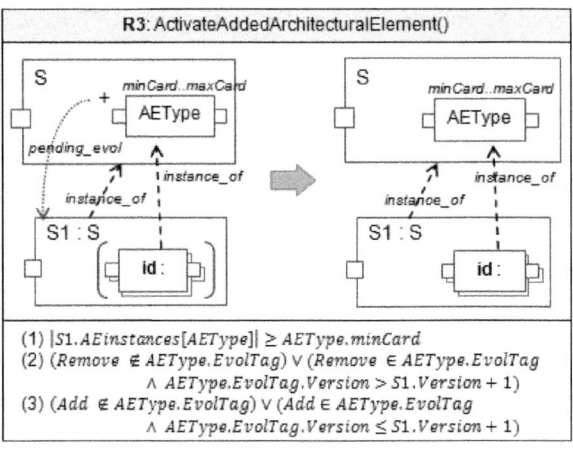

Fig. 9. Rule R3 – ActivateAddedArchitecturalElement

The activation of the quiescent instances is performed by the rule R3 (see Fig. 9). This rule is automatically triggered when any System instance (depicted as 'S1' in the figure) has so many quiescent instances of the type *AEType* as specified in the property *mincard*, defined in the System type *S* (see Fig. 9, condition 1). The conditions 2 and 3 depicted below the rule are only defined to guarantee that the System type version chosen by the rule conforms to the current version of the System instance. The execution of this rule unblocks (i.e. start the execution of) the set of instances of the type *AEType* (note that the symbols '[' and ']' have been removed from the right hand side).

Another effect of the execution of the rule is that it removes the link 'pending_evol' among the addition tag (depicted as '+' in the rule) and the System instance *S1* where the rule has been applied. This means that the System instance *S1* has integrated the new type *AEType* as specified by the System type *S*, which was introduced in its version *S1.Version+1*. In other words, the System instance now has integrated one of the evolution operations that were pending.

However, since an evolution process is made of several evolution operations, a System instance could not be promoted to a new version of the System type until all the evolution steps would be performed. This is described by the rule R4 (see Fig. 10). The left-hand side of the rule describes the condition that must be satisfied to promote a System instance to (i.e. be conformant with) the next version of the System type: a System instance *S1*, whose current version is *v*, is said to be conformant to the version *v+1* of its System type if no evolution tag with version = *v+1* exists with a 'pending_evol' link to the System *S1*.

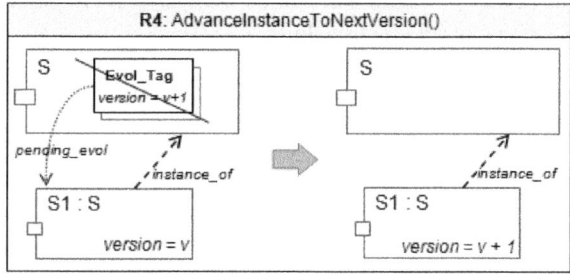

Fig. 10. Rule R4 - AdvanceInstanceToNextVersion

In addition, the 'pending_evol' link has an additional use: to monitor the evolution state of each instance at the type-level. By looking the oldest evolution tag (i.e. with the minor version) which is still linked to a System instance, it lets us know: (i) the set of evolution operations the System instance is currently involved in, and (ii) the version a System instance currently is conformant to, by decreasing the version provided by the oldest evolution tag found.

4.1.2 Updating of Architectural Types
Finally, in order to fully understand our approach, the update operation is described here. This operation replaces a type version by another and updates its instances, keeping all their existing connections and their internal state. This is formalised by two transformation rules: *UpdateAEType* and *ReplaceAE*, which act at the type-level and the instance-level respectively. The context where these rules are executed are: the System specification that contains the type to update, and the instances (or Configurations) that have imported and instantiated the type to be updated.

On the one hand, the *UpdateAEType* rule models the updating of a type version by a new one, keeping the existing interaction patterns. Let us recall that in the PRISMA model, the type-level defines the interaction patterns among types (i.e. what kind of interactions are allowed), and consequently, the interactions among their instances. When updating a type, the existing interaction patterns must be preserved, in order to guarantee the coherence of interactions at the instance-level: the instances that were connected/attached to the old, updated instance must be able to interact with the new, evolved instance.

The coherence of interactions can be only guaranteed by requiring the new type version to be syntactically and semantically compatible with existing interactions; otherwise, the update operation must not be performed. That is, compatibility evaluation when performing an update is focused on the interactions that the old type has, instead of focusing on the type itself. The reason is that a new type version, despite being incompatible with its previous version, could be semantically compatible with the types that were interacting with the previous version. This is the case when, in the context of an evolution process, the removal of *part of* the original functionality is required: a new version provides less functionality than the previous version (i.e. the new version is not compatible with the old one), and the interactions requiring this functionality have been removed from the initial set of interactions (i.e. the new version is compatible with the resulting set of interactions). If complete compatibility (or subtyping) of the new type with respect to the old type were required, we would never be allowed to remove unused functionality[4].

Since interactions among architectural types are performed through ports (i.e. they are the points of interaction), compatibility for updating is evaluated through ports. Thus, the requirement to perform an update operation is that the new type provides a set of ports syntactically and semantically compatible with the interacting ports of the old type. On the one hand, we define a port p_x as *syntactically compatible* with another port, p_y, if it provides all the services that are required from p_y, which are defined by the set of existing interactions of p_y, with exactly the same signature (i.e. name and parameters) of each service. Note that syntactic compatibility only refers to the minimum set of services *provided* by p_x: this means that a syntactically compatible port may provide additional services or require different services than the original. The goal of the updating operation is to guarantee that the updating does not break the current architecture, without taking care of the introduction of new required services. This is the responsability of other evolution rules, which will evaluate if all the required services are conveniently bound, thus guaranteeing the consistency of the architecture.

On the other hand, we define a port p_x as *semantically compatible* with another, p_y, if it provides the same observable behaviour as other elements expected from p_y. That is, the execution of the services of p_x must produce the same expected results, or sequence of traces, that p_y produces. However, the evaluation of semantic compatibility is challenging and its integration in the evolution model is not trivial. Since this is not the goal of this paper, the reader can refer to other works [18,19] in order to get more

---

[4] In fact, we could remove unused functionality from a type by removing the old version and adding the new, reduced one. However, in this case, the state migration of its instances would not be performed.

details about how some issues have been addressed from a formal perspective. For the sake of simplicity, we have abstracted semantic evaluation this way: each port is provided with a function, *CanInteractWith*, which evaluates if another port satisfies a certain contract or interaction protocol, i.e. that the observed behaviour is the required. For instance, in a port that requires compression and decompression services, a very simple function could evaluate that the compression of a sample data is decompressed correctly to the original data. That is, each port has the responsability of validating that their required services are provided correctly. Thus, we could say that a port p_x is *semantically compatible* with another, p_y, if all the ports that are connected to p_y, and request services from p_y, can interact seamlessly with p_x. In case of semantic incompatibility, a factible solution is the use of dynamically generated adaptors [9].

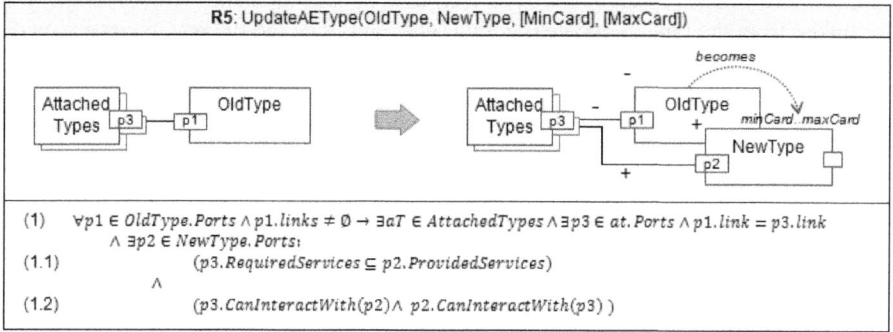

Fig. 11. Rule R5 - UpdateAEType

Syntactic and semantic compatibility is reflected in the *UpdateAEType* rule by means of condition (1) (see Fig. 11): the execution of the rule is only performed if the ports of the new type version (i.e. parameter *NewType*) are syntactically (see condition 1.1) and semantically (see condition 1.2) compatible with the ports of the old type version (i.e. parameter *OldType*). The set of types that interact with *OldType* are modelled by means of a multiobject called *AttachedTypes*. The ports of the types that interact with *OldType* (i.e. variable *p3*) are used to evaluate the syntactic and semantic compatibility of the ports provided by *NewType* (i.e. variable *p2*). Moreover, since *AttachedTypes* includes all the types connected to *OldType*, in case *OldType* were connected to itself, *AttachedTypes* would include OldType in its set. This guarantees that the updating of a self-interacting type is made consistently: the ports of the new version must be semantically compatible with the ports of the old version. Self-interacting types are common in self-organised systems: different (distributed) instances of the same type (e.g. agents) interact themselves, sharing a common interpretation of the environment. In this case, semantic compatibility guarantees that instances of the old version can interact consistently with instances of the new version.

The execution of this rule introduces the new type version tagged for addition (i.e. see the symbol '+' near the *NewType* component), and tags the old type version for removal (i.e. see the symbol '-' near the *OldType* component). The use of the addition and removal tags activates or constrains the behaviour of other instance-level rules.

For instance: the rule R2 (see Fig. 8) not only will create instances of the new type version, but will also avoid the creation of instances of the *old* type version. Another example: the rule R3 (see Fig. 9) will only allow the activation of instances of the new type version if and only if the minimum cardinality is satisfied.

However, in order to distinguish an update operation (which requires state migration) from an addition or removal operation (which also introduce addition/removal tags but do not preserve the instance state), this rule also introduces a relationship among the old type and the new type: the relationship "becomes". It specifically models which type is going to replace the older version, and activates the rule *ReplaceAE* (which is described below) for performing the migration of the instances to the new type version.

With respect to interactions, the existing connections (i.e. links to *AttachedTypes*, a multiobject which represent the set of types that are interacting with *OldType*) are unlinked from the type to update (i.e. *OldType*) and linked to the new type version (i.e. *NewType*). This is modelled by tagging the old links with the symbol '-' (i.e. a removal tag) and the new ones with the symbol '+' (i.e. an addition tag). These tags will avoid the creation of new attachments at the instance-level among instances of *OldType* and *AttachedTypes*, promoting instead the creation of attachments with instances of *New-Type*. This way, *OldType* will be progressively removed from the System.

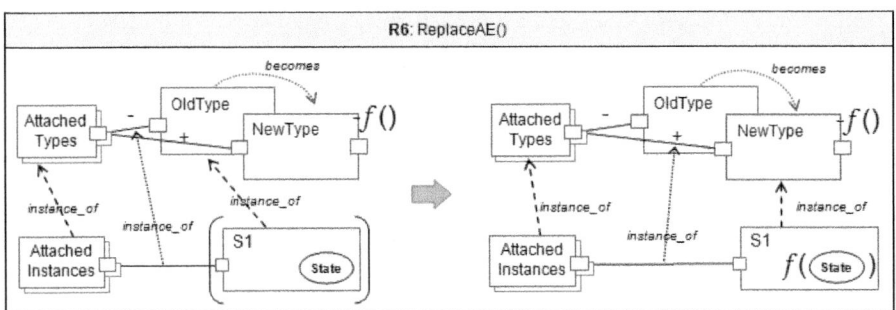

Fig. 12. Rule R6 - ReplaceAE

On the other hand, the *ReplaceAE* rule models the replacement of an instance by a new one, migrating its previous internal state and updating its existing connections (see R6, Fig. 12). This rule is activated if and only if a component instance is detected in a running System which matches the following conditions: (1) the type of such instance (which matches in the rule with *OldType*) has a "becomes" relationship with another type (i.e. *NewType* in the rule); and (2) such instance has reached a quiescent status (represented in the rule by the symbol "[ ]" around the instance *S1*). The first condition detects that the matching type has been updated. The second condition ensures that the interactions of the instance to migrate are stopped, and that the instance state is consistent, ready to be migrated. This is guaranteed by the quiescent status [25], which is only achieved when there are no running and/or pending transactions.

The result of the execution of the *ReplaceAE* rule is the migration (or transformation) of an instance of the old type to an instance of the new type. This migration is modelled by means of the modification of the *instance_of* relationship and the transformation of the internal state. An instance of a type that has been updated (i.e. *S1* is an *instance_of* the type *OldType*, which will *becomes NewType*) is transformed to an instance of another type (i.e. in the right hand part of the rule R6, *S1* is now an *instance_of* the type *NewType*). This results in that the internal state of the instance (i.e. the *State* element inside the *S1* instance, in the left hand part of the rule R6) is transformed to another (i.e. *f(State)*, in the right hand part of R6), by means of a state migration function. This function is modelled as *f()* and defines the mappings from the old data structures to the new ones.

There are two conditions to enable the state migration of instances: (i) the accessibility of their internal state, and (ii) the availability of a state migration function. On the one hand, if the old component type does not make accessible somehow the internal state of its instances, obviously we cannot migrate their state. For this reason, one of the conditions is that the old type provides a mechanism to get the internal state of its instances at runtime: this can be achieved by means of reflection, or by specialized functions that return the internal state to authorized requests (e.g. the migration function of a new version of the same type). This condition is explicitly included in the rule R6: the *State* of *S1* is visible, at least in the context of the rule. However, the rule does not reflect to *whom* it is accessible. It will depend on the specific implementation.

On the other hand, the new component type must provide a function to create new instances from the data structures of a previous type version. Otherwise, the state of an old instance could not be introduced into a new instance. The implementation of this function can be provided by means of a specialized constructor of the new type version which creates a new instance from the state of an instance of the previous type version. This condition is also explicitly included in the rule R6: *NewType* provides a function *f* , which results in the transformation of the original state of *S1* after the execution of the rule. The specific details for the creation of state migration functions are outside the scope of this paper. This rule only models the presence of such functions and what is the result obtained after the execution of the rule. If a state migration function is not provided (or the state of the instance is not accessible), then the old state cannot be migrated and is simply lost. However, the reader can refer to the works of Ritzau & Andersson [37], or Vandewoude & Berbers [46] for further details about how to automate the creation of state migration functions.

Another result of the execution of the *ReplaceAE* rule is the updating of the links of the instance to evolve. The set of instances that are interacting are represented by a multiobject called *AttachedInstances*, and their corresponding types by another multiobject, *AttachedTypes*. Note that, as a result of the execution of the type-level updating rule, the interacting types (i.e. *AttachedTypes*) are semantically compatible with the updated type (i.e. *NewType*). As a consequence, their instances (i.e. *AttachedInstances*) will be also semantically compatible (i.e. they could interact correctly) with the instance after evolution. The updating of links is modelled by means of the modification of *instance_of* relationships (which link the instance-level with the type-level): the attachments (i.e. links among instances) belong to different *attachment types* (i.e. patterns of interaction) before and after the execution of the rule. This means that, when the rule is executed, the existing links with other instances are deleted and replaced by new ones, but which point to the updated instance instead of the old one. Then, all the elements could start interacting.

Finally, note that rule R6 is abstract enough to model two kind of update approaches: *type substitution* and *type transformation*. One the one hand, in type substitution approaches, which has been commonly used in dynamic updating approaches (e.g. [26, 37, 40]), updating is performed through the *replacement* of instances of the old version by instances of the new version. Old instances are completely stopped and their state is migrated to new, updated instances. On the other hand, in type transformation approaches (e.g. [13]), the updating is performed by the *transformation* of the internal structure of old type instances to accommodate the elements introduced by the new type. From outside, an evolved instance keeps the original boundaries, links and state, but also integrates the behaviour added by the new type. From inside, only the elements that have been affected by the change are modified: their state is migrated to new ones, whereas the state of non-changed elements is kept intact. Type transformation approaches are better suited to partially change large architectural types (e.g. servers), because they do not require stopping the entire architectural instance, but only the required parts of the instance.

This is the focus that have been used in our proposal to model recursively the evolution of PRISMA Systems. From outside, the evolution of a System is perceived as a type substitution: its instances are *replaced* by new versions having their state migrated. However, from inside the evolution is performed as a type transformation: the differences among the new type specification and the current specification are used to incrementally change the original instances, by means of a set of evolution operations (e.g. see Fig. 5). This strategy can be recursively applied, until (i) the decomposition of a type is not advisable (i.e. more than the 60% of type structures are going to be changed), or (ii) the internal composition is neither available or modifiable (e.g. COTS).

4.2   Graph-Based Abstract Description of the Evolution Semantics

The semantic rules described in the previous section are formalised by mapping their architecture-based concrete description to a graph-based abstract description. Then, the graph-based abstract rules can be entered in a graph transformation tool, such as AGG [1], to simulate and validate their execution. This will allow us to analyze some properties and dependencies from an high abstraction level (e.g. interaction dependencies, instance-level implications, etc.). In order to do this, the first step is the definition of a *type graph*, i.e., a graph representing a metamodel to define the concepts and relations in the domain. Instances of this metamodel are called instance graphs. They represent the runtime states of the system and are subject to modification by transformation rules: given a valid instance graph and a rule, the rule could be applied to the graph to produce a new graph. If this graph satisfies the constraints of the type graph (i.e. the metamodel) it constitutes the successor state in the runtime model of the system. Note that elements of type graphs are not types in the architectural sense, but metamodel elements. In order to model the evolution of both types and instances, an instance graph must include not only instances (e.g. Configurations), but also the types that provide the behaviour of these instances (e.g. Systems). Therefore, a type graph must describe the ontological metamodel [4] of the concepts that are going to be subject to evolution. That is, the type graph must describe both the meta-types (i.e. the properties and relations among types) and the meta-instances (i.e. the properties and relations among instances). In addition, this type graph should also include the concepts required to describe how the concepts evolve over time: the evolution tags.

Fig. 13 shows the type graph that includes the concepts of the PRISMA metamodel and the concepts related to the evolution of types and instances. The *ArchitecturalElement* concept represents simple PRISMA architectural elements (i.e. not composed elements), which provide or request services through a set of *Ports* (see the *Port* concept in the metamodel). The *System* concept represents composed PRISMA architectural elements: they are composed of several architectural elements (*AEType*), which are connected by means of attachments (*AttachmentType*) and/or bindings (*BindingType*). Since a System is also an architectural element (e.g. it has ports), it inherits its behaviour from the ArchitecturalElement concept. The *AEType* concept provides an alias for an architectural element that is imported into a System and the allowed cardinality in such System. There are three kind of evolution tags: *Added*, *Removed*, and *Update*. However, the latter is only used by stateful entities, i.e. architectural elements. The reason is that the update of stateless entities (like attachments, bindings and ports) can be reflected by means of a removal tag and an addition tag. In the case of stateful entities, this is not enough, since the state must be migrated from the old entity to the new one. For this reason, the *updated* concept has a link (called *replaced_by*) to the architectural element that is going to replace the tagged element. All the evolution tags (see the concept *Evol_Tag*) have a link to the System instances (i.e. Configurations) that are pending to be evolved. Finally, the metamodel defines what are the elements of the instance-level, their properties and relationships (see Fig. 13, meta-instances). Note how the quiescent status has been added to the concept *AEInstance*, in order to reflect when an instance is ready to be evolved.

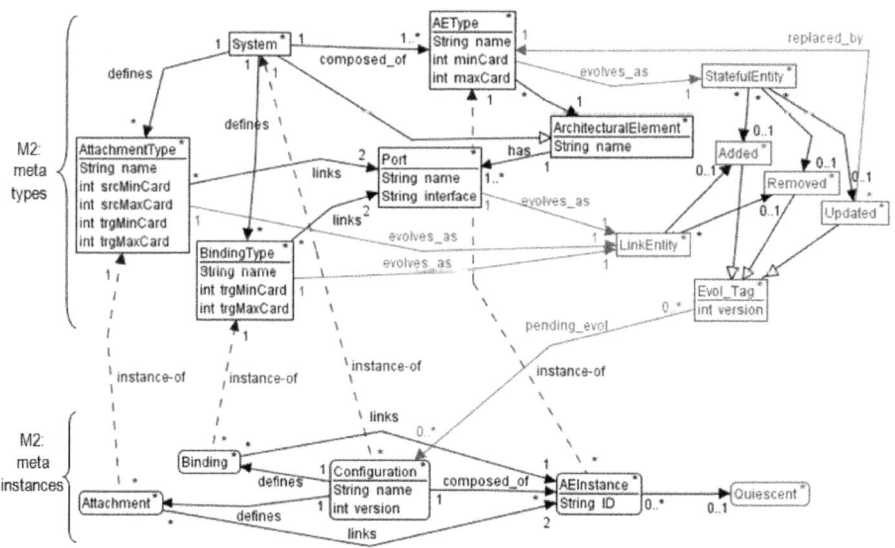

Fig. 13. Type Graph of PRISMA with Evolution Tags

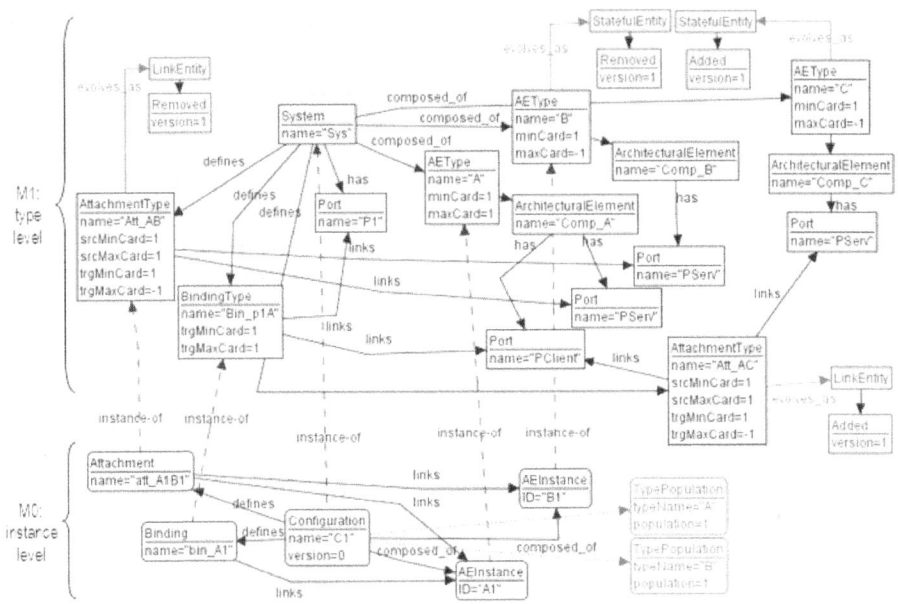

Fig. 14. Instance graph of the Sys type (at version 1) with the C1 instance (at version 0)

Fig. 14 illustrates how an instance graph looks like: it includes type-level concepts and instance-level concepts. In particular, the type-level concepts included in this graph are related to the System *Sys*, after the type-level execution of the evolution process "E_0:*Sys.v$_0$*→*Sys.v$_1$*": it removes the architectural type *B* and adds the type *C* (see Fig. 5). The System *Sys* is modelled as a graph, and each one of its structural elements (ports, architectural types, and connections) are modelled as nodes of this graph, according to the type graph described above. For instance, the type *A* is modelled with two graph nodes: (i) an *ArchitecturalElement* node named "Comp_A" which defines the behaviour of the type and which is linked to the *Port* nodes named "PClient" and "PServ"; and (ii) an AEType node named "A" which defines the usage restrictions of such type in *Sys*: a minimum and maximum cardinality of 1. Note how the elements that have been affected by the evolution process (i.e. the attachment types *Att_AB*, *Att_AC*, and the architectural elements *B*, *C*) have been tagged with removal or addition tags. Finally, the instance graph shows some elements of the instance-level (see M0): the Configuration *C1*, which is still conforming to version 0 of *Sys*.

This graph is a clear example of the benefits of using an architecture-based concrete description instead of a graph-based abstract description to describe the evolution semantics: the former is more concise than the latter. However, the former cannot be automatically validated and/or formally analysed to detect inconsistencies.

An example of the mapping of an evolution rule to a graph transformation rule is shown in Fig. 15. This figure shows how the reconfiguration operation *CreateArchitecturalElement* has been modelled in AGG, renamed as *CreateAE*. The execution of this rule requires three input parameters: *configName*, the name of the Configuration where a new instance is going to be created; *typeToInstantiate*, the name of the type to instantiate; and *newID,* the identifier of the new instance to create. The left-hand side of the rule describes the initial matching (see Fig. 15, center) required to execute the rule. The instance graph must contain a Configuration node whose attribute "name" equals to *configName.* This node must be linked to a System node (by an *instance-of* relationship), which in turn is linked to an AEType node with an attribute "name"=*typeTo Instantiate.* This checks that the type to instantiate is declared in the Configuration type (i.e. the System node *sysName* linked to the Configuration). The right-hand side of the rule describes the result of the transformation rule: a new AEInstance node, whose attribute "ID" is *newID,* has been created and linked to the selected Configuration node (i.e. *2:Configuration*). In order to reflect the fact that the new instance is in a quiescent status, it is also linked to a Quiescent node.

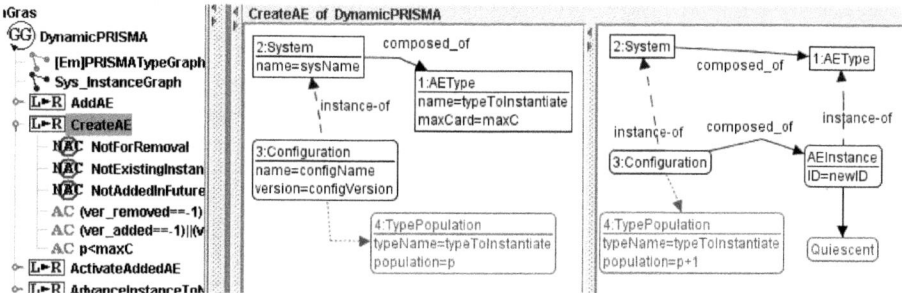

Fig. 15. CreateAE: mapping of rule R2 in the AGG [1] tool

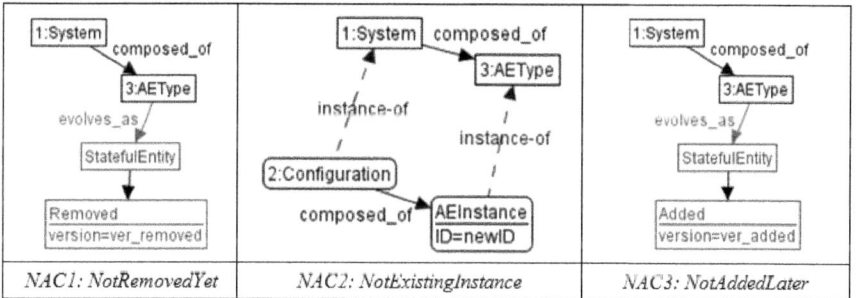

Fig. 16. NACs of the graph transformation rule *CreateAE*

The complex conditions defined by the rule R2 (see Fig. 8) have been modelled by a set of NACs (Negative Applicable Conditions) and attribute conditions (see Fig. 15, left). These conditions allow the execution of a graph transformation rule only if: (i) none of the defined NACs matches with the instance graph, and (ii) all of the attribute conditions are true. Here is described how complex R2 execution conditions have been modelled in AGG:

(i) If *typeToInstantiate* has been removed from the System type (i.e. it has been tagged for deletion), it must have been removed in a System version greater than the current Configuration version. This is checked by a NAC and an attribute condition.

NAC1 (see Fig. 16) checks if the type to instantiate (i.e. the node *3:AEType*) is tagged for deletion (i.e. it is linked to a Removed node). If true, the variable *var_removed* will contain this tag version number; otherwise (the type has not been deleted yet), it will contain the value -1. This variable is compared with the current Configuration version (see the first attribute condition in Fig. 17), checking that it is greater. If it is evaluated to false, then the rule *CreateAE* cannot be executed.

(ii) The identifier of the new instance, *newID*, must not have been used previously. This is checked by NAC2 (see Fig. 16): if the selected configuration (i.e. the node *2:Configuration*) is linked to an AEInstance node with an ID=*newID*, then the NAC condition is true and the rule *CreateAE* cannot be executed.

(iii) If *typeToInstantiate* has been added at runtime (i.e. it is tagged with an addition tag), it must have been added in a System version less than the current Configuration version. This is similarly checked as with the removal case. See NAC3 in Fig. 16 and the second attribute condition in Fig. 17.

Conditions	
Expression	OK
(ver_removed==-1)||(ver_removed>configVersion+1)	☑
(ver_added==-1)||(ver_added<=configVersion+1)	☑
p<maxC	☑
	☐

Fig. 17. Attribute conditions of the graph transformation rule *CreateAE*

(iv) The number of existing *typeToInstantiate* instances in the Configuration must be lower than the maximum cardinality defined in the System type. This is checked with the help of an auxiliary node, TypePopulation (see Fig. 15), which keeps the number of instances (attribute *population*) of a type (attribute *typeName*) that have been instantiated in the Configuration it is linked to. Then, an attribute condition (see the third attribute condition in Fig. 17) checks that the population, p, of the selected TypePopulation node (i.e. this which attribute *typeName* equals to the type to instantiate, *typeToInstantiate*) is less than the maximum cardinality, *maxC*, of the selected *AEType* node.

5 Discussion

The main goal behind the use of self-adaptive/self-organizing systems is the realization of quality attributes (e.g. performance, reliability, dependability, etc.). These quality attributes are measured at runtime to prevent their degradation, and in such a case, corrective actions are automatically performed by the system [23,32]. These corrective actions may involve runtime adaptations or re-organizations of the system, following some rules or guidelines introduced at design-time. Therefore, in such systems, runtime adaptations/reorganizations are part of the system behaviour. Dynamic evolution is beneficial in this kind of systems not only for introducing unforeseen changes to the functional concerns of the system, but also for changing the self-adaptive/self-organizing concerns of the system, like the adaptation rules or the algorithms that guide the local organization of system nodes.

However, the question here is to what extent a dynamic evolution process may impact, or be in conflict with, existing self-adaptive/self-organizing behaviour. For instance, the dynamic update of a subsystem may affect (at least temporarily) a quality attribute (e.g. performance) of the whole system. This could trigger the activation of self-adaptive behaviour, which may be in conflict with the dynamic evolution process. In order to address these possible conflicts, two possible scenarios must be considered. On the one hand, if self-adaptive behaviour is triggered by the instance (or subsystem) that is being evolved, the dynamic evolution process must prevail: external, ad-hoc changes upon a type should be prioritary instead of programmed management operations. The reason is that incorrect or conflicting self-adaptive behaviour may prevent the installation of updates, leading an instance to be indefinitely out-of-date. This does not mean that current executing processes must be aborted, but that new processes must be postponed in benefit of the execution of a dynamic evolution process. That is, when a self-adaptive action is going to be performed, the dynamic type change should be executed first, and then the need for the self-adaptive action should be reevaluated. This is reflected in our approach: there may be conflicts among the execution of a dynamic reconfiguration rule (based on an outdated type) and the integration of a new type version (e.g. see rules R2 and R3). The transformation rules take this into account: in presence of a new type version, the dynamic reconfiguration behaviour is constrained to fulfill the changes imposed by the new, evolved type, instead of the current type.

On the other hand, if self-adaptive behaviour is raised by other subsystems (i.e. not being the subject of an evolution process), the presence of dynamic evolution processes must be notified to these subsystems to avoid conflicts. For instance, if a dynamic evolution process is applied on a server component instance, then the performance of the processing rate of requests will invariably decrease. Then, if a self-adaptive subsystem detected this decrease on performance, it might trigger a dynamic reconfiguration policy for creating new instances of the "delayed" server component. In our approach, this decision would not be a problem: since a type is the first on being evolved, the new instances of the server component would be conformant to the new type version instead to the old one. And when the server instance that was evolving would finish its evolution, it could continue serving requests (probably better, since it has been updated/corrected). However, if the self-adaptive subsystem had decided to disconnect all the server component instances of the type that is being evolved and use a different server type, then the dynamic evolution of the server type would have been useless. A way to overcome these unpredicted situations is to generate an event to notify that an instance is temporarily unavailable due to an evolution process. This event should be notified to the neighbours of the evolving instance and/or to the self-adaptive subsystems. This way, a self-adaptive subsystem could take this information into account to postpone the evaluation of the quality attributes that are affected by the evolving instance. We have modelled this event in our approach by using the quiescent status: when an instance achieves this status is because it is engaged in an evolution process. Then, this information can be used by other subsystems (like a self-adaptive subsystem) to avoid reconfiguration conflicts.

In general, the execution of a dynamic evolution process will have an impact on some quality attributes of the instance to evolve and its context (mainly performance). In most cases, this impact will be temporary, since it is related to the execution of an evolution process and this evolution process has a limited duration. In fact, after the finalization of the evolution process, generally the quality attributes of the system will recover its previous values (or better values, if the installed updates benefit these attributes). However, in the worst case, it may happen that the installed updates degrade the overall quality of the system (e.g. some interrelations have not been considered by the architect). In this case, the dynamic updates can also be reverted by using the version management we have described in this paper: since a type specification keeps all the information of previous versions, in case anything fails, the changes could be reverted.

The approach presented in this paper captures the dynamic evolution process at a very high level of abstraction, as a sequence of gradual, asynchronous changes on both type and instance-level. We have chosen graph transformations to describe the evolution process because they naturally model both the system itself (as a graph of types and configurations) and the asynchronous nature of its evolution (by individual application of rules without global control). As opposed to logic-based formalisations [43] or process calculi approaches [10, 15], the use of graphs allows us to represent the system and its runtime state at a high level of abstraction. The direct mapping between graphs and visual models resembles that between UML diagrams and their metamodel-based abstract representations. We have shown that graph transformations allow to describe precisely and concisely the principles and mechanisms of dynamic evolution: (i) how, in presence of change, will the involved type and instances react, and (ii) how will their interactions with other elements be managed.

However, the details of how the quiescence status is achieved by instances have been explicitly excluded from the model. The reason is that this would require modelling also the instance execution model: for modelling the safe stopping of running transactions and the blocking of new service requests, the model should also include how instances process and execute service requests. This would add excessive complexity on the evolution model, thus eliminating the benefits of a concise description. We have decided to simply model quiescence as an attribute of an instance, which describes when this status is reached. The concrete semantics have been left to the infrastructure (i.e. the middleware). More details of the semantics can be found in [25, 47].

Finally, one consideration remains concerning the management of inactive type versions (ie. old versions without instances). Version inactivity (as described in section 2) must be considered *within the context of a System*. Every System has a local copy of a type, and each local copy can integrate the updatings (i.e. versions) at different times: a version may be inactive in one System (because it has been evolved), but still be active in another System (because due to semantic incompatibility the new version cannot be introduced). For this reason, the accessibility of inactive versions must be guaranteed, by storing them in a database or filesystem for future reference.

6 Related Work

Several works have addressed the dynamic evolution of software systems. Segal and Frieder [40] reviewed the first approaches (around the late 1970s and earlier 1980s) to support *dynamic program updating*. Such earlier works were mainly based on supporting the evolution of procedure-oriented programs, with few emphasis on the concepts of types and instances. However, Fabry's work [20] investigated the updating of (abstract) data types and the migration of their instances. His work provided deferred updates, using version attributes to distinguish the out-of-date instances. The main mechanism used was the use of indirections at code segments: all the calls to the old code were updated with the address of the new code segment. Thus, running processes could continue executing the old code, whereas new processes would start executing the new code. However, this means that not only the old code segment is modified, but also the calling programs, which results in invasive evolutions. The benefit of current architectural approaches, as our work, is that interactions among processes are made explicit and separated from the functional behaviour, thus facilitating their evolution. Another limitation of Fabry's work is that external interfaces could not be modified.

With the expansion of object-oriented languages and frameworks, the distinction among types and instances (e.g. classes and objects) become evident, and also the need for dynamically updating both. JDRUMS [2,37] provides transparent dynamic updating features to Java programs, by means of a modified Java Virtual Machine. It extends the Java class loader (i.e. the Class and Object Java meta-types) to include a link to the class (or object, respectively) that replaces it. When a class is dereferenced, JDRUMS checks if it has been updated and then returns the last version. Object updates are performed when they are dereferenced; their old, internal state is converted to the structure of the new class. Thus, different versions of the same class, as well as their objects, can coexist simultaneously: asynchronous evolution is supported. Other works also address the dynamic evolution of Java programs by extending the default class loader, such as Malabarba et al. [26] and Wang et al. [48]. However, both approaches perform a synchronised dynamic evolution, not suitable for distributed systems.

Most of the techniques for supporting dynamic evolution only differ on how the indirection is managed. An interesting review of the wide range of techniques proposed can be found at [27]. For instance, PROSE [31] supports the dynamic weaving of new concerns (which are implemented as methods) to the existing code. It uses code interception and redirection to introduce the new code. However, the use of multiple type versions is not considered, neither the migration of running instances. In general, the existing works describing how to support dynamic evolution are focused on the implementation details. The contribution of our work is that we provide a high-level model for describing the evolution.

Dynamic evolution has also raised the interest in the area of software architecture [7,10,15,32]. However, the interest has been mainly focused on the description and/or support of the dynamic (self)reconfiguration of a single system instance (i.e. a composite component instance), according to an overall architectural specification, style or pattern [12,14,24]. However, the evolution of a type specification has not explicitly taken into account. For instance, focusing on graph-based approaches, the works of Hirsh et al [22], Wermelinger et al. [49], or Bucchiarone et al. [8] use typed graph grammars to describe an architectural type (i.e. a style or a pattern) and graph instances to describe architecture instances. Then, typed graph transformation rules are used to model the dynamic reconfiguration, keeping the architectural type intact. We have also used typed graph grammars as the basis for the formalization of the dynamic evolution process. However, we have used typed graph grammars to represent a software architecture metamodel (i.e. PRISMA) and graph instances to represent both an architecture type and the set of its instances. Thus, graph transformations can change the type and its instances, while taking into account both type constraints (defined in the left-hand side of rules) and the metamodel constraints (defined in the typed graph).

Other approaches that have addressed dynamic evolution from an high-level of abstraction are those based on reflective concepts, such as [3,5,11,15]. Reflection is a powerful concept to describe the ability of a system to reason about itself and act upon itself. However, these approaches do not perform asynchronous evolution as covered in this work.

7   Conclusion and Further Work

In this paper we have introduced the semantics of the asynchronous evolution of types in the context of software architecture. This is an important feature for the design and development of large systems with a long-time usage, since it allows the introduction of concurrent changes at runtime without waiting to sequence all changes, which would delay the introduction of needed capabilities or problem fixes. When considering the use of dynamic evolution support, its benefits (i.e. supporting the correction or improvement of executing software artifacts) should be evaluated against its disadvantages (i.e. a negative impact on the system quality attributes during the execution of an evolution process).

This work has been focused on the dynamic evolution of the structural view of systems (i.e. the architecture of their subsystems). It has been addressed from a white-box perspective: the internal structure of the architectural instances are evolved gradually, by adding or removing its elements at runtime, connecting/disconnecting them, etc. From a black box perspective, this is perceived as a replacement of the entire architecture and the migration of their state. However, internally, only some parts have been changed, while keeping the state of the other elements.

The evolution semantics described here allows the concurrent, but ordered, dynamic evolution of both the type and its instances. The evolution is concurrent because both the type and its instances evolve asynchronously, independently of each other. But the evolution is also ordered because instances evolve towards the last updated version of the type, while preserving the order in which different evolution processes (i.e. those that create new type versions) were introduced. Version management is carried out by means of evolution tags, which allow keeping evolution traceability, so that each instance can follow the evolution of its type across the time. Thus, although a type would evolve several times, at the end each instance would converge to the last version of the evolved type.

The evaluation of the evolution semantics, by simulating all the rules in a tool like AGG, is an ongoing work. However, there are issues that have not been covered, such as the description of how runtime faults are managed. Since rules only describe pre-conditions and post-conditions of actions, it is not addressed what happens in the middle of such actions, for instance, if an evolution operation cannot be finished. Another issue is the correct evaluation of semantic compatibility: our model would have to be complemented with a mechanism to establish compatibility among types. In addition, further work remains, such as the definition of a fully distributed and decentralized asynchronous evolution model. The model presented is decentralized at the instance-level: each instance evolves at different times, without requiring to be located at the same node where the evolution started. However, the model relies on a centralized type which receives the evolution requests and propagates the changes to its (distributed) instances. In a fully decentralized model, a type may also be distributed among different nodes. In this case, when an evolutionary change is requested on a type, it should be propagated properly to the other nodes. Then, some issues arise like: (i) how to keep (distributed) type specifications synchronized; (ii) how to manage concurrent type evolution requests and avoid type version branching. These issues are similar to those dealt in versioning management approaches [16,45]. This is an ongoing work which has not been covered in this paper.

References

1. AGG: Attributed Graph Grammar System Tool,
    http://user.cs.tu-berlin.de/~gragra/agg/
2. Andersson, J., Comstedt, M., Ritzau, T.: Run-time support for dynamic Java architectures. In: ECOOP 1998 workshop on Object-Oriented Software Architectures (WOOSA 1998), Brussels (1998)
3. Andersson, J., De Lemos, R., Malek, S., Weyns, D.: Reflecting on Self-Adaptive Software Systems. In: ICSE workshop on Software Engineering for Adaptive and Self-Managing Systems (SEAMS 2009), Vancouver, Canada (2009)
4. Atkinson, C., Kühne, T.: Model-Driven Development: A Metamodeling Foundation. IEEE Software 20(5), 36–41 (2003)
5. Bencomo, N., Blair, G.S., Flores-Cortés, C.A., Sawyer, P.: Reflective Component-based Technologies to Support Dynamic Variability. In: 2nd Int. Workshop on Variability Modelling of Software-Intensive Systems (VaMoS'08), Universität Duisburg-Essen, Germany (2008)
6. Beydeda, S., Book, M., Gruhn, V.: Model-Driven Software Development. Springer, Heidelberg (2005)
7. Bradbury, J.S., Cordy, J.R., Dingel, J., Wermelinger, M.: A Survey of Self-Management in Dynamic Software Architecture Specifications. In: Workshop on Self-Managed Systems (WOSS 2004), Newport Beach, CA (2004)
8. Bucchiarone, A., Melgratti, H., Gnesi, S., Bruni, R.: Modelling Dynamic Software Architectures using Typed Graph Grammars. In: Graph Transformations for Verification and Concurrency. ENTCS, vol. 213(1), pp. 39–53. Elsevier, Amsterdam (2007)
9. Cámara, J., Salaün, G., Canal, C.: Composition and Run-time Adaptation of Mismatching Behavioural Interfaces. J. of Universal Computer Science 14(13), 2182–2211 (2008)
10. Canal, C., Pimentel, E., Troya, J.M.: Specification and Refinement of Dynamic Software Architectures. In: Working IFIP Conference on Software Architecture (WICSA 1999), San Antonio, Texas, USA (1999)
11. Cazzola, W., Ghoneim, A., Saake, G.: Software Evolution through Dynamic Adaptation of Its OO Design. In: Ryan, M.D., Meyer, J.-J.C., Ehrich, H.-D. (eds.) Objects, Agents, and Features. LNCS, vol. 2975, pp. 67–80. Springer, Heidelberg (2004)
12. Cheng, S., Garlan, D., Schmerl, B.R.: Making Self-Adaptation an Engineering Reality. In: Babaoğlu, Ö., Jelasity, M., Montresor, A., Fetzer, C., Leonardi, S., van Moorsel, A., van Steen, M. (eds.) SELF-STAR 2004. LNCS, vol. 3460, pp. 158–173. Springer, Heidelberg (2005)
13. Costa-Soria, C., Hervás-Muñoz, D., Pérez, J., Carsí, J.A.: A Reflective Approach for Supporting the Dynamic Evolution of Component Types. In: 14th IEEE International

Conference on Engineering of Complex Computer Systems (ICECCS 2009). Potsdam, Germany (2009)

14. Costa-Soria, C., Pérez, J., Carsí, J.A.: An Aspect-Oriented Approach for Supporting Autonomic Reconfiguration of Software Architectures. In: 2nd Workshop on Autonomic and SELF-adaptive Systems (WASELF 2009), San Sebastián, Spain (2009)

15. Cuesta, C.E., Romay, P., de la Fuente, P., Barrio-Solórzano, M.: Reflection-Based, Aspect-Oriented Software Architecture. In: Oquendo, F., Warboys, B.C., Morrison, R. (eds.) EWSA 2004. LNCS, vol. 3047, pp. 43–56. Springer, Heidelberg (2004)

16. De Lucia, A., Deufemia, V., Gravino, C., Risi, M.: Behavioral Pattern Identification through Visual Language Parsing and Code Instrumentation. In: 13th European Conference on Software Maintenance and Reengineering (CSMR 2009), Kaiserslautern, Germany (2009)

17. Ehrig, H., Ehrig, K., Prange, U., Taentzer, G.: Fundamentals of Algebraic Graph Transformation (Monographs in Theoretical Computer Science. An EATCS Series). Springer, Heidelberg (2006)

18. Engels, G., Heckel, R., Küster, J.M., Groenewegen, L.: Consistency-Preserving Model Evolution through Transformations. In: Jézéquel, J.-M., Hußmann, H., Cook, S. (eds.) UML 2002. LNCS, vol. 2460, pp. 212–227. Springer, Heidelberg (2002)

19. Engels, G., Heckel, R., Küster, J.M.: Rule-Based Specification of Behavioral Consistency Based on the UML Meta-model. In: Gogolla, M., Kobryn, C. (eds.) UML 2001. LNCS, vol. 2185, pp. 272–286. Springer, Heidelberg (2001)

20. Fabry, R.S.: How to design a system in which modules can be changed on the fly. In: 2nd International Conference on Software Engineering (ICSE 1976), San Francisco, California, USA (1976)

21. Heckel, R.: Graph Transformation in a Nutshell. In: School on Foundations of Visual Modelling Techniques (FoVMT 2004). ENTCS, vol. 148(1), pp. 187–198. Elsevier, Amsterdam (2006)

22. Hirsch, D., Inverardi, P., Montanari, U.: Graph grammars and constraint solving for software architecture styles. In: 3rd Int. Software Architecture Workshop (ISAW-3). ACM Press, New York (1998)

23. Kephart, J.O., Chess, D.M.: The Vision of Autonomic Computing. IEEE Computer 36(1), 41–51 (2003)

24. Kramer, J., Magee, J.: Self-managed systems: an architectural challenge. In: ICSE - Future of Software Engineering (FOSE 2007). IEEE, Los Alamitos (2007)

25. Kramer, J., Magee, J.: The Evolving Philosophers Problem: Dynamic Change Management. IEEE Transactions on Software Engineering 16(11), 1293–1306 (1990)

26. Malabarba, S., Pandey, R., Gragg, J., Barr, E., Barnes, J.F.: Runtime support for typesafe dynamic Java classes. In: Bertino, E. (ed.) ECOOP 2000. LNCS, vol. 1850, pp. 337–361. Springer, Heidelberg (2000)

27. McKinley, P.K., Sadjadi, S.M., Kasten, E.P., Cheng, B.H.C.: Composing Adaptive Software. IEEE Computer 37(7), 56–64 (2004)

28. Medvidovic, N., Taylor, R.N.: A Classification and Comparison Framework for Software Architecture Description Languages. IEEE Transactions on Software Engineering 26(1), 70–93 (2000)

29. Mens, T., Wermelinger, M.: Separation of concerns for software evolution. J. of Software Maintenance and Evolution 14(5), 311–315 (2002)

30. Milner, R.: The Polyadic π-Calculus: A Tutorial. Laboratory for Foundations of Computer Science Department, University of Edinburgh (1993)

31. Nicoara, A., Alonso, G., Roscoe, T.: Controlled, Systematic, and Efficient Code Replacement for Running Java Programs. SIGOPS Operating Systems Review 42(4), 233–246 (2008)

32. Oreizy, P., Gorlick, M., Taylor, R.N., et al.: An Architecture-Based Approach to Self-Adaptive Software. IEEE Intelligent Systems 14(3), 54–62 (1999)
33. Pérez, J., Ali, N., Carsí, J.A., Ramos, I., et al.: Integrating aspects in software architectures: PRISMA applied to robotic tele-operated systems. Information & Software Technology 50(9-10), 969–990 (2008)
34. Pérez, J., Ali, N., Carsí, J.A., Ramos, I.: Designing Software Architectures with an Aspect-Oriented Architecture Description Language. In: Gorton, I., Heineman, G.T., Crnković, I., Schmidt, H.W., Stafford, J.A., Szyperski, C., Wallnau, K. (eds.) CBSE 2006. LNCS, vol. 4063, pp. 123–138. Springer, Heidelberg (2006)
35. Pérez, J.: PRISMA: Aspect-Oriented Software Architectures. PhD Thesis, Universidad Politécnica de Valencia (2006)
36. Perry, D.E., Wolf, A.L.: Foundations for the Study of Software Architecture. SIGSOFT Software Engineering Notes 17(4), 40–52 (1992)
37. Ritzau, T., Andersson, J.: Dynamic Deployment of Java Applications. In: Java for Embedded Systems, London (2000)
38. Rogers, A., Jennings, N.R., Farinelli, A.: Self-Organising Sensors for Wide Area Surveillance using the Max-Sum Algorithm. In: WICSA/ECSA Workshop on Self-Organizing Architectures (SOAR 2009), Cambridge, UK (2009)
39. Rombach, H.D.: Design for Maintenance - Use of Engineering Principles and Product Line Technology. In: 13th European Conf. on Software Maintenance and Reengineering (CSMR 2009), Kaiserslautern, Germany (2009)
40. Segal, M.E., Frieder, O.: On-the-Fly Program Modification: Systems for Dynamic Updating. IEEE Software 10(2), 53–65 (1993)
41. Serugendo, G.D.M., Gleizes, M.P., Karageorgos, A.: Self-organisation and emergence in MAS: An Overview. Informatica (Slovenia) 30, 45–54 (2006)
42. Software Engineering Institute: Ultra-Large-Scale Systems: Software Challenge of the Future. Technical Report, Carnegie Mellon University, Pittsburgh, USA (2006)
43. Stirling, C.: Modal and Temporal Logics. In: Handbook of Logic in Computer Science, vol. II. Clarendon Press, Oxford (1992)
44. Taylor, R.N., Medvidovic, N., Dashofy, E.M.: Software Architecture: Foundations, Theory and Practice. Wiley, Chichester (2009)
45. Thao, C., Munson, E.V., Nguyen, T.N.: Software Configuration Management for Product Derivation in Software Product Families. In: 15th Int. Conf. on Engineering of Computer Based Systems (ECBS 2008), Belfast, Northern Ireland (2008)
46. Vandewoude, Y., Berbers, Y.: Component state mapping for runtime evolution. In: Int. Conf. on Programming Languages and Compilers, Las Vegas, Nevada, USA (2005)
47. Vandewoude, Y., Ebraert, P., et al.: Tranquillity: A low Disruptive Alternative to Quiescence for Ensuring Safe Dynamic Updates. IEEE Transactions on Software Engineering 33(12), 856–868 (2007)
48. Wang, Q., Shen, J., Wang, X., Mei, H.: A Component-Based Approach to Online Software Evolution. Journal of Software Maintenance and Evolution 18(3), 181–205 (2006)
49. Wermelinger, M., Lopes, A., Fiadeiro, J.L.: A graph based architectural (re)configuration language. SIGSOFT Software Engineering Notes 26(5), 21–32 (2001)

A Self-organizing Architecture for Traffic Management

Rym Zalila-Wenkstern, Travis Steel, and Gary Leask

*MAVs Lab, Department of Computer Science,
University of Texas at Dallas
Richardson, TX, USA
{rmili,steel,gml013000}@utdallas.edu
http://mavs.utdallas.edu

Abstract. In this paper we discuss the use of self-organizing architectures for traffic management systems. We briefly introduce Soteria, a multi-layered, integrated, infrastructure for traffic safety enhancement and congestion reduction. We highlight Soteria's use of micro- and macro-level models and its hybrid top-down/bottom-up strategy for traffic management. We then present a generic architecture that can be used to develop simulation systems for real world self-organizing systems. Lastly, we describe how this generic architecture can be instantiated to create the architecture of Matisse, a tailor made distributed simulation system for Soteria.

Keywords: multi-agent systems; traffic management; simulation.

1 Introduction

Traffic congestion is a major problem in most large cities throughout the world and several strategies have been defined to address this problem [1]. Transportation technologies known as Intelligent Transportation Systems (ITS) have been considered as possible solutions. ITS are defined as "the application of advanced sensor, computer, electronics, and communication technologies and management strategies in an integrated manner to increase the safety and efficiency of the surface transportation system" [19].

In this paper we briefly discuss Soteria, an ITS based multi-layered, integrated, infrastructure for safety enhancement and congestion reduction. Soteria's architecture was recently developed by a team of researchers at the University of Texas at Dallas [8]. At this stage the architecture has been baselined, and several components are in the process of being implemented. Soteria is based on the following premises a) the traffic model consists of micro- and macro-level components; b) traffic is a bottom-up phenomenon that is the result of individual decisions at the micro-level (vehicles, traffic control devices, etc); c) traffic management is a top-down activity that is the result of decisions taken at the

---

* The MAVs Lab is supported by Rockwell Collins under grant number 5-25143.

D. Weyns et al. (Eds.): SOAR 2009, LNCS 6090, pp. 230–250, 2010.

macro-level. At the micro-level, vehicles adapt to evolving traffic conditions, affecting the state of the system. At the macro-level, traffic control devices monitor the state of the system and adapt the system conditions to maintain the overall system stability.

The purpose of this paper is to discuss the architecture of Matisse (Multi-Agent based TraffIc Safety Simulation systEm), a tailor-made large scale distributed multi-agent based simulation system designed to support Soteria by simulating traffic phenomena. In Matisse, micro and macro-level entities are modeled as software *agents*. The macro-level components called *controllers* inform, monitor, and guide micro-level agents to ensure that, at some level, micro-level behaviors and interactions are consistent with the global system behavior.

Existing traffic simulation systems model scenarios using either a microscopic or a macroscopic approach. Matisse is designed to model scenarios using a hybrid approach, allowing simulation users to observe the effects of micro-level behaviors on a global scale and macro-level behavior on individual decisions.

The rest of this paper is organized as follows: in Section 2, we give an overview of Soteria. In Section 3, we describe a generic architecture for simulation systems designed to model self-organizing systems. In Section 4, we discuss Matisse's high level architecture. In Section 5, we discuss a model execution using a case study, and in Section 6 we describe some related works.

2   Components of a Self-organizing Cyber-Physical System

Soteria is a novel, large-scale, decentralized cyber-physical infrastructure [29,8] for improving safety and reducing congestion on roads and highways. The term cyber-physical systems refers to the tight conjoining of and coordination between computational and physical resources [5]. Soteria aims to enforce communication, interaction, and collaboration between computational and physical resources at the micro- and macro-levels.

As shown in Fig. 1, our proposed infrastructure is a conglomerate of infrastructures each having individual concerns and objectives. The purpose of Soteria is to combine and coordinate the efforts of these sub-infrastructures while maintaining the individual integrity of the sub-infrastructures. In this sense, Soteria is a system of systems that addresses larger traffic management objectives in a top-down manner while allowing sub-systems to handle more detailed aspects of traffic management in a bottom-up fashion.

The proposed infrastructure is based upon two underlying concepts.

– In order to manage the traffic environment efficiently, it is necessary to partition the physical space into smaller defined areas called *cells*. The cell partitioning is specified for a particular environment and provided to the system.
– Each cell contains a device, called a *cell controller*, that is physically deployed within that cell. Each cell controller is networked with neighboring

cell controllers to form the cell controller infrastructure. A cell controller is
responsible for 1) autonomously managing and controlling a portion of the
physical environment (i.e., cell) including vehicles and traffic lights, and 2)
notifying other controllers of changes that may affect their cells.

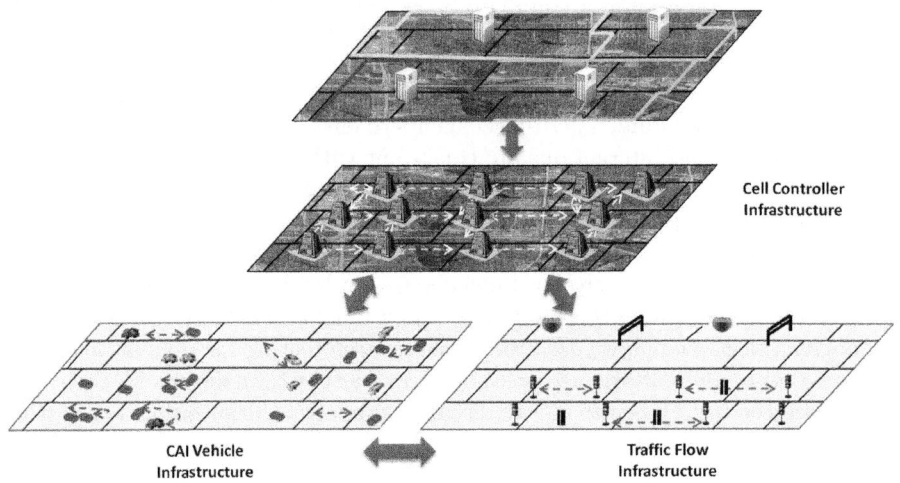

Fig. 1. The Soteria Infrastructure

Cell Controller Infrastructure. The purpose of this infrastructure is to keep
the Vehicle and Traffic Flow Infrastructures up to date with respect to traffic
and safety information. It consists of cell controllers, i.e., interactive devices
responsible for keeping the vehicles and selected traffic devices aware of local
traffic information. As such, a cell controller is required to be: 1) autonomous;
2) aware of the vehicles and the traffic devices located in its defined physical
area; 3) able to communicate with the vehicles and the traffic devices; 4) able to
communicate with other cell controllers to inform them of external events and
their effect.

In order to manage traffic information efficiently, it is necessary to organize
the cell controller infrastructure as a hierarchy (see Fig. 2). This hierarchical
organization allows cell controllers to communicate quickly with distant cell controllers. The depth of the hierarchy depends on the size and traffic density of
the area being managed. This approach is particularly important for congestion
caused by an accident, when micro-level information is insufficient for vehicles
to determine the best alternative route. In this case, a cell controller (e.g., C11)
will communicate with a higher level controller (e.g., C1) to obtain a broader
image of the traffic situation, determine the best vehicle routing scheme, and
pass the information on to its local vehicles.

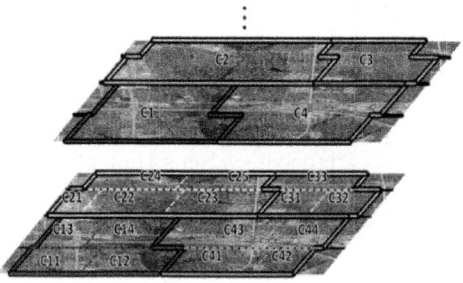

Fig. 2. Cell Controller Hierarchy

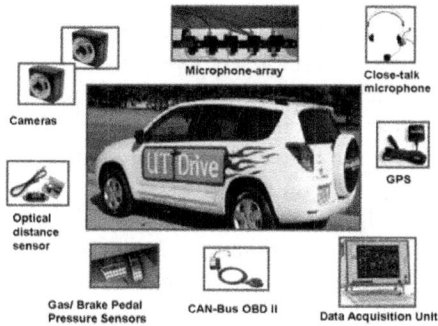

Fig. 3. UTD's Context Aware Intelligent Vehicle

Context-Aware Intelligent Vehicle Infrastructure. This component consists of vehicles equipped with devices that allow them to 1) monitor the driver's behavior in order to prevent possible accidents, 2) communicate with other vehicles, and 3) interact with cell controllers to obtain traffic information and guidance in real time. Figure 3 shows the UTD Context-Aware Intelligent Vehicle (CA-IV). Currently, the CA-IV is equipped with two main systems. The *Driver Status Monitoring system* determines the driver's maneuvers and distraction levels through state of the art computer vision systems (for head and eye tracking) and powerful processing modules that record and analyze CAN-Bus signals, gas/brake pressure, and audio [9,26]. The *Environment Monitoring system* uses a forward-looking monocular camera as well as computer vision systems to track vehicle and lane positions and identify road signs [10]. Additional radar and LIDAR technologies will soon be introduced to enhance vehicle tracking in the vicinity.

Traffic Flow Infrastructure. The purpose of this infrastructure is to improve safety and traffic flow on roads and highways by providing information about the physical traffic infrastructure and congestion condition. This infrastructure consists of three types of stationary traffic devices: traffic lights, traffic collection devices, and relay units.

Traffic lights are equipped with systems that allow them to 1) interact with cell controllers to obtain traffic information in real time, and 2) communicate with other traffic light controllers to improve traffic flow when necessary. They are also equipped with autonomous adaptive systems that allow them to independently determine the best course of action when unexpected events occur.

Traffic collection devices are used on highways (e.g., toll units) to collect information about traffic, and communicate the information to cell controllers for further analysis (e.g., identification of a drunk driver on the highway).

Relay units are used to pass on information between the various communicating entities when the physical distance is too great.

At the micro-level, each component is governed by a set of goals which it tries to achieve through communication and collaboration with other micro-level entities. For example, a vehicle's goal may be to travel from point A to point B in a reasonable amount of time. Vehicles will collaborate with other vehicles and traffic devices to achieve their goals. If for any reason this goal is not achievable (e.g., driver falls asleep), the vehicle will inform the macro-level component (i.e., the cell controller) in a bottom-up fashion that there is an emergency condition that may have a widespread effect on traffic. At the macro-level, cell controllers will communicate and collaborate to determine the best course of actions required to maintain the system's stability as required by their goal. This macro-level decision is then decomposed and delegated to the micro-level components in a top-down fashion.

As mentioned in Section 1, Soteria's architecture has been baselined, and the components are at various stages of development. The systems onboard the CA-IV are being enhanced; various communications and networking options (wireless cellular and 802.11p for wireless link and access, Internet Protocol (IP) for end-to-end networking, fast addressing scheme for mobile nodes (i.e., vehicles)) are being investigated, and prototypes developed; and a decentralized distributed agent-based logic programming approach for traffic management is being developed.

In order to validate the concepts underlying the large scale cyber-physical system, it was necessary to develop a simulation system. A thorough analysis of the problem helped us identify abstract patterns that can be used to design simulation systems for other self-organizing systems. These patterns are discussed in the next section.

3 Components of a Generic Architecture for Simulating Self-organizing Systems

Software architectures for simulating self-organizing systems, such as those used for traffic management, must accommodate bottom-up and top-down objectives, and should therefore incorporate the concepts of micro- and macro-level components. At first it seems natural to model these components as *agents*. However, a thorough analysis reveals a clear distinction between the roles and responsibilities of micro and macro-level agents, hence the definition of two design patterns[1]. In the remainder of this paper and for the sake of simplicity, we will refer to micro-level agents as "agents", and macro-level agents as "cell controllers".

3.1   The Micro-level Agent Pattern

The micro-level agent pattern [23,22] is used to design and implement autonomous entities that have individual goals, perceive and have knowledge of their environment, exhibit organizational behavior through collaboration with other agents, and influence their environment as a result of their decisions. The micro-level agent pattern establishes the general software architecture of an agent and consists of *interaction modules, information module*, a *task module*, and a *planning and control module* (see Fig. 4).

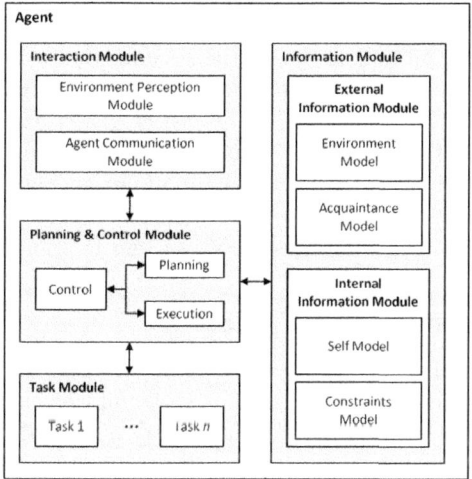

Fig. 4. The agent architecture consists of an Interaction Module, an Information Module, a Task Module, and a Planning and Control Module. Together, these components satisfy the needs of autonomous agents.

Interaction Module. The interaction module handles an agent's interaction with external entities and separates environment interaction from agent interaction. Environment interaction is handled by the *environment perception module* which allows an agent to perceive its environment. Agent interaction is handled by the *agent communication module* which allows agents to communicate with each other.

---

[1] In this section the word pattern refers to a "description of communicating objects and classes that are customized to solve a general design problem in a particular context" [15].

Information Module. This is partitioned into the External Information Module (EIM) and Internal Information Module (IIM).

- The EIM serves as the portion of the agent's memory that is dedicated to maintaining knowledge about entities external to the agent. It consists of the *Environment Model* and the *Acquaintance Model*. The Environment Model is maintained according to the agent's perception of its surroundings while the Acquaintance Model is maintained according to the agent's collaboration with other agents.
- The IIM serves as the portion of the agent's memory that is dedicated for keeping information that the agent knows about itself. This module consists of the agent's *Self Model* and the *Constraint Model*. The Self Model maintains the essential properties of the agent (e.g., position, heading, velocity) while the Constraint Model maintains the agent's physical limitations (e.g., maximum velocity, maximum deceleration) and collaborative limitations (e.g., priority of requests, ability to broadcast messages).

Task Module. This module manages the specification of atomic tasks that the agent can perform in the domain in which it is being deployed.

Planning and Control Module. This serves as the brain of the agent. It uses information provided by the other agent modules to plan, execute tasks, and make decisions.
Detailed information regarding these modules can be found in [23].

3.2   The Macro-level Agent Pattern

The macro-level agent pattern also called cell controller pattern [23,22] applies to autonomous entities that manage and store information about a dedicated region of the *environment*, participate in satisfying global system goals, inform local agents about changes in their surroundings, and exhibit organizational behavior through interaction with neighboring cell controllers.

As depicted in Fig. 5, the cell controller pattern consists of *interaction modules*, *information module*, a *task module*, and the *planning and control module*. The description of the task module and planning and control modules are identical to those in the Micro-Level Pattern discussed in Section 3.1.

Interaction Modules. These modules handle asynchronous communication among cell controllers as well as synchronous communication between cell controllers and agents. Cell controller-to-agent interaction must be synchronous in order to ensure that agents receive a consistent state of the environment. Since cell controller-to-cell controller interaction is involves high-level self-adaptation and has no bearing on the consistency of the state of the simulated environment, these interactions occur asynchronously.

Information Module. This module contains the data a controller needs to function. It is composed of the

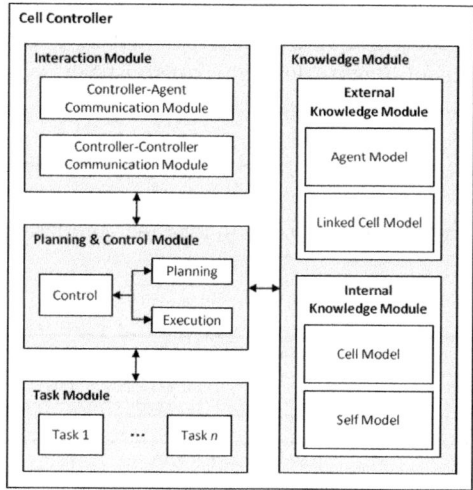

Fig. 5. The cell controller architecture closely resembles the agent architecture

– Agent Model. This model contains minimal information, such as identification and location, about the agents within the cell's environment region.
– Adjacent Cell Model. This model maintains a list of neighboring cells, that is, those which share a border with the cell. The cell controller uses this information to handle events that occur near boundaries and potentially affect adjacent cells.
– Self Model. This model contains information regarding the essential characteristics of the cell assigned to this cell controller such as its identifier and region boundaries.
– Object Model. This model includes information detailing physical entities that are situated within the cell region but are not actual agents (e.g., barricades, buildings, road signs).

3.3   Reference Architecture for Homogeneous Micro-level Agent Platform

Agents of a self-organizing software system can typically be classified into types or species based on similarities in perception, behavior, knowledge, abilities, and constraints. Since agents of the same type often interact differently with other agents of their own type than with agents of another type, separating homogeneous and heterogeneous interaction at an architectural level allows greater control over intra-agent collaborative behavior. Hence, software architectures for self-organizing systems should exploit this characteristic by managing agents of the same type with a single homogeneous agent platform.

The homogeneous agent platform reference architecture is applied for each type of agent in the system and manages only that type of agent. It consists of an

Agent Management System and an *Agent-Agent Message Transport Service.* The Agent Management System is responsible for creating and managing agents of a specific type. The Agent-Agent Message Transport Service provides a dedicated communication medium between agents of the same type, allowing intra-species collaboration. The Agent-Agent Message Transport Service is exclusively limited to agents of the same type, therefore no outside communications are sent via this service.

Instances of this reference architecture for CAI vehicles and traffic devices are shown in Fig. 6.

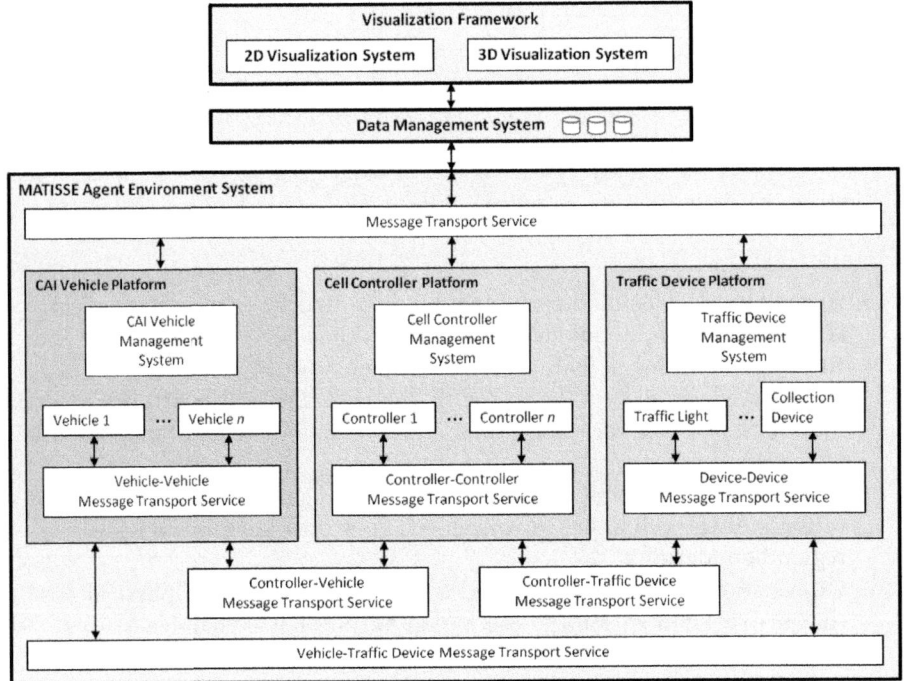

Fig. 6. Matisse's High Level Architecture

3.4   Reference Architecture for Cell Controller Platform

Cell controllers are managed through their own platform. This platform consists of a *Cell Controller Management System* and a *Controller-Controller Message Transport Service.* The Cell Controller Management System (CCMS) not only creates and manages cell controllers, but organizes them into a hierarchy. The CCMS dynamically assigns cell controllers to manage corresponding regions of the environment. In large environments it becomes difficult to maintain the level of integrity necessary to achieve macro-level objectives due to the increased amount of required collaboration between cell controllers. In this case, the CCMS creates several parent cell controllers to coordinate the efforts of the lower-level controllers. As the environment scales, the CCMS maintains a hierarchy suitable for achieving global system goals. The hierarchy, which is conceptually a tree, allows decomposition of global objectives, distributing the responsibility of achieving those objectives to cell controllers in a divide-and-conquer fashion.

An instance of this reference architecture for cell controllers is shown in Fig. 6.

3.5   Communication between Heterogeneous Platforms

A software architecture for self-organizing systems should allow communication between the various platforms to accommodate interaction and collaboration among different types of agents and cell controllers. A dedicated *Message Transport Service* (MTS) exists for each pair of platforms that have the need to communicate directly. Since the means of communication between platforms may be specialized, they are separated according to the platforms they service.

4   Matisse

As mentioned in Section 1, Matisse (Multi-Agent based Traffic Safety Simulation systEm) is a "tailor made" simulation framework designed to specify and execute simulation models for Soteria. More precisely, it allows the simulation of various traffic safety improvement and congestion reduction scenarios under nominal and hypothetical conditions.

The models developed for Matisse are used to validate the concepts introduced by Soteria. The simulation framework contains a software model of each of Soteria's components. Each model simulates the behavior of the actual Soteria component it represents, providing a means to evaluate the models in the context of the larger system. Additionally, since each of Soteria's sub-systems can exist as a stand-alone system, their corresponding models can also be evaluated in an isolated manner.

The rest of the section describes the details of the Matisse architecture starting with a high level view of the system, followed by more specific details regarding the implementation. Finally, we explain how the components interact in a self-organizing manner.

4.1   Matisse's High Level Architecture

At a high level, Matisse contains several subsystems (see Fig. 6). The main constituent is the *Agent-Environment System* (AES) which creates and executes simulation instances. The *Data Management System* (DMS) stores and processes simulation data collected from the AES for use during performance and behavioral analysis. The *Visualization Framework* receives information from the DMS to create 2D and 3D images of the simultion and also allows users to interact with the simulation at run-time. The data analysis and visualization mechanisms increase the effectiveness of Matisse as a validation tool. For the sake of conciseness, we have depicted instances of the various references architectures described in Section 3 in Fig. 6.

Message Transport Services. Soteria's cyper-physical systems utilize physical mediums (wired and wireless) to support communication among system entities (such as agents, cell controllers, and other devices). Since Matisse is a pure software simulation system, it must also simulate Soteria's communication infrastructure for the simulated software models. Furthermore, Matisse must be deployed in a distributed manner to satisfy scalability requirements of large, complex traffic simulations. Thus, Matisse uses several Message Transport Services (MTS's) to allow for distributed communication among simulated components. The MTS's are implemented using the Java Message Service (JMS) [21] and provide distributed, encrypted, reliable, and asynchronous messaging services for Matisse simulations.

Agent-Environment System. Matisse's Agent Environment System (AES) creates and executes simulation instances by instantiating the patterns and reference architectures described in Section 3. It creates a platform for each type of agent as well as the MTS's needed for these platforms to communicate. Each of the agent platforms represents an individual sub-infrastructure.

4.2   Matisse's Micro-level Agent Platforms

Context-Aware Intelligent Vehicle (CAI) Platform. This is an instance of the reference architecture discussed in Section 3.3. This platform manages mobile agents that represent vehicles. It is important to note that we are not modeling human drivers, but rather the effect of the driver's actions on the vehicle. These agents, have individual goals (e.g., safely arriving at some destination in a reasonably short time), influence other agents (e.g., braking and changing lanes), are governed by environmental norms and constraints (e.g., speed limits and traffic signals), and may not be directly aware of global objectives. Within this platform, vehicle agents are created by the *Vehicle Agent Management System* and communication is handled through the *Vehicle-Vehicle Message Transport Service*.

Traffic Device Platform. This is an instance of the reference architecture discussed in Section 3.3. This homogeneous agent platform manages stationary agents that represent traffic lights and information collection devices. The *Traffic Device Agent Management System* creates traffic device agents within the simulation while the *Device-Device Message Transport Service* handles communication between traffic lights, collection devices, and other stationary traffic agents.

4.3   Matisse's Cell Controller Platform

This is an instance of the reference architecture discussed in Section 3.4 In addition to the agent platforms, the AES also contains the *Cell Controller Platform*. The *Cell Controller Platform* uses the reference architecture described in Section 3.4 to create and manage cell controllers. The *Cell Controller Management*

System creates cell controllers, assigns them to a dedicated region of the environment, and additionally maintains the cell controller hierarchy for the simulation. The *Controller-Controller Message Transport Service* handles communication between cell controllers.

4.4   Interactions in Matisse

The interaction between platforms contained by the AES must be coordinated to enable a self-organizing architecture. This coordination is handled by a discrete time influence-reaction model. This model maintains synchrony among the distributed platforms using a discrete *tick* that governs when certain interactions may take place. These interactions are classified based on the types of entities at the sending and receiving end of the interaction.

Agent-Agent Interaction. Vehicle and Traffic Device agents often have need to communicate with each other. For instance, a vehicle may inform a traffic device that the vehicle is in a state of emergency, or a traffic device may need to communicate with other traffic devices to ensure traffic flows steadily through a busy part of town. In each of these instances, the agents utilize the *Agent-Agent MTS* created for such interaction. Agent-Agent messages are received by the agent's *Interaction Module* and sent to the *Planning and Control Module* for processing.

Environment-Environment Interaction
Environment-Environment interaction occurs in two forms: vertical and horizontal. The form used to send a message is determined by the urgency, proximity, and level of detail of the communication. Horizontal interaction is sufficient for frequent, low-level communication between cell controllers of adjacent cells. Vertical interaction is reserved for less frequent, high priority communication that may need to reach distance cell controllers.

Horizontal Interactions. The horizontal form of environment-environment interaction deals with cell controllers and their immediate neighbors. This type of interaction is frequently used by cell controllers when they must inform adjacent cell controllers of events that will propagate the cell's boundary [28]. Since each cell is directly aware of its neighbors via the *Adjacent cell model* (see Fig. 5) located in each cell controller's *Information Module*, there is no need to invoke the cell controller's parent. Cell controllers handle interaction with adjacent cell controllers by sending messages directly to them via the *Controller-Controller MTS*. When the message arrives the adjacent cell controller it is sent to the *Planning and Control Module* where the message is acted upon.

Figure 7 shows a scenario that would require this horizontal interaction. An event originating in cell 1 (controlled by cell controller C1) at time t_0 eventually propagates into cells 2, 3, and 4. At time t_3, C1 realizes that the event will propagate into C2's cell by t_4. C1 directly notifies C2 about the event and the notification is received by t_4. Once C2 is notified of the event, it is able to

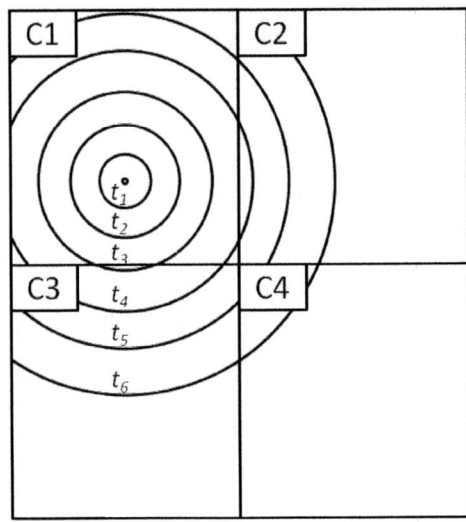

Fig. 7. Event Propagation: Horizontal Environment-Environment Interaction

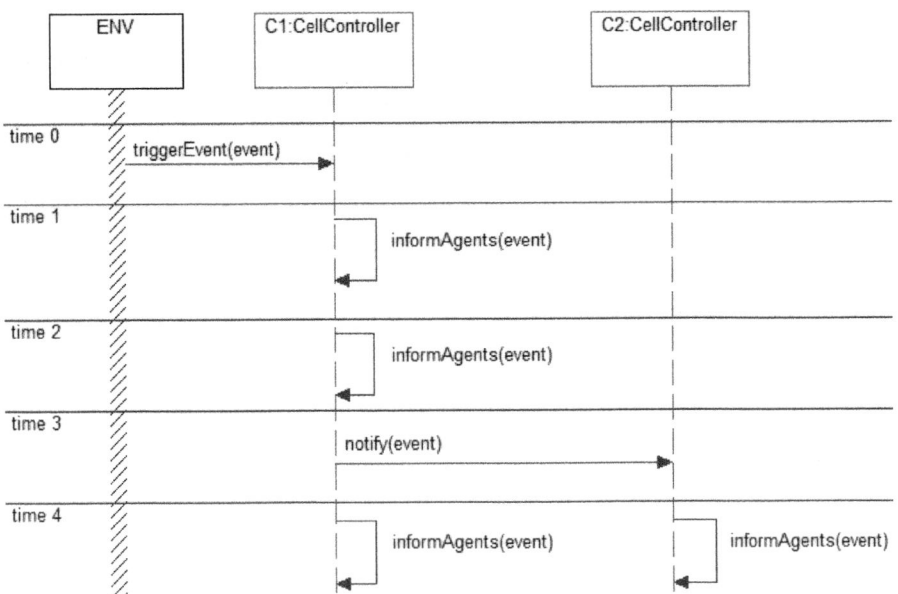

Fig. 8. UML Sequence Diagram: Event Propagation

inform the agents situated in cell 2 about the event. In such situations, horizontal environment-environment interaction is sufficient for delivering messages among adjacent cell controllers.

This scenario is represented using a UML sequence diagram in Fig. 8. Additional scenarios involving environment-environment interaction can be found in [28].

Vertical Interactions. Cell controllers will often interact to send and gather information to and from remote cell controllers. This is the vertical form of environment-environment interaction. Depending on the proximity of the cells, cell controllers may need to employ the use of the cell controller hierarchy in order to make long distance communication more efficient. Cell controllers do so by using the *Controller-Controller MTS* contained in the *Cell Controller Platform*. If a request is made by one cell controllers that must be addressed by a cell controller situated many cells away, the requesting cell controller will immediately direct the request to its parent cell controller via the *Controller-Controller MTS*. The parent cell controller will send the request to its parent and so on, until the request makes it to a cell controller that contains the addressee. The parent directs the request to the recipient who responds and sends the replay in a similar manner. Lesser levels in the hierarchy contain cell controllers with limited but detailed awareness of the area they manage. In contrast, higher levels contain cell controllers with greater but less detailed awareness of the areas they manage. This organization is key to achieving both top-down and bottom-up objectives.

Agent-Environment Interaction. The final type of interaction involves agents and their environment. Unlike the other types of interaction, the agent-environment interaction model must ensure that the agents act realistically with respect to their target environment. For example, a model allowing agents to move freely throughout the environment without respecting the physical laws of the environment would produce invalid results. Vehicle agents in the physical world are subject to collision if their paths intersect at the same time. This and similar considerations must be handled by the simulation system. Matisse uses the influence-reaction model [14,20] to handle such temporal and spatial aspects of the simulation. Details of the influence-reaction used in this paper can be found in [27].

Agents must be constantly informed about changes in their own state and the state of their surroundings. Similarly, the environment must react to the influences imposed upon it by agents.

When agents execute actions, they produce influences that are synchronously communicated to the environment (i.e., the controller agent managing the cell in which the agent is situated). The environment interprets and combine these agent influences, reacts to them, and updates the state of the environment. The new state is then synchronously communicated back to the agents (see Fig. 9).

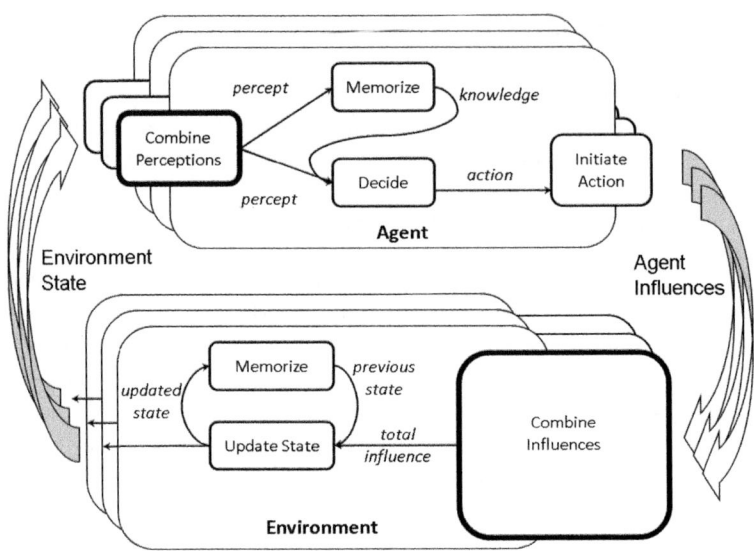

Fig. 9. Matisse's Influence-Reaction Model

5   Case Study - Model Execution

In this section, we consider two scenarios that illustrate self-organizing characteristics of Matisse.

5.1   Safety Enhancement

The first scenario highlights self-organization in response to a potentially unsafe real-world situation. A drunk driver begins to lose control of the vehicle. The CAI Vehicle System takes over control of the vehicle, contacts a law enforcement dispatcher, and attempts to safely maneuver the vehicle out of harm's way. The following description provides details for a simulation of this scenario is realized using Matisse. This scenario begins when the CAI Vehicle agent has taken over control of its simulated vehicle. Figure 10 is a UML sequence diagram that illustrates the scenario.

1. The CAI Vehicle system sends a law enforcement dispatch request to the cell controller of the cell in which the vehicle is currently situated.
2. The cell controller determines the location of the nearest available law enforcement officer and forwards the dispatch request to the law enforcement vehicle. For this example, we assume the nearest officer is several cells away.
3. When the cell controller is unable to locate an available law enforcement officer within its cell, it sends a request to neighboring cells via the Controller-Controller MTS.

4. After the dispatch request is sent, the CAI Vehicle agent contacts nearby vehicles via the Vehicle-Vehicle Message Transport Service, informing them of the vehicle agent's desire to safely pull into the shoulder of the roadway.
5. Upon receipt of this desire, nearby vehicle agents may or may not choose to honor the request based on individual objectives or other requests.
6. Assuming the drivers of nearby vehicles respect the request, the CAI Vehicle system brings the vehicle to a stop in the shoulder of the road (or other safe location).
7. Until the law enforcement officer arrives on the scene, the CAI Vehicle agent continues to send updates to the law enforcement vehicle.

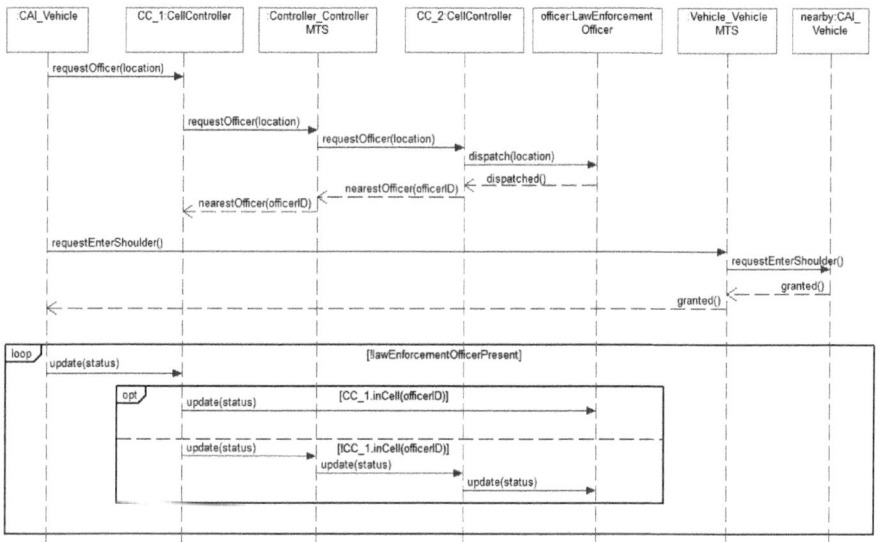

Fig. 10. UML Sequence Diagram: Safety Enhancement

5.2   Congestion Reduction

Considers the accident scenario depicted in Fig. 11. An accident (shown as a star) has occurred in cell 12, and the vehicle onboard Collision Avoidance System informs cell controller CC_12 of the accident. Figure 12 is a UML communication diagram showing the interactions involving cell controller CC_12.

Under this scenario, CC_12 immediately performs the following steps:

1. Informs all the vehicles in the cell of the accident. CC_12 achieves this by broadcasting an accident notification over the cell controller-to-vehicle MTS to all vehicle agents currently stored in the cell's vehicle agent model.
2. Informs the traffic light controllers of the accident. CC_12 broadcasts an accident notification over the cell controller-to-traffic device MTS to all traffic

light controller agents currently stored in the cell's traffic light controller agent model.

3. Informs adjacent cell controllers of the accident. CC_12 sends accident notifications over the cell controller-to-cell controller MTS to neighboring cell stored in the linked cell model.

4. Communicates with higher level controllers to obtain broader traffic information. This information is used to determine general directions as to the best exit routes for vehicles (to avoid creating congestion on secondary streets, and maintain the global system stability).

All vehicles in the cell make use of the broader traffic information and directions to avoid the accident. Traffic Lights communicate with other traffic lights to optimize traffic flow (e.g., approaching cars are not allowed to enter the cell) and may decide to turn green to allow traffic to flow. The adjacent cell controllers, upon receipt of the accident notification, act in a similar manner as CC_12 to notify vehicles, traffic light controllers, and adjacent cell controllers of the accident.

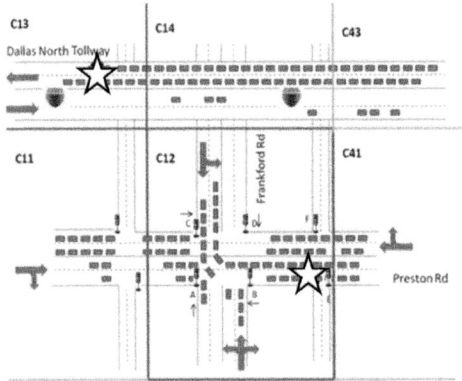

Fig. 11. Accident Scenario

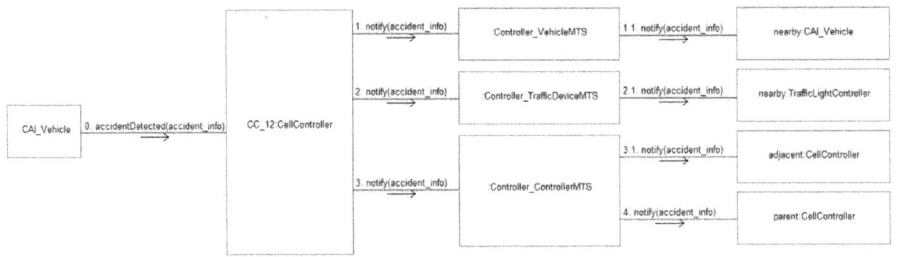

Fig. 12. UML Communication Diagram: Congestion Reduction

6 Related Work

6.1 Cyber-Physical Infrastructures

There have been several ITS initiatives and large-scale projects to address the congestion problem. MIT Intelligent Transportation Systems Program has developed a real-time computer system, DynaMIT [3], to effectively support Advanced Traveler Information System and Advanced Traffic Management Systems. DynaMIT is able to estimate network conditions and generate traveler information to guide drivers. DynaSMART-X [4] offers the same capabilities as well as interactions with multiple sources of real-time information (e.g., vehicle probes, roadside sensors). Although very successful, neither DynaMIT nor DynaSMART-X are able to handle non-recurrent congestion, i.e., congestion due to an unpredicted accident. Both systems approach traffic management in a top-down fashion although traffic information is collected bottom-up. Research and Innovative Technology Administration (RITA) has several major initiatives, including the Cooperative Intersection Collision Avoidance Systems (CICAS), the Integrated Corridor Management Systems (ICMS), and IntelliDrive [6]. CICAS includes autonomous vehicles and cooperative communication systems aimed at intersection collision problem. ICMS targets underused parallel routes (train, metro, bus) for corridor management rather than intra-network optimization. IntelliDrive combines advanced wireless communications, on-board computer processing, advanced vehicle sensors, and GPS to provide the capability for vehicles to identify threats. Each of these systems addresses a particular aspect of the traffic management problem, and follow a bottom-up approach. The unique features of Soteria are its integrated, multi-layered structure as well as its hybrid top-down/bottom-up strategy for traffic management.

6.2 Traffic Simulation Systems

Two models have emerged to simulate traffic scenarios. Macroscopic models [18,11,16,17,2] use a top-down approach, focusing on the overall behavior of the system. The systems are seeded with initial conditions where traffic entities follow predefined behavioral models. In order to simulate traffic congestion, these models use metrics such as: average speed, traffic density, probability distributions, traffic light frequency, and ramp-merge messages. By varying the models, the overall behavior can be analyzed and optimal traffic control can be implemented.

Microscopic models [24,13,25,12,7] use a bottom-up approach where every entity in the simulation is modeled as an agent and given a set of behavioral characteristics. This includes vehicles, roads, traffic signals, and signs which all interact. Agents control their own destiny thereby giving a realistic behavior to the simulation.

Our research enhances the conventional traffic simulation as we present the design of a hybrid system that implements both a macroscopic and microscopic approach. From a macroscopic view point, the cell controllers define initial conditions, monitor, and guide the global system behavior in a decentralized fashion.

At a microscopic level, each autonomous agent influences and adapts to changes while interacting, cooperating, and coordinating actions with other agents. This results in a simulation where micro-level agent behaviors are connected with the macro-level system behavior.

7 Conclusion

In this paper we discussed Soteria, a multi-layered, integrated, infrastructure for safety enhancement and congestion reduction. Soteria's unique features are its use of micro- and macro-level models, and its hybrid top-down/bottom-up strategy for traffic management. We presented a generic architecture that can be used to develop simulation systems for real world self-organizing systems, and finally described how this generic architecture can be instantiated to create the architecture of Matisse, a tailor made distributed simulation system for Soteria. Our initial observations show that Matisse's ability to model Soteria's hybrid strategy for traffic management will enable it to support the development of Soteria in the near future.

Matisse is in its early stage of development. Distributed versions of AES components have been implemented with basic functionality, including all message transport services. Working prototypes for the distributed Visualization Framework and Data Management Systems have been developed and integrated with the AES. Future efforts will include development of a distributed cell controller hierarchy, enhancement of the Visualization Framework, and further definition of agent behavior.

Acknowledgments. We would like to thank Professors John Hansen, Ovidiu Daescu, Andrea Fumagalli, Klaus Trumper and Drs. Pinar Boyraz and Marco Tacca for their contribution to the definition of the Soteria cyber-physical infrastructure.

References

1. The congestion problem, http://www.ops.fhwa.dot.gov/publications/bnprimer/02conge-stion_prob.htm (accessed August 2009)
2. Contram: Continuous traffic assignment model, http://www.contram.com/ (accessed August 2009)
3. Dynamit, http://mit.edu/its/dynamit.html (accessed August 2009)
4. Dynasmart-x, http://mctrans.ce.ufl.edu/featured/dynasmart/ (accessed August 2009)
5. NSF cyber-physical systems, http://www.nsf.gov/pubs/2008/nsf08611/nsf08611.htm (accessed January 14, 2010)
6. Rita, http://www.rita.dot.gov/ (accessed August 2009)
7. Bazzan, A.L.C.: A distributed approach for coordination of traffic signal agents. Autonomous Agents and Multi-Agent Systems 10(2), 131–164 (2005)

8. Boyraz, P., Daescu, O., Fumagalli, A., Hansen, J., Trumper, K., Wenkstern, R.:
   Soteria: An integrated macro-micro transportation super-infrastructure system for
   management and safety. Technical report, Erik Jonsson School of Engineering and
   Computer Science, University of Texas at Dallas, Dallas, USA (2009)
9. Boyraz, P., Kerr, D., Acar, M.: Signal modelling and hidden markov models for
   driving manoeuvre recognition and driver fault diagnosis in an urban road scenario.
   In: Proceedings of the 2007 IEEE Intelligent Vehicle Symposium, Istanbul, Turkey,
   June 13-15, pp. 897–992 (2007)
10. Boyraz, P., Yang, X., Sathyanarayana, A., Hansen, J.: Computer vision systems
    for context-aware active vehicle safety and driver assistance. In: Proceedings of
    the 21st International Technical Conference on the Enhanced Safety of Vehicles.
    Stuttgart, Germany, June 15-18 (2009)
11. Broucke, M., Varaiya, P.: A theory of traffic flow in automated highway systems.
    Transportation Research, Part C: Emerging Technologies 4, 181–210 (1995)
12. Dresner, K., Stone, P.: Multi-agent traffic management: An improved intersection
    control mechanism. In: Proceedings of the Fourth International Joint Conference
    on Autonomous Agents and MultiAgent Systems, Utrecht, The Netherland (July
    2005)
13. Erol, K., Levy, R., Wentworth, J.: Application of agent technology to traffic simu-
    lation. Technical report, Department of Transportation, Federal Highway Admin-
    istration, USA (1998)
14. Ferber, J., Müller, J.: Influences and reaction: a model of situated multi-agent sys-
    tems. In: Proceedings of the 2nd International Conference on Multi-agent Systems
    (ICMAS 1996), December 10-13, pp. 72–79. The AAAI Press, Menlo Park (1996)
15. Gamma, E., Helm, R., Johnson, R., Vlissides, J.: Design Patterns: Elements of
    Reusable Object-Oriented Software. Addison Wesley, Reading (1998)
16. Helbin, D., Hennecke, A., Shvetsov, V., Treiber, M.: Micro- and macro simulation
    of freeway traffic. Mathematical and Computer Modelling 35(6), 517–547 (2002)
17. Jin, W., Zhang, H.M.: The information and structure of vehicle clusters in the
    payne-whitham traffic flow model. Transportation Research, Part B: Methodolog-
    ical 37, 207–223 (2003)
18. Lieu, H., Santiago, A.J., Kanaan, A.: Corflo: an integrated traffic simulation system
    for corridors. In: Proceedings of the Engineering Foundation Conference, Palm
    Coast, Florida, April 1-6 (1991)
19. Meyer, M.: A tool box for alleviating traffic congestion and enhamcing mobility.
    Technical report, Institute of Transportation Engineers, US Department of Trans-
    portation, Federal Highway Administration (1997)
20. Michel, F.: The IRMS4S model: The influence/reaction principle for multi-agent
    based simulation. In: Proceedings of Conference on Autonomous Agents and Multi-
    Agent Systems (AAMAS 2007), Honolulu, Hawaii (May 2007)
21. Sun Microsystems, Java message service, jms (2002),
    http://java.sun.com/products/jms/ (Accessed January 2010)
22. Mili, R.Z., Oladimeji, E., Steiner, R.: Architecture of the DIVAs simulation system.
    In: Proceedings of Agent-Directed Simulation Symposium ADS 2006, Huntsville,
    Alabama (April 2006)
23. Mili, R.Z., Steiner, R., Oladimeji, E.: DIVAs: Illustrating an abstract architecture
    for agent-environment simulation systems. Multiagent and Grid Systems, Special
    Issue on Agent-oriented Software Development Methodologies 2(4), 505–525 (2006)

250   R. Zalila-Wenkstern, T. Steel, and G. Leask

24. Rasche, R., Naumann, R., Tacken, J., Tahedl, C.: Validation and simulation of de-centralized intersection collision avoidance algorithm. In: Proceedings of IEEE Conference on Intelligent Transportation Systems (ITSC 1997), Boston, Massachusetts (1997)
25. Roozemond, D.A.: Using intelligent agents for urban traffic control systems. In: Proceedings of International Conference on Artificial Intelligence in Transportation Systems and Science (1999)
26. Sathyanarayana, A., Boyraz, P., Hansen, J.: Driver behaviour analysis and route recognition by hidden markov models. In: Proceedings of the 2008 IEEE International Conference on Vehicular Electronics and Safety, Ohio, USA, September 22-24, pp. 276–281 (2008)
27. Steel, T., Kuiper, D., Wenkstern, R.: Context-aware virtual agents in open environments. In: Sixth International Conference on Autonomic and Autonomous Systems, Cancun, Mexico, March 7-13, pp. 1160–1168 (2010)
28. Steel, T., Wenkstern, R.: Simulated event propagation in distributed, open environments. In: Agent Directed Simulation Symposium (ADS 2010), Orlando, Florida, USA, April 12-14 (2010)
29. Wenkstern, R.Z., Steel, T., Daescu, O., Hansen, J., Boyraz, P.: MATISSE: A large scale multi-agent system for simulating traffic safety scenarios. In: Proceedings of IEEE 4th Biennial Workshop on DSP for In-Vehicle Systems and Safety, Dallas, USA, June 25-27 (2009)

On the Modeling, Refinement and Integration of Decentralized Agent Coordination
A Case Study on Dissemination Processes in Networks

Jan Sudeikat* and Wolfgang Renz

Multimedia Systems Laboratory (MMLab),
Faculty of Engineering and Computer Science,
Hamburg University of Applied Sciences,
Berliner Tor 7, 20099 Hamburg, Germany
Tel.: +49-40-42875-8304
{jan.sudeikat,wolfgang.renz}@haw-hamburg.de

Abstract. The integration of self-organizing processes in distributed software systems allows to equip applications with adaptive features that rise from the (inter-)actions of agents. In this paper, we address the systematic development of decentralized inter-agent processes in Multi-Agent Systems (MAS) that give rise to self-organizing properties. A *systemic* programming model is utilized that comprises a modeling approach to the conception of feedback loop structures and a reference architecture for the automated enactment of process prescriptions. A key design criterion for this tool set is the ability to supplement decentralized adaptations to already constructed software systems. Here, we present and exemplify the systematic design of decentralized coordination. The refinement of coupled feedback loop structures is presented as a design activity that is part of the systematic integration of coordination processes. This development approach is exemplified in two simulation scenarios, where the epidemic dissemination of information is integrated.

1 Introduction

The increasing size and complexity of today's distributed software systems demands novel engineering approaches. The utilization of self-organizing processes has been proposed to enable the adaptiveness of inherently decentralized system architectures [1]. *Self-Organization* describes concerted phenomena, where the coaction of autonomous system entities gives rise to global structures. These phenomena can be found in *physical, biological, social* as well as *artificial* systems [2] where structures are established and adapted.

---

* Jan Sudeikat is doctoral candidate at the Distributed Systems and Information Systems (VSIS) group, Department of Informatics, Faculty of Mathematics, Informatics and Natural Sciences, University of Hamburg, Vogt–Kölln–Str. 30, Germany, jan.sudeikat@informatik.uni-hamburg.de

D. Weyns et al. (Eds.): SOAR 2009, LNCS 6090, pp. 251–274, 2010.
© Springer-Verlag Berlin Heidelberg 2010

In natural and artificial self-organizing systems, the activities/configurations of individual system elements, e.g. particles, cells, and software elements, influence each other. The exhibited interdependencies form feedback loops that steer the collective, localized adjustment of system elements [3]. Agent-based software development techniques facilitate the construction of self-organizing systems. Software applications are conceived as sets of autonomous actors, i.e. *Multi-Agent Systems* (MAS) [4]. Interactions of agents with other agents [4] and the system environment(s) [5] are conceptually supported.

The construction of collective phenomena, e.g. self-organizing processes, faces the fundamental challenge to revise the designs of system entities in order to enforce the rise of global system properties. The major tools to bridging this gap are simulation-based development processes [6], design patterns [7,8] and formal modeling techniques [9] (see Section 2). In earlier works, we revised tool support that allows to model coordinating processes as networks of mutual interdependencies between system elements. A *systemic* modeling level has been proposed [10], that supplements agent-oriented modeling techniques with an orthogonal description-level. This level focuses on the interdependencies of system elements to express decentralized self-organization. A corresponding programming model allows to prescribe and enact self-organizing inter-agent processes in MAS. It comprises a configuration language [10] and a reference architecture [11]. This tool set is based on the externalization of decentralized coordination processes. These are self-contained design artifacts that can be shared, reused and incrementally revised.

Here, we want to stimulate the software-technological utilizations of self-organization in application development. The externalization of process models is beneficial as it enforces the separation of concerns and enhances the customizability of processes. Particularly, the supplementation of self-adaptive system properties [12] to existing software systems, based on self-organization principles, is prepared by this development approach. Adding self-organization allows to adjust the dynamic behavior of MAS. A key challenge is the conceptions of the process that is to be supplemented. One approach is the utilization of nature-inspired template processes, e.g. reviewed in [8].

Alternatively, it is desirable to deliberately conceive processes that allow to enhance the behavior of a MAS. In this paper, we discuss how the systematic conception of decentralized adaptation process among autonomous agents. This conception is presented as a structured development activity that is part of the systematic integration of inter-agent coordination. First, the *problematic*, unintended system behavior is modeled and these descriptions are incrementally refined to a detailed coordination process description that can be enacted automatically, i.e. without modifying the original agent implementations. Based on the separation of agent and coordination development [13], we discuss the incremental refinement of externalized coordination models. This refinement process is exemplified in a case study, where the epidemic dissemination of information is integrated in agent-based application designs. These MAS are used to simulate application behaviors.

The remainder of this paper is structured as follows. In the next section, we discuss related work, i.e. concepts and tools for the development of self-organizing applications. In the following Section 3 the here utilized, systemic programming model is outlined and interrelated to the foundational activities of software development. Part of this discussions is the iterative development of self-organizing processes that match the dynamics of the application context (see 3.3). The systematic conception and integration of inter-agent process is exemplified in Section 4, before we conclude and give prospects for future work.

2   Related Work

Decentralized, self-organizing processes are powerful means for the coordination of large-scale applications [1]. The properties of self-organizing processes (see e.g. [8]) and the specific mechanisms to their realization (see e.g. [7]) are well understood. However, their software technological utilization, e.g. share and reuse, found limited consideration. Development practices are dominated by techniques that blend system elements with algorithms that are highly specialized for the application context. The foundational aspects of the development of these algorithms are the *modeling* of the intended dynamics, the *integration* of coordination mechanisms in software agents, and the *systematic* utilization of the former two aspects in MAS development.

2.1   Modeling Self-organization

The modeling of self-organizing dynamics in MAS concerns the abstraction of the operation of individual agents to describe how agent mutually influence and give rise to collective phenomena that emerge from the agent (inter-)actions. These approaches describe the stochastic processes that underly the operation of agents by analyzing the possible transitions between microscopic agent states or agent configurations. Examples are the characterization of *averaged* transitions with differential equations (e.g. [14]) or explicit *stochastic* models, e.g. [9].

While it is well-understood that distributed feedbacks are imperative for the rise of self-organizing phenomena (e.g. see [3]) their *explicit* modeling is commonly not exercised in MAS development. One approach for the visualization of networks of feedbacks is given in Section 3.1.

2.2   Construction of Self-organization

The computational mechanisms that allow to construct self-organizing processes in MAS can be subdivided into two categories [15]. *Interaction*-level mechanisms, e.g. discussed in [7], provide interaction modes for agents and are means for the realization of information flows among agents. Interactions are either based on *direct* communications among agents, e.g. using public visible tags or market-based negotiations [7], or are *mediated* by the systems environment, e.g. using stigmergy principles [16]. Practical development tools are coordination frameworks (e.g. [17]) that encapsulate the details of interaction modes. Agents are decoupled as the agent developer can delegate interactions to a dedicated infrastructure [18].

Secondly, *adaptation*-level mechanisms, a subset of the mechanisms that are cataloged in [1], control how the perceived information cause agent-intern adjustments. Approaches to separate these mechanism from the agent functionality are either based on specialized agent architectures (e.g. [19]) or additional programming language extensions [20]. An orthogonal approach is the control of element adjustments by remote system elements, e.g. services [21].

A comprehensive approach to the integration of both types of coordination mechanisms is discussed in Section 3.2. Interaction infrastructures and the separation of the adjustment-logic are consolidated in a middleware layer. This enforces the encapsulation of coordination-related activities.

2.3   Systematic Design of Self-organization

The systematic construction of self-organizing systems faces the fundamental challenge to revise the interactions of system elements (*microscopic*-level) in order to provoke system-wide structuring (*macroscopic*-level), i.e. the autonomous ordering of system aspects. Two fundamental concerns are the *construction* of a MAS and the *adaptation* of implementation parameters that influence the rise of the intended phenomena [6]. Due to the non-linearity of these phenomena, e.g. the exhibition of phase transitions, often system simulations are required to estimate system-level properties. Thus the development alternates between the construction of a software system (*top-down*) and the validation of the intended system-level behaviors by system simulations (*bottom-up*).

A successful design strategy is the resembling of nature-inspire processes [8]. Particularity interaction-level mechanisms have been proposed as means to resemble the self-organization in natural systems. Approaches to design processes from scratch are rare and this is thematized in Section 3.3.

In addition, system designers typically decide early for the utilized coordination mechanisms. Coordination and agent function are blended and later adjustments of the coordination techniques or strategies imply redesigns of the agents. The programming model in the in the following section enables the specification of inter-agent coordination in externalized design models. This outsourcing can be used to supplement processes, as systematized in Section 3.3.

3   A Systemic Programming Model for Decentralized Agent Coordination

The expertize and effort that are required to conceive self-organizing coordination (cf. Section 2) justifies our interest in fostering the software-technological utilization, i.e. the systematic conception, share and reuse. In previous works, we advocated that the coordination strategy itself should be considered as a first

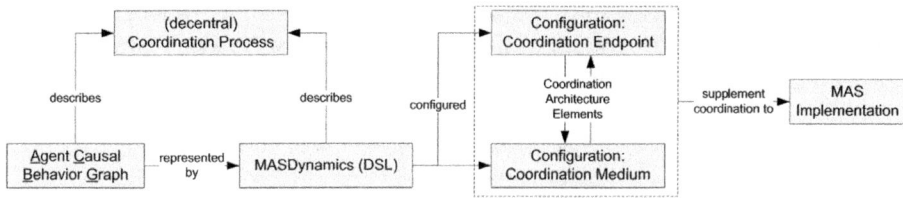

Fig. 1. The systemic programming model

class design element. The systemic programming model, proposed in [10,11], allows developers to describe coordination in externalized, application-independent formats and automate their enactment in MAS.

The constituent parts of the programming model are summarized in Figure 1. A *Coordination Process* describes the collective behavior of agents that results from the interactions of the individual agents. Here, we consider decentralized process that manifest self-organization. These processes structure aspects of a MAS, e.g. the agent configurations, the organizational structure, or the systems environment. A graph-based modeling approach (*Agent Causal Behavior Graph*) allows to describe how agent activities casually influence each other [10,13]. This modeling stance particularly highlights feedback structures among agents and is the semantic foundation for a configuration language (*MASDynamics*), which can be used to configure the enactment of interactions among agents. The enactments is carried out by an architecture that provides an event-based middleware layer for the establishment of inter-agent influences. The two foundational building blocks are *Coordination Endpoints* and *Coordination Media*. Endpoints are specialized agent modules that contain the adaptation logic of individual components and automate coordination-related activities. Media provide communication infrastructures that mediate influences among endpoints. Using this architecture, self-organizing process, which are composed of influences, can be supplemented to *MAS implementations*.

The ability to supplement coordination to MAS is a key design criterion for this tool set. The behavior of a conventionally developed MAS can be adjusted by integrating decentralized coordinating processes. This approach requires the strict separation of the activities that are conceptually related to coordination from the functioning of the agents. Thus this framework opposes established development approaches that focus on blending the agent logic with coordination issues (see Section 2.2).

3.1 Systemic Modeling and Configuration

The mainspring of self-organizing processes is the *distributed* feedback among system elements that governs the concerting of individual adjustments. The visualization of these feedbacks is addressed by a graph-based modeling approach [10] that refines *System Dynamics* (SD) [22] modeling concepts to describe the dynamics of agent-based software systems. The macroscopic state of an application is characterized by a set of interrelated system variables. These variables describe the number of agents that exhibit a specific behavior, e.g. a play a *role* [23]. Relations among these variables are illustrated by links and indicate how behaviors influence each other. Mathematically speaking, these relations describe the rates of changes of variable values.

The MAS-specific refinement, is coined *Agent Causal Behavior Graph* [10,13] $ACBG :=< V_{acbg}, E_{acbg} >$, where V_{acbg} is a set of nodes and E_{acbg} is a set of directed edges ($E_{acbg} \subseteq V_{acbg} \times V_{acbg}$). Nodes represent the state variables. Nodes are typed to enable the distinction between different design concepts of agent-behaviors, e.g. the adoption of *roles* [23]. The connecting relations denote either additive/subtractive *influences* or positive/negative causal *interdependencies*. Influences describe that changes in originating variables effectuates the increase or decrease of connected variables. An example is a hypothetical set of manufacturing agents. Their individual productive behavior (*producing*) increases the amount of workpieces in a *stock* (cf. Figure 2, I). The production of items adds elements in a stock variable and when the agents are not showing their productive behavior, the stock is not affected.Interdependencies describe causal relations, where connected variables change in the same (positive) or opposite (negative) directions. An example are client-server relations among agents. *Requesting* agents are related by a positive interdependency to service providers (*providing*). When the number of active (asynchronous) requesters in the system increases, the activations of services, i.e. their invocations, increase as well (cf. Figure 2, II). When the number of requesters decreases, the number of the corresponding invocations decreases as well. Relations are typically associated with a time delay and can be described by a macroscopic rate that averages the microscopic interactions of agents.

Graphs can be abridged by *stacking* nodes of the same type that participate in conjoint relations. By default, incoming links to these stacks affect all member variables and affected nodes can be constrained by annotating links with set definitions. Hierarchies of system variables support the *traceability* of refinements and are expressed by placing (sub-)variables within super-variables that are composed of the values of enclosed variables. Links to the super-variable affect all enclosed links and links from a super-variable can transmit changes of any enclosed variable. Besides, links are also allowed to connect enclosed variables directly. These constructs are utilized in the Figures 6, 9, and 10.

In [10], an ACBG-based configuration language for is presented. This language allows to configure the agent-behaviors and influences in an ACBG. These configurations prepare the automated enactment (cf. Section 3.2) as they contain the configurations and parameterizations of Endpoints and Media. Conceptually, the language allows to map the abstract elements of a process definition to the concrete realizations of agents coordination media. This mapping is systematized and exemplified in [13]. Coordination Media are simply parameterized and the set of possible parameters depends on the concrete Medium realizations. The configuration of endpoints concerns when agents show the behaviors that

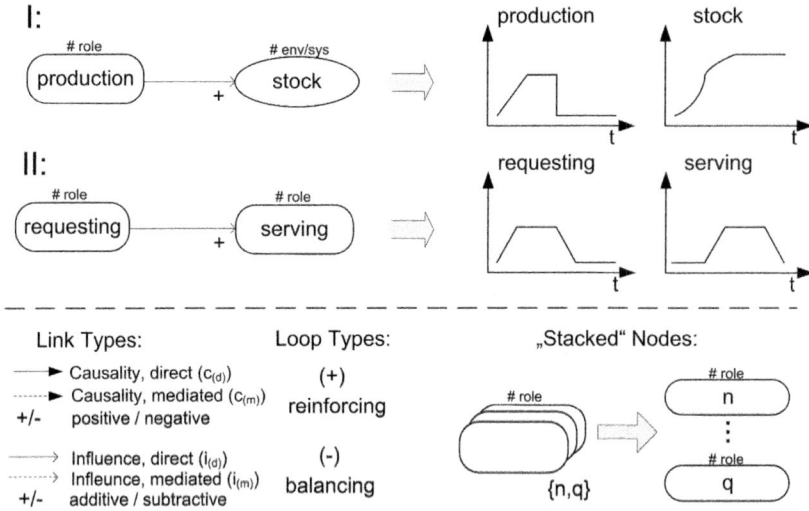

Fig. 2. The graphical ACBG notation

are part of a process model. These configuration is specific for the utilized agent platform and has been exeplified for BDI agents in [10].

3.2   A Reference Architecture for the Integration of Decentralized Coordination in Distributed Systems

Here, we outline a *reference* architecture, i.e. a blueprint for families of application architectures [24], that automates the enactment of decentralized coordination processes (cf. Section 3.1) [13]. Figure 3 (I) illustrates the conceptual, layered structure of this architecture [13]. The *Application Layer* realizes the system functionality and comprises autonomous agents. The coordination of these agents is encapsulated by an underlying *Coordination Layer*. This layer comprises these agents and a set of *Coordination Media* (CM) that are realized by *coordination infrastructures* (e.g. [5]). The Media encapsulate interaction models, e.g. *digital pheromones* or *token distribution algorithms* etc. [7], and thus provide abstractions that *enable* and *govern* interactions. The architecture encapsulates these infrastructures and separates activities that concern the coordination, e.g. communication, from agent models.

Coordination is enacted by equipping agent instances with a *Coordination Endpoint*, i.e. an agent module that is able to observe and influence the agent execution (cf. Figure 3, II 1). Using this decentralized approach, coordination is provided by a middleware service that is accessed via CE-modules. CEs establish information flows between agents as these modules inform each other about the observations of behavior adjustments by exchanging *Coordination Information*

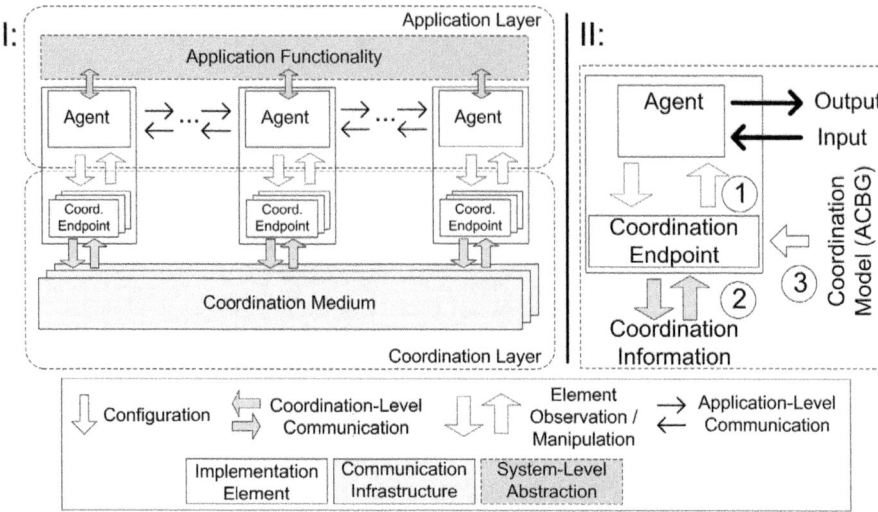

Fig. 3. DeCoMAS operating principle [13]

(CI, cf. figure 3, II 2). Perceptions of CI are interpreted by CEs that influence the agent execution by initializing agent behaviors. Behaviors are triggered by the dispatching of agent reasoning events, e.g. goal adoptions or belief modification. It is to note that these influences respect agent autonomy, since the dispatching of events suggests local adjustments and agents have the authority to drop inappropriate behaviors. The observed and suggested behavior adjustments are defined in terms of agent reasoning events that are specified by sets of *Systemic Coordination Specification* (cf. Figure 3 II 3), e.g. ACBGs. These provide a generic interface between the enactment architecture and agent models.

The interplay of the exchange of CI and the local adjustments of agent behaviors manifests the feedback loop structure that is specified by *Systemic Coordination Specification* (cf. figure 3 II, 3). Their automated enactment is configured by specifying the data that are exchanged between CEs, parameterizing the Medium realizations to control the dynamics of the information exchanges and specifying how CEs adjust the local activities of agents [10]. The adjustments are encapsulated in *Adaptivity Components* that contain the procedural logic to observe and modify agent states. This architecture has been prototypically realized [11] on top of the *Jadex*[1] agent platform.

The feedback that is enacted is *distributed* among the participating system entities. Endpoints observe when influences are to be effectuated and the receiving endpoints respond by checking whether behavior adjustments should be proposed to the host agent. The relation to conventional feedback-oriented designs of self-adaptive applications is discussed in Section 3.4.1.

---

[1] http://jadex.informatik.uni-hamburg.de/bin/view/About/Overview

3.3   Systematic Conception of Self-organizing Processes

The previously described development tools allow to externalize the description of a self-organizing algorithm from the MAS implementation. The utilization of this externalization enables to principal development strategies. First, the development of the software system from scratch can utilize the externalization. In this context, agents are understood as autonomous, self-contained providers of functionalities and the concerting of these activities is outsourced in an explicit process description. This approach is particularly attractive for development context where frequent changes in the inter-agent coordination are likely, e.g. when prototyping an application. The customization of the process structure and configuration is facilitated (see Section 3.1) and the need to modify agent realizations, when trying-out alternative coordination strategies, is minimized. An alternative approach is the supplementation of self-organization during or after the development of agents. Agents are designed as interactive system components, thus when not explicitly outsourced, the agents will show coordination in their interplay. The adding of self-organization can be used to enhance the operation of software systems and ad additional aspects of adaptivity to a functioning software. This approach requires the analysis of the already present system behavior in order to identify an appropriate supplement. The here discussed development procedure allows for both approaches.

In [25] it has been argued that the development of (externalized) coordination processes can be separated from the construction of the coordinated application elements. A five-step development procedure is prosed that can be applied in parallel to the conventional development of MAS, e.g. discussed in [26]. The contained development activities extend MAS development with the systematic conception and integration of self-organizing processes. This procedure (cf. Figure 4) realizes an alternation between *top-down* development and *bottom-up* validations activities (see Section 2.3).

The process description follows the *Software Process Engineering Meta-Model*[2] (SPEM) [27] a standard from the *Object Management Group*[3] (OMG) that, among others, provides concepts for the modeling and partitioning of software engineering procedures. This standard has been applied to describe agent-oriented development methodologies (e.g. [28]). Software development procedures are composed of *Activities* that create and modify *Artifacts*. Artifacts describe the work products of a software engineering process. These comprise design models and executable software elements. Activities are composed of *Tasks*, i.e. assignable units of work. Tasks can be subdivides in development *Steps*.

The *Adaptivity Requirements* Activity guides the elicitation of the intended adaptivity. Based on the description of the application *domain* and initial models of the planned MAS (*Organizational / MAS Environment*), these requirements (*Adaptation Requirement*) are derived [29]. Subsequently, requirement models and system descriptions are (top-down) refined (*Coordination Process Definition*) to a proposal of an appropriate structure of agent-interdependencies

---

[2] http://www.omg.org/technology/documents/formal/spem.htm
[3] http://www.omg.org/

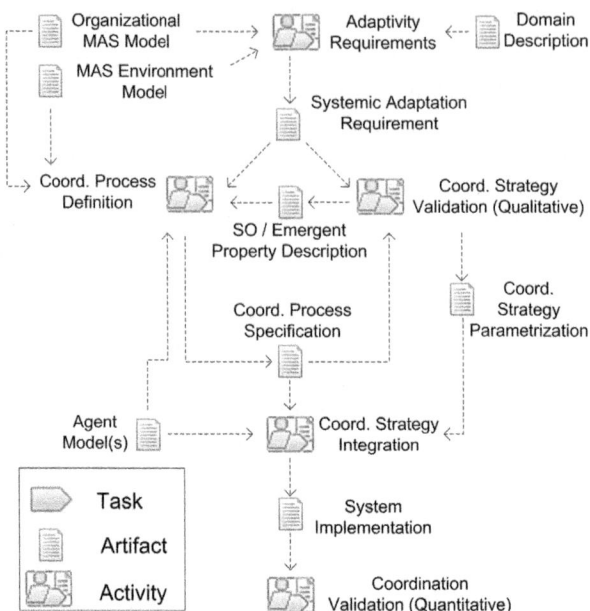

Fig. 4. The systematic development of externalized coordination models that combines top-down modeling and bottom-up analysis, following [25]

(*Coord. Process Specification*). Prior to its implementation, this model is (bottom-up) analyzed to validate the dynamical properties that arise from the proposed element interdependencies [25]. Two approaches are the formal modeling of emergents, e.g. [9] and the behavioral modeling of agents by stochastic simulation techniques [30]. After the applicability of the coordination model has been shown, developers embed it in an actual MAS (*Coord. Strategy Integration*) [13]. This Activity comprises the detailing of process configurations and results in an executable MAS that is augmented with coordination (*System Implementation*). Simulations of these realizations check the system behavior and enable the parametrization of implementations of system elements.

The construction of the coordination process (*Coordination Process Definition*, see Figure 4) is composed of four fundamental Tasks. First, the descriptions of the application domain and the conceived MAS, e.g. initially specified as a set of interdependent actors [31], is transferred to a systemic description (see Section 3.1) of the system and its application context. Subsequently, the resulting model is refined to a coordination process (*Systemic Coordination Refinement*). The refinement indicates additional agent states and activities. A following Tasks checks whether these require the modifications of the agent models or can be added within Adaptivity Components (cf. Section 3.2). Finally, the interaction mechanisms [7] are identified and configured. These Tasks partition the incremental refinement of coordination strategies. In incremental development

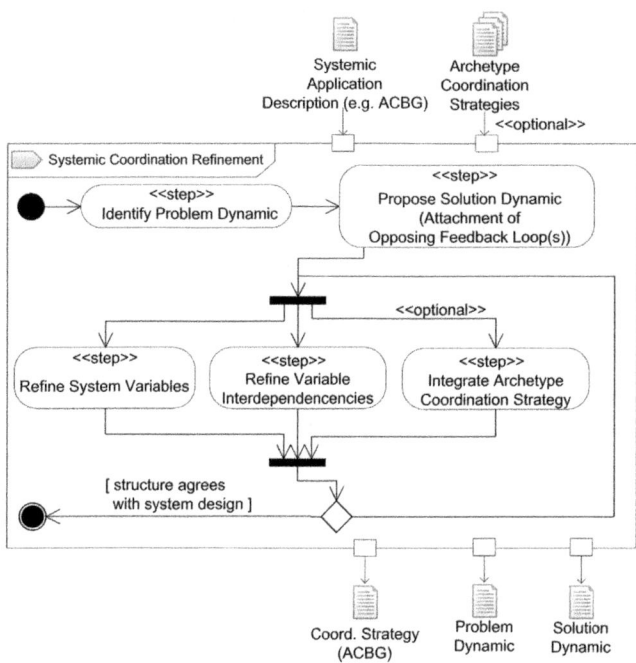

Fig. 5. The refinement of a systemic coordination model, described by a stereotyped UML Activity Diagram

processes, these Tasks are iterated in order to refine coordination, e.g. when agent models are detailed or the system requirements are adjusted.

The Task *Systemic Coordination Refinement* is illustrated in Figure 5. Inputs to the refinement process is a model of the application dynamics (*Systemic Application Description*). The refinement particularly addresses the incremental detailing of ACBGs. Additionally, development teams may maintain a library of field-tested coordination strategies (*Archetype Coordination Strategies*). These provide reusable structures, where the dynamic properties are known in advance [8,32].

Initially, the model of the application dynamic is examined to identify the original, e.g. un-coordinated, dynamical properties of the future application (*Identify Problem Dynamic*). This activity leads to a model of the *Problem Dynamic*, i.e. the unintended and problematic system behavior that is to be adjusted by the application realization. This model may describe processes of the system environment or process within a given MAS design that are to be modified. Secondly, a so-called *Solution Dynamic* is derived. Additional feedback loops are introduced as to counterbalance the unintended behaviors. In section 4.2, this is exemplified for a scenario where an inevitable, but unsolicited s-shaped grows is balanced by the introduction of an balancing feedback loop. In the subsequent Steps, system state variables and their interdependencies are refined. Optionally, well-known coordination pattern (e.g. [8]) may be integrated and/or combined, in order to enable the realization of the Solution Dynamic, as exemplified in [32]. The detailing of this dynamic leads to the definition of the *Coordination Strategy* model.

3.4   Discussion

3.4.1   The Relation to Self-adaptive Software Architectures

The driving force for distributed, adaptive software systems is the presence of feedback loops that enforce adjustments within system components to respond to changes in the application context [3]. Feedbacks can be utilized in two fundamental modes. First, the configurations of individual system elements are managed by feedbacks that *decide* the adjustments that are to be made. These approaches transfer control theoretic concepts to software architectures. The principal parts of these control loops are the *collection* of information, the *analysis* of these information, the *planning* of an adjustment, and its *actuation*. Sophisticated system architectures are available to realize managing system entities that impose these feedbacks on sets of system elements. Examples are the reference architectures that are proposed by the *autonomic* [33] and *organic* [34] computing initiatives. One successful strategy to the realization of these feedbacks is the *architecture-based self-adaptation* which utilizes explicit models of the managed software system, e.g. [35]. These models constrain the selection of valid adjustments and make the key properties explicit that are to be maintained by the imposed feedback.

Alternatively, *decentralized* feedbacks can be embedded among system entities when the activities of system elements influence other elements. Cyclic propagations of influences are an alternative mode for the control of the system adaptivity that is prevalent in natural self-organizing systems. System entities use signaling techniques to inform each other about their configurations and adjust due to perceptions. Examples are discussed in [8], e.g. the nest construction in termite colonies where building block distribute a scent that actuate other termites to reinforce the build structure.

The here presented design approach concerns the latter type of feedback. The systemic modeling approach facilitates the planning for collective effects that result from decentralized feedbacks within an agent society (see Section 3.1). The corresponding architecture (see Section 3.2) reflects the enacted feedback as it provides a middleware layer that is distributed among system elements and automates the accessing of shared communication infrastructures. This middleware can be used to integrate sets of mutual influences among agents and the agents are equally managed by locally associated endpoints. Endpoints respect the agent autonomy and propose adjustments. The agents however have the authority to deny proposals, based on their current execution context. This tool set is a prerequisite for the supplementation of self-organizing features to MAS (see Section 3.3).

The design of self-adaptive and self-organizing systems is typically approached by non-intersecting research communities and only few studies compare the applied design principles, e.g. [3]. The design of self-organization concerns the effects among elements within the same abstraction level while self-adaptive architectures guide the integration of managing entities that adjust subordinates. The integration of both development stances allows for an comprehensive development approach as it has been recognized that multiple, concurrent self-adaptations can lead to unwanted, emergent effects [3,33,2].

Enhancing the Programming of Self-Organization. The here discussed development approach distinguishes from other development techniques as two development aspects gain center stage. A long term goal for the revisions of self-adaptive architectures is the ability to supplement adaptivity to conventionally developed software systems, e.g. described in [12]. This is attractive as development can first focus on the functionality of the system components and the system behavior can be adjusted afterwards. Self-organizing systems are typically build from scratch and the exhibition of self-organizing phenomena is an initial design criterion (see Section 2.2). In this respect, the clear separation of self-organization aspects and agent functionality enables a comparative approach. The dynamic behavior of a MAS is studied and problematic dynamics are counterbalanced by the integration of additional feedbacks that tune the collective behavior of the system elements (see Section 3.3).

Secondly, the treatment of coordination models as first class design elements is attractive as it supports the customization of the supplemented feedbacks. The practicability of this externalization demands the minimization of the effort to integrate coordination processes. The major importance of providing the context information that are necessary for the individual, local adaptation has been noticed, e.g. in [36] where an *architectural strategy* has been proposed that supports the design of self-adaptive applications by *situated* autonomous agents. The systems environment becomes a self contained design element [37] that, amongst others, controls the interactions of agents. Sophisticated coordination/environment frameworks [5] decouple communication partners and encapsulate the interaction logic, but developers have to manually integrate the interactions into component designs. An example is the the *Agents & A* approach [38] that structures the environment of agents with artifacts, i.e. passive environment elements. These elements can encapsulate interaction techniques, e.g. protocols. However, the control flow within agents is blended with the control of the artifact-based interactions, e.g. when to invoke artifacts and how to respond to interactions [18]. Approaches to automate the use of agent-interaction techniques for agent coordination, found yet minor attention. Examples are *Interaction-Oriented Programming* (IOP) [39], that has been proposed to automate externalized coordination models and coordination service architectures, e.g. [40], that control the agent execution by observing and dispatching agent-internal events.

The here presented coordination framework extends these works. This tool-set allows developers to separate coordination, i.e. the establishment of information flows and the local adaptivity of agent models, from the core agent functionality. The proposed coordination layer allows to encapsulate established coordination infrastructures and interaction techniques, e.g. [5,38], in Coordination Media.

The accessing of these infrastructures, e.g. the initiation of an interaction, is automated and deliberatively configured. In addition, the effects of interactions are encapsulated in Endpoints as well. Consequently, the inter-agent coordination is a design element that is clearly separated from the agent models. The technical aspect of this separation is reflected and exploited by the methodical development processes (cf. Section 3.3) that guides the supplementation of coordination to an application.

4   Embedding Epidemic Information Dissemination

Epidemic Algorithms provide mechanisms to the dissemination of information in distributed systems [41]. These algorithms mimic the spread of infections in populations to ensure the propagation of data among large numbers of system elements.

Here, we exemplify the utilization of the systemic programming model and the systematic refinement of coordination processes by supplementing MAS with epidemic algorithms. The algorithms are described by systemic coordination models (cf. Section 3.1) and are enacted by a prototype realization of the proposed coordination architecture (cf. Section 3.2). First, the embedding of an externalized coordination model is exemplified (cf. Section 4.1). Secondly, the conception of a counterbalancing process, to work against a problematic system behavior, is described. The former cases study illustrates the supplementation of an inter-agent processes that is independent form the application domain. The second example demonstrates how problematic self-organizing phenomena can be resolved by appending a counteracting self-organizing process.

4.1   Convention Emergence

Epidemic algorithms can be used to address agreement problems in MAS (e.g. discussed in [42,43]). In these settings, agents start with arbitrary local settings and have to agree on a particular value. These values can *emerge* in agent populations when (1) individuals mutually inform each other about their local values and (2) the individuals adjust their local configuration, based in the exchanged information. If the size of the system and/or the communication overhead permits the broadcast of local values, agents can be arranged in (overlay) networks that constrain communications. Local adjustment policies model the *suggestibility* of individuals, i.e. the likelihood that perceptions cause adjustments. Several reasoning techniques are appropriate, but often combinations of majority rules and bounded memories of past communications are applied (e.g. see [42,43]). When both modeling levels are well-matched, the initially diverse configurations converge.

Figure 6 illustrates the causal structure, i.e. ACBG, of this convergence process. External factors actuate agents to execute arbitrary *Activities*. A side-effect of these activities is the communication of local convention value to connected agents (*inform convention value*). Perceptions trigger individual reconfigurations

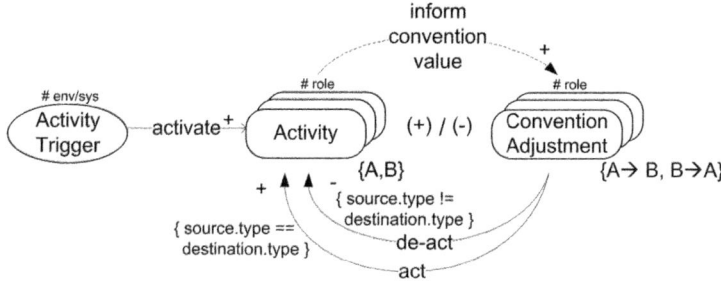

Fig. 6. Convention Emergence: Externalized coordination model

(*Convention Adjustment*), according to the local adjustment policies. The local configuration feeds back to the agent activities, e.g. the configurations may control how activities are proceeded or constrain individuals. These causal relationships form a reinforcing (+) and a balancing (-) feedback loop. The system history decides which value dominates. In the realized simulation model, two configuration types are available (*A, B*) and reconfigurations alternate between these (*A → B, B → A*). The local configuration (*source.type*) can be used to control the available activities (*destination.type*).

This decentralized agreement strategy has been realized with the systemic programming model (cf. Section 3). Agents are equipped with *Adaptivity Components* that, based on a fixed memory size of 9 communications, adjust the local value when the majority of the perceptions differs from the local setting. A *Coordination Medium* enforces an overlay network topology. The *Coordination Endpoints* (CEs) register themselves at the Medium when associated agents are initialized and the Medium provides the addresses of virtually connected agents that the CEs are allowed to inform. The Medium can be configured to exhibit different network topologies.[4] Figure 7 exemplifies the exhibited dynamics and visualizes the self-organized structuring of agent-configurations. 300 agents are randomly assigned an initial, local configuration. These agents are arranged in a random graph and the communication of configuration values is randomly triggered (cf. Figure 6]).

The self-organizing process controls the structuring of the agent values, i.e. the convergence. The convergence time is a measurement that characterizes this structuring. It depends on several parameters, most importantly the network topology and the local communicativeness of the agents. Here, we briefly discuss their impact. The stochastic properties of complex graphs have been studied in literature. Of particular interest are randomly generated and *power law* distributed graphs [44]. In our simulations, random graphs follow the *Erdős-Rényi* binomial model[4] (cf. [44], p. 8) that connects each pair of N nodes with a fixed probability p. The expected number of edges is $E(N) = p\frac{N(N-1)}{2}$. Power law

---

[4] Using: Java Universal Network/Graph Framework (JUNG):
http://jung.sourceforge.net/

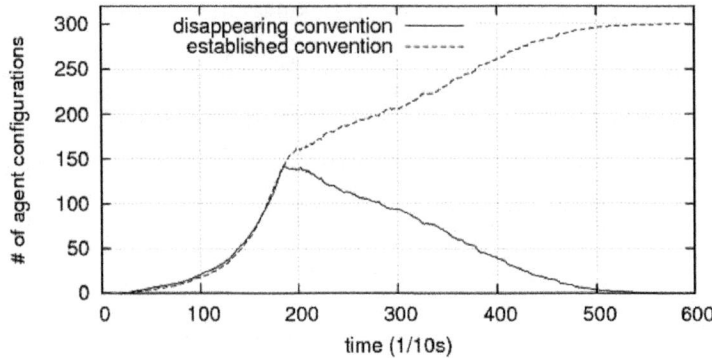

Fig. 7. Sample Convention Emergence. 300 agents reach agreement (averaged over 10 runs) in a random graph (p = 0.2).

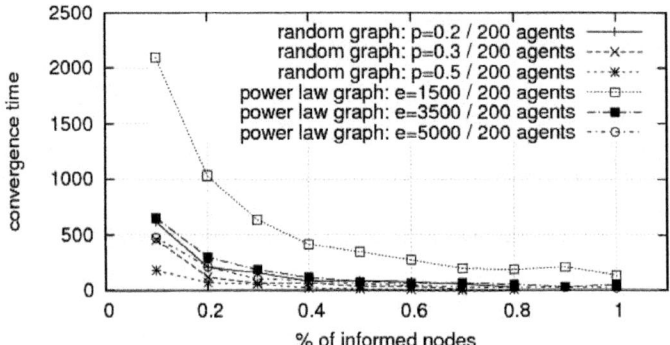

Fig. 8. Comparison of the convergence time that is exhibited by 200 agents in different graph topologies, in each case averaged by 10 runs

distributed graphs were generated[3] according to [45]. The *node degree*, i.e. the number of edges of a node, is distributed by a function that follows a power law, i.e. the probability that a node has k edges follows $P(k) \sim k^{-\gamma}$.

In Figure 8, we compare the effects of these network types on the exhibited convergence times. The y-axis denotes the time needed till 95% of the agents agreed on a local value. The x-axis shows *communicativeness* of the CEs, i.e. the percentrage of connected nodes that are informed. It can be seen that both graph types behave comparatively. Convergence is quickly achieved for mid-level values and increasing the number of informed agents, i.e. the commmunicativeness, above 70% rewards minor impact.

However, comparing the number of edges between the nodes (cf. Table 1), it can be seen that power law distributed topologies more effectively disseminate the information that is needed to reach convergence. These graphs are set to

Table 1. The numbers of edges in the simulated graph topologies

Graph Type	# of Edges
Random (p=0.2 / 200 nodes)	approx. 3992.7 [5]
Random (p=0.3 / 200 nodes)	approx. 5975.9 [5]
Random (p=0.5 / 200 nodes)	approx. 9949.5 [5]
Power Law (200 nodes)	set to fixed values

comprise less edges (between 3500 and 5000, cf. figure 8) and therefore the immanent communication overhead is reduced.

This scenario demonstrates the coordination of agents by the enactment of an externalized coordination model. The implemented model corresponds to a well known class of algorithms [42,43]. In the following section, we exemplify the systematic conception of coordination models, that is driven by the adjustment of problematic dynamics (cf. Section 3.3).

4.2   Supplementing a Patching Process

In a hypothetical scenario, we assume a network of agents to be subject to malicious *infections*. Infections randomly spread among the agents and the design objective is to realize an opposing process that contains these virtual epidemics. The initial step for the conception of a coordinating process is the identification of the *Problem Dynamic* (cf. Section 3.3). Here, the problematic behavior is the unrestricted spreading of infections (cf. Figure 9) that drives the system to a configuration where all agents are infected. To model this scenario we assumes that agents play two roles. Agents are either *susceptible* to infections or *infectious*, as they have already been infected. The local interactions between these roles are characterized by a fluctuating *infection* rate. This rate reduces the number of susceptible agents (negative link) and increases the number of infectious agents (positive link). It depends on the availability of susceptible and infectious agents (positive links) and declines when the number of susceptible and/or infectious agents is reduced. These causalities form a pair of feedback loops. A balancing feedback gradually reduces the amount of susceptible agents and a reinforcing feedback increases the number of infected agents. Infected agents do not recover by themselves, thus the system tends to a steady state where all agents are infected.

The model of the infection process illustrates that the presence of susceptible and infectious agents enforces the rate of further infections. Therefore, the proposal of an initial Solution Dynamic (cf. the second step in Figure 5) concerns the attachment of feedbacks that limit these reinforcing factors. The proposed *Solution Dynamic* (cf. Section 3.3) supplements two balancing feedback loops that reduce the number of infectious and susceptible agents. Two processes *recover* infected agents and *patch* susceptible agents. The affected agents enter

---

[5] Averaged over 100 generated graphs.

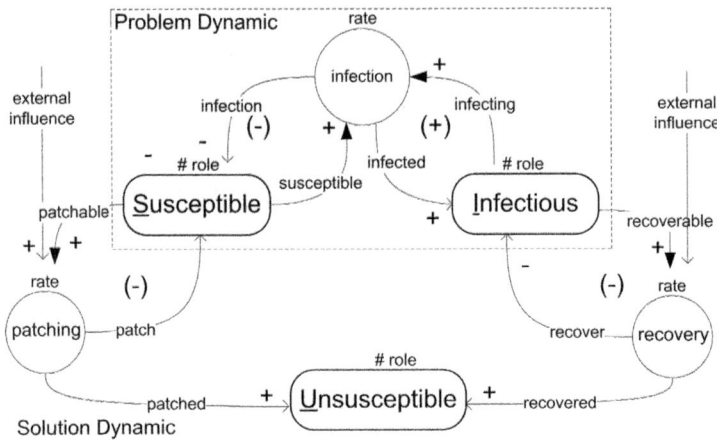

Fig. 9. The problem model is supplemented with an opposing feedback

a *Unsusceptible* state and are immune for futurse infections. This is an early conception of the intended dynamics and their realization is detailed by the subsequent refinement of variables and their interdependencies (cf. the second step in Figure 5).

Figure 10 illustrates the finally enacted coordination strategy that details the enactment of the additional feedback loops within the Solution Dynamic. Starting point for the derivation of this model is the refinement of the agent behaviors to activities that can be integrated using the enactment architecture (cf. Section 3.2). Subsequently, the influences among these behaviors are detailed to complete the intended feedbacks. Two agent roles have been added that realize the *patching* of susceptible agents and the *recovery* of infected agents. These are realized as Adaptivity Components (cf. Section 3.2) in the agent programming language and are capable to modify the agent state. In the simulation model, these mimic the patching of agent instances at run-time. The *Unsusceptible* agent role has been refined to two roles that denote how the agent has been rendered unsusceptible (*U by recovery*, *U by patching*). The recovery process of infectious agents is initialized by the manual recovery of an infected agent (*Manual Recovery (administration)*). This describes administrative interventions that trigger the process to recover (role: *recovery*) an infected agent. This agent will enter the corresponding unsusceptible state (*U by recovery*) and all Unsusceptible agents forward the received patch (*forward patch* links), i.e. they inform other agents about the update and the reception of this update triggers the local conversion of agents (*patching,recovery*) to the unsusceptible role. These causalities form two reinforcing feedback loops that increase the numbers of recovered and patched agents. Therefore, the infection process, that reinforces infections, is countered with two processes that reinforce unsusceptibility. The parameterization of the local agent activities and the communication Medium control which feedback loops dominate.

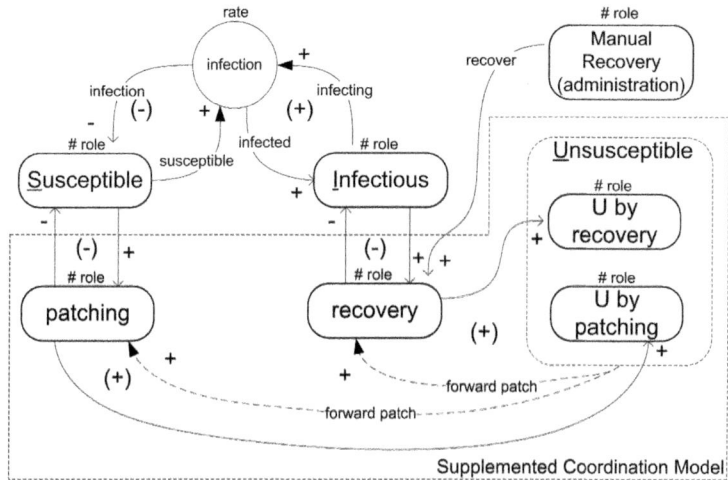

Fig. 10. Coordination model: The enacted ACBG that supplements a MAS with the Solution Dynamic, i.e. the recovery / patching of agents

In our simulations, all agents are initially susceptible and a randomly selected agent is infected immediately after the initialization of the system. Infected agents randomly infect a fixed number of other population members. After a fixed time delay, the initially infected agent is recovered to mimic a manual intervention. Subsequently, this agent spreads the patch, according to the coordination process. The recovery of an infected agent is set to take twice as long as the patching of an unsusceptible agents. Figure 11 shows a sample simulation run in a scenario, where the infection is quickly spreading. The patch distribution then opposes the epidemic by converting infected nodes.

A measure for the effectiveness of the supplemented patching process is its ability to limit the number of infected agents. As the conversion times are fixed, a basic parameter to influence the rate of conversions is the communicativeness of unsusceptible agents, i.e. the number of neighboring agents that patches are communicated to. Following the examination of the first case study, power law distributed graph support the effective dissemination of messages and this configuration is used as an overlay topology among Endpoints. Figure 12 (left) shows the plot of the maximum number of infected agents against the communicativeness of endpoints.[6]

Simulations of the conceived patching process indicate that the behavior of the process can range between two regimes. In a *subsequent* regime, the majority of the conversions to unsusceptibility is based on recovering infected agents. Consequently, the distribution of patches follows the infections. This regime is imposed when the rate of infections is larger then the rate of conversions. Secondly, the patching behaves *competitive* if the majority of conversions are

---

[6] Each measurements is averaged over 10 simulation runs.

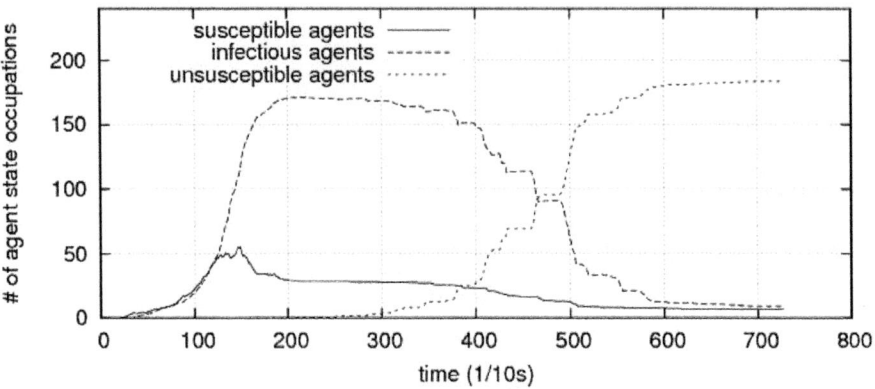

Fig. 11. Patching dynamics: Sample simulation run. 200 agents (averaged over 10 runs) are patched by a power law distributed graph (5000 edges, r = 10000).

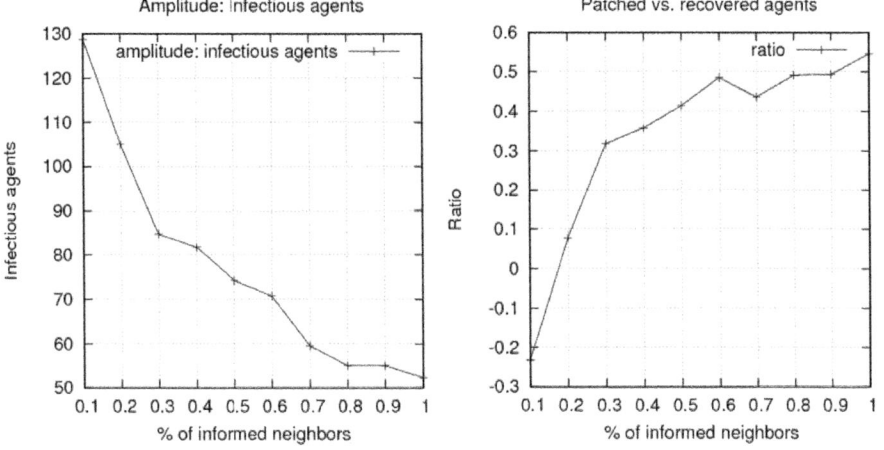

Fig. 12. Patching dynamics: Simulation results

based on the patching of susceptible agents. This regime is desirable since both processes compete for a common resource, i.e. the amount of susceptible agents, thus the infections are effectively eliminated. The exhibition of the regimes can be measured by the ratio

$$\rho = \frac{max_p - max_r}{|U|}, \tag{1}$$

where max_p and max_r are the maxima of patched, respectively recovered agents in a single simulation run. $|U|$ is the cardinality of the agent set that is made unsusceptible. Since $max_p + max_r = |U|$, measurements range from -1 to $+1$. negative values indicate subsequent regimes and positive values signify the competitive behavior of the processes. Figure 12 (right) plots the measurements of the ratio against the communicativeness of Endpoints.[5] Due to the effective dissemination of patches, subsequent regimes are only shown for small values and quickly approach the competitive regime.

5 Conclusions

In this paper, the utilization of a systemic programming model for the integration of self-organizing dynamics in MAS is discussed. This development approach is based on a middleware layer that serves as a reference architecture of the integration of decentralized coordination in MAS. This architecture is inherently interrelated with its systematic utilization, i.e. the corresponding development methodology that guides the supplementation of inter-agent processes to bring by the intended system behaviors.

In the discussion of the corresponding development procedure, we particularly focused on a refinement phase (cf. Section 3.3) that guides the planning of decentralized coordination process within MAS. This refinement is based on a systemic modeling approach to decentralized agent coordination and a layered coordination architecture that automates the enactment of coordination models. This systematic procedure has been defined in a standard software engineering notation and has been exemplified in a case study where coordinating processes, e.g. epidemic information dissemination, are integrated in MAS.

Future work addresses the analysis of ACBG-based coordination models, e.g. by the (semi-)automation of required simulations. In addition, we plan to detail the integration of the proposed coordination development in established agent-oriented development methodologies and to further compare the presented modeling level to designs of self-adaptive applications.

Acknowledgment

We would like to thank the *Distributed Systems and Information Systems* (VSIS) group at Hamburg University, particularly Winfried Lamersdorf, Lars Braubach and Alexander Pokahr for helpful discussion.

References

1. Serugendo, G.D.M., Gleizes, M.P., Karageorgos, A.: Self-organisation and emergence in mas: An overview. In: Informatica, vol. 30, pp. 45–54 (2006)
2. Mogul, J.C.: Emergent (mis)behavior vs. complex software systems. Technical Report HPL-2006-2, HP Laboratories (2005)
3. Brun, Y., Marzo Serugendo, G., Gacek, C., Giese, H., Kienle, H., Litoiu, M., Müller, H., Pezzè, M., Shaw, M.: Engineering Self-Adaptive Systems through Feedback Loops. In: Cheng, B.H.C., de Lemos, R., Giese, H., Inverardi, P., Magee, J. (eds.) Software Engineering for Self-Adaptive Systems. LNCS, vol. 5525, pp. 48–70. Springer, Heidelberg (2009)

4. Jennings, N.R.: Building complex, distributed systems: the case for an agent-based approach. Comms. of the ACM 44(4), 35–41 (2001)
5. Viroli, M., Holvoet, T., Ricci, A., Schelfthout, K., Zambonelli, F.: Infrastructures for the environment of multiagent systems. Autonomous Agents and Multi-Agent Systems 14, 49–60 (2007)
6. Edmonds, B.: Using the experimental method to produce reliable self-organised systems. In: Brueckner, S.A., Di Marzo Serugendo, G., Karageorgos, A., Nagpal, R. (eds.) ESOA 2005. LNCS (LNAI), vol. 3464, pp. 84–99. Springer, Heidelberg (2005)
7. DeWolf, T., Holvoet, T.: Decentralised coordination mechanisms as design patterns for self-organising emergent systems. In: Brueckner, S.A., Hassas, S., Jelasity, M., Yamins, D. (eds.) ESOA 2006. LNCS (LNAI), vol. 4335, pp. 28–49. Springer, Heidelberg (2007)
8. Mamei, M., Menezes, R., Tolksdorf, R., Zambonelli, F.: Case studies for self-organization in computer science. J. Syst. Archit. 52, 443–460 (2006)
9. Randles, M., Zhu, H., Taleb-Bendiab, A.: A formal approach to the engineering of emergence and its recurrence. In: Proc. of the Sec. Int. Workshop on Engineering Emergence in Decentralised Autonomic Systems, EEDAS 2007 (2007)
10. Sudeikat, J., Renz, W.: MASDynamics: Toward systemic modeling of decentralized agent coordination. In: Kommunikation in Verteilten Systemen, pp. 79–90 (2009)
11. Sudeikat, J., Renz, W.: Decomas: An architecture for supplementing mas with systemic models of decentralized agent coordination. In: Proc. of the 2009 IEEE/WIC/ACM Int. Conf. on Intelligent Agent Technology (2009) (accepted)
12. Cheng, S.W., Garlan, D., Schmerl, B.R.: Making self-adaptation an engineering reality. In: Babaoğlu, Ö., Jelasity, M., Montresor, A., Fetzer, C., Leonardi, S., van Moorsel, A., van Steen, M. (eds.) SELF-STAR 2004. LNCS, vol. 3460, pp. 158–173. Springer, Heidelberg (2005)
13. Sudeikat, J., Renz, W.: Programming adaptivity by complementing agent function with agent coordination: A systemic programming model and development methodology integration. Communications of SIWN 7, 91–102 (2009)
14. Lerman, K., Galstyan, A.: Automatically modeling group behavior of simple agents. In: Modeling Other Agents from Observations (MOO 2004) at the International Joint Conference on Autonomous Agents and Multi-Agent Systems, New York (2004)
15. Sudeikat, J., Renz, W.: Building Complex Adaptive Systems: On Engineering Self-Organizing Multi-Agent Systems. In: Applications of Complex Adaptive Systems. IGI Global, pp. 229–256 (2008)
16. Brueckner, S., Czap, H.: Organization, self-organization, autonomy and emergence: Status and challenges. International Transactions on Systems Science and Applications 2, 1–9 (2006)
17. Viroli, M., Casadei, M., Omicini, A.: A framework for modelling and implementing self-organising coordination. In: SAC 2009: Proceedings of the 2009 ACM symposium on Applied Computing, pp. 1353–1360. ACM, New York (2009)
18. Ricci, A., Piunti, M., Acay, L.D., Bordini, R.H., Hübner, J.F., Dastani, M.: Integrating heterogeneous agent programming platforms within artifact-based environments. In: AAMAS 2008: Proceedings of the 7th international joint conference on Autonomous agents and multiagent systems, Richland, SC, International Foundation for Autonomous Agents and Multiagent Systems, pp. 225–232 (2008)
19. Shabtay, A., Rabinovich, Z., Rosenschein, J.S.: Behaviosites: A novel paradigm for affecting distributed behavior. In: Brueckner, S.A., Hassas, S., Jelasity, M., Yamins,

D. (eds.) ESOA 2006. LNCS (LNAI), vol. 4335, pp. 82–98. Springer, Heidelberg (2007)

20. Seiter, L.M., Palmer, D.W., Kirschenbaum, M.: An aspect-oriented approach for modeling self-organizing emergent structures. In: SELMAS '06: Proceedings of the 2006 international workshop on Software engineering for large-scale multi-agent systems, pp. 56–59. ACM Press, New York (2006)

21. Serugendo, G.D.M., Fitzgerald, J.: Designing and controlling trustworthy self-organising systems. Perada Magazine (2009)

22. Sterman, J.D.: Business Dynamics - Systems Thinking and Modeling for a Complex World. McGraw-Hill, New York (2000)

23. Mao, X., Yu, E.: Organizational and social concepts in agent oriented software engineering. In: Odell, J.J., Giorgini, P., Müller, J.P. (eds.) AOSE 2004. LNCS, vol. 3382, pp. 1–15. Springer, Heidelberg (2005)

24. Kruchten, P.: The Rational Unified Process: An Introduction. The Addison-Wesley Object Technology Series. Addison Wesley Professional, Reading (2003)

25. Sudeikat, J., Randles, M., Renz, W., Taleb-Bendiab, A.: A hybrid modeling approach for self-organizing systems development. Comms. of SIWN 7, 127–134 (2009)

26. Henderson-Sellers, B., Giorgini, P. (eds.): Agent-oriented Methodologies. Idea Group Publishing (2005), ISBN: 1591405815

27. Object Management Group: Software & systems process engineering meta-model specification version 2.0 (2008), http://www.omg.org/spec/SPEM/2.0/PDF

28. Cossentino, M., Gaglio, S., Garro, A., Seidita, V.: Method fragments for agent design methodologies: from standardisation to research. Int. J. Agent-Oriented Software Engineering 1, 91–121 (2007)

29. Sudeikat, J., Renz, W.: On expressing and validating requirements for the adaptivity of self–organizing multi–agent systems. System and Information Sciences Notes 2, 14–19 (2007)

30. Gardelli, L., Viroli, M., Omicini, A.: On the role of simulations in engineering self-organising mas: The case of an intrusion detection system in tucson. In: Brueckner, S.A., Di Marzo Serugendo, G., Hales, D., Zambonelli, F. (eds.) ESOA 2005. LNCS (LNAI), vol. 3910, pp. 153–166. Springer, Heidelberg (2006)

31. Bresciani, P., Giorgini, P., Giunchiglia, F., Mylopoulos, J., Perini, A.: Tropos: An agent-oriented software development methodology. Journal of Autonomous Agents and Multi Agent Systems 8, 203–236 (2004)

32. Sudeikat, J., Renz, W.: Toward systemic mas development: Enforcing decentralized self-organization by composition and refinement of archetype dynamics. In: Proc. of Engineering Environment-Mediated Multiagent Systems, pp. 39–57 (2008)

33. Huebscher, M.C., McCann, J.A.: A survey of autonomic computing—degrees, models, and applications. ACM Comput. Surv. 40, 1–28 (2008)

34. Richter, U., Mnif, M., Branke, J., Müller-Schloer, C., Schmeck, H.: Towards a generic observer/controller architecture for organic computing. In: INFORMATIK 2006 – Informatik für Menschen, GI-Edition, Lecture Notes in Informatics, vol. P-93, pp. 112–119. Köllen Verlag (2006)

35. Garlan, D., Cheng, S.W., Huang, A.C., Schmerl, B., Steenkiste, P.: Rainbow: architecture-based self-adaptation with reusable infrastructure. Computer 37, 46–54 (2004)

36. Weyns, D., Holvoet, T.: An architectural strategy for self-adapting systems. In: SEAMS 2007: Proceedings of the 2007 International Workshop on Software Engineering for Adaptive and Self-Managing Systems (2007)

37. Weyns, D., Omicini, A., Odell, J.: Environment as a first class abstraction in multiagent systems. Autonom. Agents and Multi-Agent Systems 14, 5–30 (2007)
38. Ricci, A., Viroli, M., Omicini, A.: Give agents their artifacts: the a & a approach for engineering working environments in mas. In: AAMAS '07: Proceedings of the 6th international joint conference on Autonomous agents and multiagent systems, pp. 1–3. ACM, New York (2007)
39. Singh, M.P.: Conceptual modeling for multiagent systems: Applying interaction-oriented programming? In: Conceptual modeling, pp. 195–210 (1999)
40. Singh, M.P.: A customizable coordination service for autonomous agents. In: Rao, A., Singh, M.P., Wooldridge, M.J. (eds.) ATAL 1997. LNCS, vol. 1365, pp. 93–106. Springer, Heidelberg (1998)
41. Eugster, P.T., Guerraoui, R., Kermarrec, A.M., Massoulieacute, L.: Epidemic information dissemination in distributed systems. Computer 37, 60–67 (2004)
42. Lakkaraju, K., Gasser, L.: Improving performance in multi-agent agreement problems with scale-free networks. In: Proceedings of 3rd International Workshop on Emergent Intelligence on Networked Agents, WEIN 2009 (2009)
43. Villatoro, D., Malone, N., Sen, S.: Effects of interaction history and network topology on rate of convention emergence. In: Proc. of 3rd Int. Workshop on Emergent Intelligence on Networked Agents, WEIN 2009 (2009)
44. Albert, R., Barabasi, A.L.: Statistical mechanics of complex networks. Rev. of Modern Physics 74 (2002)
45. Eppstein, D., Wang, J.: A steady state model for graph power laws. CoRR cs.DM/0204001 (2002)

A Self-organizing Architecture for Pervasive Ecosystems

Cynthia Villalba, Marco Mamei, and Franco Zambonelli

Dipartimento di Scienze e Metodi dell'Ingegneria,
University of Modena and Reggio Emilia
Via Amendola 2, 42100 Reggio Emilia, Italy
{cynthia.villalba,mamei.marco,franco.zambonelli}@unimore.it

Abstract. It is getting increasingly recognized that the models and
tools of standard service-oriented architectures are not adequate to tackle
the decentralized, pervasive, and very dynamic scenarios of modern ICT
(Information and Communication Technologies) systems, and that innovative and flexible software architectures have to be identified. This paper discusses how these architectures could get inspiration from natural
systems, so as to enforce those features of self-adaptability and evolvability that are inherent in natural systems. In particular, we propose
to get inspiration from ecological systems to model and deploy services
as autonomous individuals, spatially-situated in an ecosystem of other
services, data sources and pervasive devices. Services will be able to self-
organize their interaction patterns according to a sort of "food web" and
in respect of a limited set of interaction laws. Accordingly, the paper introduces a general reference architecture to frame the key concepts of our
ecological approach, details its characteristics, and also with the help of a
case study, discusses its implementation and presents simulation results
to show the effectiveness of the approach.

1 Introduction

It is getting increasingly recognized that the engineering of modern ICT systems
and services – being increasingly complex, dynamic, pervasive, and situated –
can no longer be faced with traditional models and tools, such as those promoted
by service-oriented architectures [1]. Current service-oriented architectures consider services as simple functional entities, whose activities are orchestrated by
static and rigid patterns of interactions, and with the support of heavyweight
middleware services such as discovery, routing, and context services. In fact, the
complexity and dynamics of modern systems calls for integrating within systems
features of self-adaptability, self-management, and long-lasting evolvability. However, these features can hardly fit the nature of service-oriented architectures,
which tend to enforce pre-defined patterns of interactions across components.
There, capabilities of autonomous adaptation and evolution can only be enforced
via ad-hoc solutions and/or additional middleware services, with the result of a
notable increase in complexity.

D. Weyns et al. (Eds.): SOAR 2009, LNCS 6090, pp. 275–300, 2010.

In this context, innovative approaches have to be identified to effectively accommodate the emerging needs of modern ICT systems [2,3]. In particular, there is a need of re-thinking the architecture of service systems so as make them inherently exhibit features of self-adaptation and self-management [4], as well as flexibly tolerate evolutions over time without forcing significant re-engineering to incorporate innovations and changing needs [5,6]. In addition, these innovative architectures should effectively fit the increasingly situated and spatial nature of modern pervasive computing services, by effectively integrating spatial concepts and interactions [7,8].

Recently, a lot of effort has been devoted to the study of self-adaptive architectures [9,10,11], as a mean to enforce self-adaptability and self-management in software systems. Typically, self-adaptive architectures define means and rules by which components and their interaction patterns can dynamically change upon specific contingencies. That is, most work on self-adaptive architectures adopt a top-down approach in which the possible patterns of self-adaptivity and self-management have to be identified and engineered a priori. Unfortunately, such an approach is too rigid to be suitable for the very dynamic and unpredictable scenarios of modern pervasive services, where it is very hard – if not impossible – to foresee all possible contingencies requiring self-adaptation and self-management. Also, most architectural approaches do not properly account the inherent spatiality of modern scenarios.

Accordingly, in this paper, we argue we should look for more flexible and fluid architectural approaches, where the connections between components dynamically emerge in a bottom-up way via self-organization. In other words, we think that a proper architecture for self-adaptive and self-managing services, while clearly prescribing the role and characteristics of each components, should be very fluid as far as the interactions among components is concerned. That is, it should leave interaction patterns spontaneously self-organize based on the current needs and status of components, and on the status of their context/environment.

With this regard, a very promising direction is that of taking inspiration from the architecture of natural systems, towards the identification of a sound and flexible architecture for self-organizing service systems, capable of inherently accommodating features of self-adaptation, self-management, and evolvability, and inherently accommodating spatiality as a primary concept.

We are aware that nature-inspired solutions have already been extensively exploited in the area of distributed computing for the implementation of specific middleware solutions or of specific distributed services [12,13]. Similarly, we are aware that natural and ecological metaphors have been adopted to characterize the complexity of modern ICT and service system [14,15]. But here we go further, we argue that natural metaphors can act as a reference architecture around which to conceive, model, and develop a fully-fledged adaptive pervasive service framework and all the components within.

In general, natural systems can be considered at different abstraction levels e.g., at the physical [16], chemical [17], biological [13], or social/ecological

[18] levels. Whatever the case, from the conceptual viewpoint, all these perspectives reflect the same architecture: above a spatial environmental substrate, autonomous individuals (i.e., agents) of different kinds interact, compete, and combine with each other in respect of the basic laws of nature. Within such general conceptual architecture, specific actual instances of nature-inspired service systems can be realized according to some specific metaphors. From our side, we consider that the ecological metaphor is one of the most promising. There, the architectural substrate is a space organized into well defined yet permeable niches each hosting an ecosystem of goal-oriented individuals whose patterns of interactions dynamically shape as a sort of self-organized food web.

Building on this background, we provide the following contributions in this paper:

- We introduce a general reference architecture for nature-inspired pervasive service ecosystems to show how ecosystem concepts can be framed into a unifying conceptual scheme. We also briefly discuss the possible approaches to realize the architecture and the related works in the area (Section 2).
- We detail the specific ecological approach that we have started investigating and implementing. The approach organizes the ecosystem into sort of niches. The components of these niches are abstracted as goal-oriented organisms, driven by laws of survival, interact with each other and self-organize their activities, according to dynamic food-web relations (Section 3).
- We introduce a representative case study in the area of pervasive display ecosystems [19] to clarify the concepts of our approach (Section 4).
- We present a number of simulation experiments that we have performed to assess the effectiveness of our approach (Section 5). The experiments, simulating a spatial scenario of display ecosystems, show that our approach is capable of effectively exhibit interesting self-organization features in a variety of different situations.

Eventually, in Section 6, we conclude and outline open research directions.

2   A Reference Architecture for Pervasive Service Ecosystems

As stated in the introduction, our goal is to identify a fluid, open and dynamic architecture inspired to natural metaphors, in which features of self-adaptation and self-management are obtained via self-organization of interaction patterns among components. To this end, in this section, we define a reference conceptual architecture around which to frame the key components of nature-inspired service systems and their role, independently of the specific natural metaphor adopted in the actual realization of the system (see Figure 1).

In such architecture, the lowest level is the physical ground on which the ecosystem will be deployed, i.e., a very dense and widely populated infrastructure (ideally, a world-wide pervasive continuum) of networked computing devices

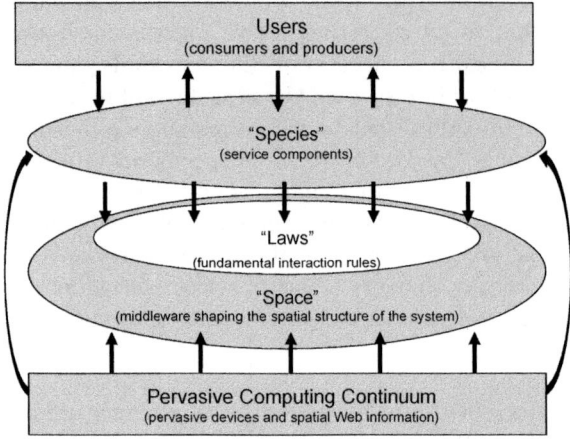

Fig. 1. A reference architecture for pervasive service ecosystems

(e.g., PDAs, tags) and information sources (Web 2.0 fragments). The highest
level is the one in which, service developers, producers and consumers of services and data access the open service framework for using/consuming data or
services, as well as for producing/deploying in the framework new services and
new data components. At both the highest and lowest levels, the architecture
exhibits a high-degree of openness and dynamics, as new devices, users, services,
data components can join and leave the system at any time. This means that the
heterogeneous components of this architecture are dynamically inserted into or
deleted from the system without requiring the restart of the system or any down-
time. Between these levels, there are the components of the pervasive ecosystem
architecture.

The "Species" level is the one in which physical and virtual devices of the pervasive system (digital and network resources of any kind, persistent and temporary knowledge/data, contextual information, events and information requests,
and of course software service components) are all abstracted as "living individuals" (or agents) of the system. Although such individuals are expected to be
modeled (and computationally rendered) in a uniform way, they will have specific characteristics very different from each other, i.e., they will be of different
"species". The dense population of devices and actors involved at the highest
and lowest levels, together with their dynamics, reflect in the presence of a very
massive and dynamically varying number of individuals and species.

The "Space" level provides the spatial fabric for supporting individuals, their
spatial activities and interactions, as well as their life-cycle. In addition, this level
helps to locate the individuals into a well defined environment, by providing the
necessary knowledge about where are they, how is their environment and how
they can change it. From a conceptual viewpoint, the "Space" level gives shape
to and defines the dynamical structure of the virtual world in which individual

lives. What the actual structure and shape could be, might depend on the specific abstractions adopted for the modeling of the ecosystem. In general, the "Space level" is devoted to support the execution and life cycle of individuals, and to enforce concepts of locality, local interactions, and mobility, coherently to a specific and dynamic structure of the space.

The way in which individuals live and interact (which may include how they should modify their own behaviors in response to the changes of the environment, how they produce and diffuse information, how they move in the environment, how they self-compose and/or self-aggregate with each others, how they can spawn new individuals, and how they decay or die) is determined by the set of fundamental "Laws" regulating the overall activities of the architecture. Such laws, or "eco-laws", are expected to act on the basis of spatial locality principles, as in real laws of nature (which is also what makes real ecosystems scalable): the enactment of the laws on individuals will typically affect and be affected by the local space around them and by the other individuals on. It is worth emphasizing that the "Space" level, with its embedded eco-laws, can somehow act as a sort of reflective substrate that, as in most of self-adaptive architectures [20], reflects the current state of the system and of its components and can indirectly influence activities at the "Species" level.

The self-organizing dynamics of the ecosystem are overall determined by the individuals. These individuals reason about their near environment and act and interact based on their own internal goals, yet being subject to the eco-laws for their actions and interactions. The fact that the way eco-laws apply may be affected by the presence and state of others individuals, provides the closing of the feedback loop which is a necessary characteristic to enable self-organization.

As far as adaptation over time and long-term evolution are concerned, the simple existence of the eco-laws can make the overall ecosystem sort of eternal, and capable of tolerating dramatic changes in the structure and behavior of the species. Simply said in ecological terms: while the basic laws of life (i.e., the basic architecture and its laws) are eternal and do not change (i.e. do not require re-engineering), the forms under which it manifests continuously evolve (i.e., the actual service and data species), naturally inducing new dynamics for the interactions between individuals and for the ecosystem as a whole.

2.1   Metaphors and Related Work

The key difference in the possible approaches that can be undertaken towards the actual implementation of the above conceptual architecture stands in the metaphor adopted to model the ecosystem, its individuals, the space in which they live, and its laws.

Physical metaphors consider that the species of the ecosystem are sort of computational particles, living in a metric space of other particles and virtual computational fields, which act as the basic interaction means. In fact, all activities of particles are driven by laws that determine how fields should diffuse and how particles should be influenced by the local gradients and shape of some computational field. Physical metaphors have been proposed to deal with several specific middleware-level aspects in dynamic network scenarios such as Proto [8] and TOTA [16] or in the area of amorphous computing [21]. However, they appear not suitable for general adoption in large-scale pervasive service ecosystems, due to the fact that they hardly tolerate diversity and evolution: the number of behaviors and interactions that can be enforced via fields is limited, unless one wants to sacrifice simplicity by increasing the number of different fields and the complexity of eco-laws.

Biological metaphors typically focus on biological systems at the small scale, i.e., at the scale of individual organisms (e.g., cells and their interactions) or of colonies of simple organisms (e.g., ant colonies). The species are therefore either simple cells or very simple (unintelligent) animals, that act on the basis of very simple reactive behaviors and that are influenced in their activities by the strength of specific chemical gradients (i.e., pheromones) in their surroundings. Biological metaphors are very similar to physical ones. Indeed, they too have been proposed to solve specific algorithmic problems in distributed network scenarios, for example the conceptual framework presented in [13] and the Digital Hormone Model (DHM) described in [22]. However, the number of patterns that can be enforced by the spread of chemical gradients and by the reactions of simple individuals seem (as it is in the physical metaphor) quite limited, and this does not match with the need for time evolution and adaptation.

Chemical metaphors consider that the species of the ecosystem are sorts of computational atoms/molecules, living in localized solutions, and with properties described by some sort of semantic descriptions which are the computational counterpart of the description of the bonding properties of physical atoms and molecules. The laws that drive the overall behavior of the ecosystem are sort of chemical laws that dictate how chemical reactions and bonding between components take place to realize self-organizing patterns and aggregations of components. Chemical metaphors have been proposed to facilitate dynamic service composition in pervasive and distributed computing systems (such it is presented in [23] and [17]), and they appear to well tolerate diversity and evolution (because they can accommodate a large number of different components with the same simple set of laws). However, a metaphor strictly adhering to the idea of a single chemical solution would hardly address the aspects of spatial distribution that characterize large-scale pervasive systems, unless properly coupled with some bio-inspired aspects, as we have recently proposed in [17].

Social and ecological metaphors focus on biological systems at the level of animal species and their finalized interactions, e.g., market-based (who needs to buy what from who [24,25,26]) or trophical (who needs to eat who [18]). The components of the ecosystem are thus goal-oriented individuals (i.e., agents) and the laws of the ecosystem determined how interactions between individuals should take place, thus determining the overall dynamics of the ecosystem. In general, the adoption of such metaphors promises to be very suitable for large-scale pervasive service ecosystems. In fact, other than supporting adaptive spatial forms of self-organization based on local economic or food-web forms of interactions, they also inherently support diversity (a large number of different agent classes can be involved in interactions) and evolution (no matter who is engaged in interactions and how the individuals evolve over time, the overall laws of the ecosystems need not to be affected).

Others approaches like [27,28] are similar to ours, and also manage similar concepts (e.g. interactions mediated and ruled by the environment). The key difference is that our approach defines the complete architecture and components for pervasive service systems, not just the environment or the individuals (see Section 2). Others approaches like [29], which exploit the biological metaphor to model pervasive service systems, differ from ours in that it lacks a deep embedding of spatial concepts and expresses a lower lever of abstraction.

Based on the above consideration, on the fact that ecological models appear to be able to properly support spatiality, adaptivity, and long-lasting resilience, and last but not least based on the fact that such models are largely unexplored (e.g., in [18] the adoption of a trophic-based interaction model is argued but not put at work), we have decided to develop, explore and experiment with one of such models.

3   The Ecological Approach

In this section, we go into more details about the key characteristics of the proposed ecological metaphor and about the modeling of the individuals accordingly. Then, we discuss how a corresponding architecture is being designed and implemented.

3.1   Key Components

Ecological metaphors generally focus on biological systems at the level of animal species and of their interactions. In our specific approach, the components of the ecosystem are sort of goal-oriented animals (i.e., agents) belonging to a species (i.e., agent classes), that are in search of "food" resources to survive and prosper (e.g., specific resources or other components matching specific criteria). That is, individuals have the ego-centric goal of surviving by finding the appropriate food and resources.

The laws of the ecosystem determine how the resulting "food web" should be realized and ruled, that is, they determine how and in which conditions animals are allowed to search food, eat, and possibly produce and reproduce, thus influencing and ruling the overall dynamics of the ecosystem and the interaction among individuals of different species.

In general, an ecological system can consider the presence of different classes of living forms (see Figure 2). As far as food-webs are concerned, we can consider the following key classes. Passive life forms (i.e., the flora system) do not actively look for food, although their existence and survival must be supported by nutrients that are in the space. Primary consumers (i.e., herbivores) need to eat vegetables to survive and prosper. Secondary consumers (i.e., carnivores) typically need to eat other animals to survive, though this does not exclude that can also act as primary consumers (eating vegetables too). The result of the metabolization of food by both primary and secondary consumers ends up in feeding lower-level "digesters" life forms (e.g., bacteria), densely spread in space, and that in their turn produce and diffuse resources and nutrients for the flora.

Let us now translate the above concepts in computational terms. Passive life forms represent the data sources and the computational/memory/communication resources of the ecosystem, which are not to be considered proactive and autonomous computational entities (i.e., they are not expected to act). Primary consumers represent those services that require to digest information to be of any use (i.e., to reach their goals), and that are computationally autonomous in pursuing such goals. Secondary consumers, instead, are those services that, to reach their goals, need the support of other computational services/agents, other than possibly of information sources. Digesters can be generally assimilated to all those background computational services that are devote to monitor the overall activities of the system, and either produce new information about or influence the existing information and resources.

From Figure 2, it is clear that the existence of such a food-web chain (or, in computational terms, of such inter-dependencies in the activities of agents) provides for producing a close food-web loop. The existence of such loops is indeed a known basic regulatory ingredient to enable adaptation and self-organization in both natural and computational systems [12]. However, unlike in more structured approaches which consider the existence of a single (or of a limited set) of regulatory loops [3], the ecological approach can make a number of related loops co-exist, depending on the number of different classes on individual and on their specific trophic relations.

As for the spatiality of the approach, the ecosystem space can be typically organized around a set of localities, i.e., ecological niches (considered as a set of local pervasive computing environments), yet enabling interactions and diffusion of species across niches. Each locality will determine how the different species organize to live, will describe how individuals of each species respond to the distribution of resources and other species, and how they alter these factors. Clearly, such way of organizing spatiality around open ecological niches provides for the existence of spatially distributed regulatory loops that ensure adaptation and self-organization at the global level, rather than at the level of individual niches.

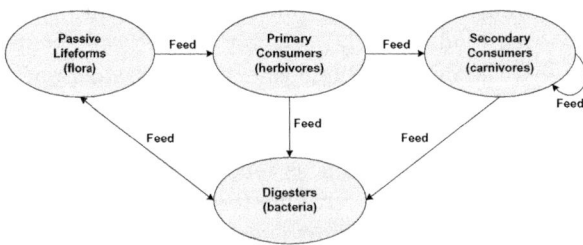

Fig. 2. Key elements of the food web for an ecological system

The ecological metaphor promises to be very suitable for local forms of spatial self-organization (think at equilibria in ecological niches), and are particularly suited for modeling and tolerating evolution over time (think at how biodiversity has increased over the course of evolution, without ever mining the health existence of life in each and every place on earth).

We are perfectly aware that understanding how to properly control the local and global equilibria of real ecological systems is a difficult task, and so it would be in their computational counterparts. Yet, and despite the challenges involved in it, we think that is an interesting and promising approach to explore.

3.2   Modeling of Individuals

Let us now go into details about the modeling of the computational elements that populate the ecosystem (i.e., the ecosystem individuals) and of the mechanisms underlying their interactions (i.e., forming the basis of the eco-laws).

In our perspective, the key components of an individual include:

- A public semantic description of its own characteristics;
- A set of "needs" expressing which other individuals it needs to interact (i.e., "eat") with;
- A "happiness" status acting as a main drive for the individual activities;
- An internal logic for the maximization of its happiness, based on its needs and it happiness status, and relying on a set of allowed actions.

Active individuals typically express all the above components, passive ones (i.e., pure data items and resources) typically express only a public semantic description.

The basic idea of our approach is that each individual express its own characteristics in a public semantic description. At the same time, each individual expresses its own needs in a sort of semantic "templates", describing the characteristics of the needed resources or individuals.

Then, based on the needs of an individual, and on the semantic descriptions of the individuals around it, a discovery process based on semantic matching takes place. Such process of matching typically occurs within a niche, though without excluding the possibility for one description to diffuse in spatially close niches and to enable cross-niches matching. The existence of a match of one individual with another needing it establishes a session of interaction between the two components, which we metaphorically abstract as an "eating" action by the individual that has found the right food. Pragmatically, we can think at a service component that has found the data/resources it needed or at a composite service that has found the needed computational partners required to fulfill its promises.

More formally, given the semantic description ind_a of an individual a and the needs $ind_b need$ of another individual b, the process assumes the existence of a matching function

$$match(ind_a, ind_b need)$$

that returns a value expressing the degree of match between individuals. The eco-laws of the ecosystem have the duty of automatically triggering possible matches in a niche (or across close niches) and of establishing a connection between matching individuals.

We do not go into details here on the specifics of such semantic representations and of the matching function. What we want to emphasize here is that both the semantic descriptions of individuals and their expressed needs are dynamic entities that can change over time depending on the current state and context of individuals.

Upon occurrence of a match, the involved individuals can, according to their own specific behaviors, start interacting with each other. The specific capability (i.e., the set of actions allowed) within individuals will determine the course of such interactions. In general, though, the internal actions of an individual are mostly driven by the need to maximize its happiness.

The happiness value of an individual expresses the satisfaction degree of each individual. The happiness value of an individual is generally a function of time, and it is affected by the previous happiness value and by the current state of the individual.

The happiness increases when the individual finds the proper resources to eat (i.e., a lot of matching individuals), because this implies it is effectively achieving the goals its has been conceived for. However, the happiness of an individual can also be affected by different situations in its niche (or in close niches), and specifically by the happiness of the other individuals in the niche.

Thus, we can distinguish between the punctual happiness of an individual, which is merely a function of the existing instantaneous matchings, i.e., for an individual i in a niche with n individuals:

$$H_i^{punctual}(t) = f(match(ind_j, ind_i need), j = 1, ..., n). \qquad (1)$$

and the current happiness of an individual, which is a function of the punctual happiness of the individual, of the previous happiness, and of the current happiness of the other individuals in the niche, i.e., for an individual i:

$$H_i^{current}(t) = f(H_i^{punctual}(t), H_i^{current}(t-1), H_j^{current}(t-1)), j = 1, ..., n). (2)$$

The set of allowed actions (usually aimed at egoistically increasing the happiness of the individual) are the final elements that define an individual. Beside specific internal computational actions and beside interactions with other individuals, additional action capabilities of an individual can include reproducing itself, moving and/or replicating across niches, or even dying. For example, an individual that does not find suitable matches in the current niche can decide to move to another niche to search for matches that can increase its happiness.

As a final note, we emphasize that – as far as happiness and actions are concerned – similar proposals have been made in the area of goal-oriented and utility-oriented agent systems [30,31,32]. These proposals share similar concepts to the happiness and how this influences into the actions of individuals. They differ in the way to define the satisfaction of individuals. In our proposal, the happiness of individuals is defined in a general way and specifically suited for our eco-inspired architecture, to be instantiated in a specific function depending on the application in witch it will be used (see Section 4.3). The peculiarity of our approach is thus, to couple happiness with the overall natural inspiration applied to pervasive computing scenario and to exploit a flexible model of dynamic composition.

3.3 Implementation Issues

Let us continue by describing how the overall proposed architecture can be implemented. We continue describing the organization of a niche and its implementation. As from the Section 3.1, we organize the ecosystem space around a set of localities: ecological niches. These local areas are delimited physical spaces defined according to specific characteristics of each device (for instance, in a wireless scenario, a local niche can be defined as the area covered by the device radio range), which supports the appropriate place for the actions and interactions between the components (i.e. the agents) executing on the niche. In the next paragraphs we describe how to implement individual and enclosing niches.

Individual niches. Figure 3 shows us the components of a single niche: tuple space, agents and rules. Each of these components will be better described at following.

The tuple space is the shared space that reifies the "space" in which agents live and that supports the actions and interactions between agents. All the data collection of the niche is stored here as tuples. The agents can insert new tuples into the tuple space or read a tuple that already have been inserted. The tuple space provides a suitable support in such way that the process of matching takes place. This process matches the corresponding tuples according to some internal rules (described later) of the niche. For the implementation of the tuple space, we used a customized version of the LighTS system [33].

A niche can host a large number of agents, each of them will belong to one specific basic class (from our ecological model: one single individual belongs to one specie). According to the food web described in Section 3.1, we have four basic classes from which will extend specialized classes of agents according to the application for which they will be deployed. Under this context, we implemented these basic classes as Java Agents, where each class of agent represents one of the species described in the food web. Figure 3 shows these four classes of agents as: *(i)* Data agents that manage all the data resources. They insert all the necessary data into the tuple space as data tuples – they are the passive life-forms in the ecological metaphor. *(ii)* The data consumer agents that need data tuples to achieve their goals – they are the primary consumers. *(iii)* The service consumer agents that typically use the results (in the forms of tuples) of other agents – they are secondary consumers. *(iv)* The monitoring agents that monitor the complete activities of the niche – they are the digesters in the ecological metaphor.

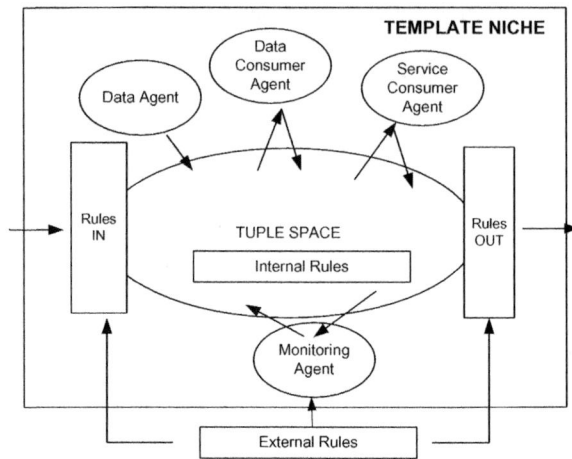

Fig. 3. An individual niche

The whole activities concerning to the niche are regulated by some rules reifying the concept of eco-laws. We define two kinds of rules: internal and external.

The internal rules orchestrate the actions and interactions between the hosted agents of the niche. For example, an internal rule can regulate the access to specific kind of data by the agents of a niche, e.g. if class.agent = "dataAgent" then cannot access to the data managed by others classes of agents.

The external rules regulate the actions and interactions between a niche and the rest of the space (other niches). They regulate the data and agents flow, intended as the data or agents that get IN to the niche (Rules IN) and the data or agents that get OUT (Rules OUT) from the niche. Here must be specified the conditions under which the agents move or migrate to another niche, as also, all the particular situations in which a data exchange between different tuples spaces can happen, e.g. if the agents' happiness =0 then moves to another niche; never receive a "*.jpg" from others niches.

In the current implementation, these rules have been implemented in Java in the form of reactive behavior of the tuple space, and plain socket programming to manage interactions among niches. Nevertheless, we plan to test in the future with larger numbers of rules and to manage them with an inference engine like Jess [*http://www.jessrules.com*] or Drools [*http://www.jboss.org/drools*].

It is important to notice that one niche can also enclose other niches. These kind of niches will be described in the next section.

Enclosing niches. The niches can be organized in a "flat" mode, where all the niches are at the same level, or in a "nested" mode, where a niche (enclosing niche) is composed by one or more smaller niches (enclosed niches). Consider for example the niche of a building, this niche is again composed by other small niches, e.g. departments, rooms, cell phones, printers, etc. Thinking in bigger, we can consider the niche of a city composed by other niches like buildings, houses, cars, etc.

Figure 4 shows how one enclosing niche can be typically organized. Each niche presented here has the same components that the individual niche previously described. We organize the niches into a hierarchical tree. For instance, niche C embeds the niches A and B. Within this hierarchical organization a dynamic composition of niches is allowed e.g. in the niche of a room new mobiles devices (like cell phones, PDAs, etc) each representing a niche in their own, can join and leave the room's niche at any time.

We assume that the components of a enclosed niche are also components of the enclosing niche, e.g. we can say that a user who is in a room, also is in a building, but we cannot say the contrary. Considering this, the interaction between a enclosed niche and the corresponding enclosing niche does not have constrains, e.g. the agents of niche A can access freely to both: the tuple space of niche A itself and of the niche C (see in Figure 4) . The others kinds of interactions are ruled according to the predefined external rules.

As we said before, the complete interaction between the components of differences niches(both niches), including data and agents, are regulated by the external rules. For example, in the external rules must be defined the policies that should be applied for the replication of data from the enclosed niche's tuple space to the enclosing niche's tuple space or to another enclosed niche's tuple space (that means that niches have the same level). In other words, the data that

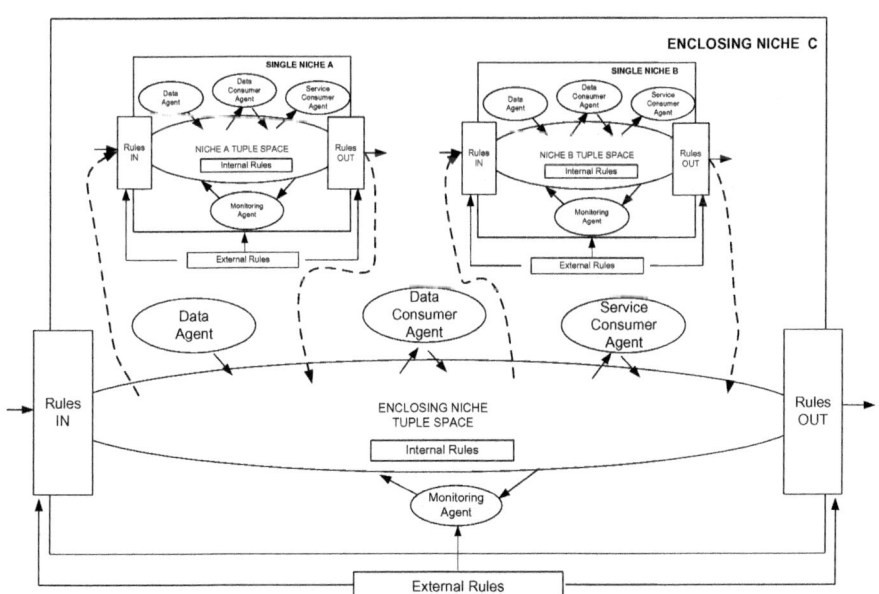

Fig. 4. An enclosing niche

is shared in a enclosed niche (e.g. niche A) can also be shared in other niches (eg. niche B or C) according to the external rules of the enclosed niche.

As a last annotation, we emphasize that the idea of enclosing niches, affects only the perception of space by agents. Instead, an enclosing niche does not perceive an enclosed one other than as a related niche (not as an agent).

4   A Case Study

To clarify the concepts of the proposed ecological approach, let us introduce a representative case study in the area of adaptive display systems, mostly inspired by the related work of Ferscha [34,35,19], and by some previous experiences of our group in the area of adaptive advertisement [36].

Consider a scenario with different kind of pervasive devices spread on it, like a thematic park or an exhibition center, and in particular densely pervaded with digital wall-mounted screens where to display information for visitors, movies, advertisements, or whatever. We can consider each of these screens (i.e., the computational resources associated with each of them) as a spatially confined ecological niche (from now on, a display's niche). The overall goal of the system is to properly satisfy in a stable and balanced way all the stakeholders involved in it. These include: visitors looking for information, advertisers that want to display commercials on the screens, and the displays themselves, that have the goal of maximizing their own exploitation (and possible revenues from advertising companies). We assume that visitors carry on PDAs with them, and then we can structure the space scenario in terms of PDA niches (enclosed niches) and display niches (enclosing niches).

4.1   PDA Niches

In this scenario, users have a wireless PDA running applications and communicating with displays nearby. In particular, each PDA runs an application that automatically detects what are the user's interests and notify screens around with such an information.

To automatically construct a profile of the users we are extending the ideas proposed in [37].

Our basic idea is to infer the interests of the user on the basis of the places where she usually spends time. We assume that each PDA includes also a GPS device, from which we obtain the list of places, e.g. restaurants, stores, visited by the user. In particular, for this application we can define the following agents: (i) the places agent that looks for the GPS track, with the goal to generate a list with the places where the user has been in a period of time; (ii) the weight agent that looks for the list of places and gives a weight to each place, according to the time spent there by the user; (iii) the user's profile agent that generates the appropriate user's preferences, accordingly to the places she visited most; (iv) the monitoring agent that look for possible errors and unlikely inference results.

We can consider each PDA as an individual niche, hosting the above classes of agents working together to infer user's profiles.

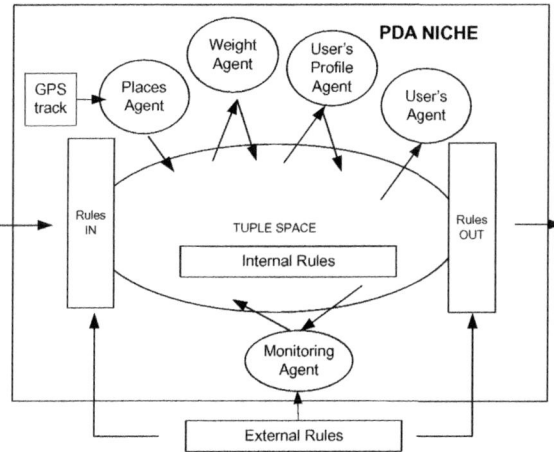

Fig. 5. PDA's niche

Making the analogy of these components with the components of the niche presented in Figure 3 we will obtain the Figure 5. In this context, each agent presented here should extend from its corresponding agent class presented in Section 3.3, e.g. the "places agents" should be an extension of the data agent class, etc. Internal rules and externals rules should be defined in order to regulate the actions and interaction between the agents and other niches.

Concurrently, we can think of sort of "user agents" that looks for specific information for users, according to the changing users' profiles. These agents run in the PDA niches but interact with the display niches.

4.2 Display Niches

Display niches host PDA niches and four kinds of agents: information, user, advertiser and display agents. Each one of these agents will be described in the next paragraphs.

Several types of "information agents" are in charge of controlling the display of useful information in the screens. Each information agent will roam across the network of screens and settle once a suitable screen is found. Once settled, the agent try to acquire the control of the screen and to display some information content.

User agents run on PDA niches but interact with the display niches. These agents, once in the proximity of a screen, (i.e. while finding themselves into that specific display niche) start looking for specific information to eat (i.e. to have it displayed) in the display niche. User agents would thus act as primary consumers and would be an specialization of the data consumer agent class (defined in Section 3.3), with the goal to find the information preferred by users (according to the inferred users' profiles) and maximizing the users' happiness.

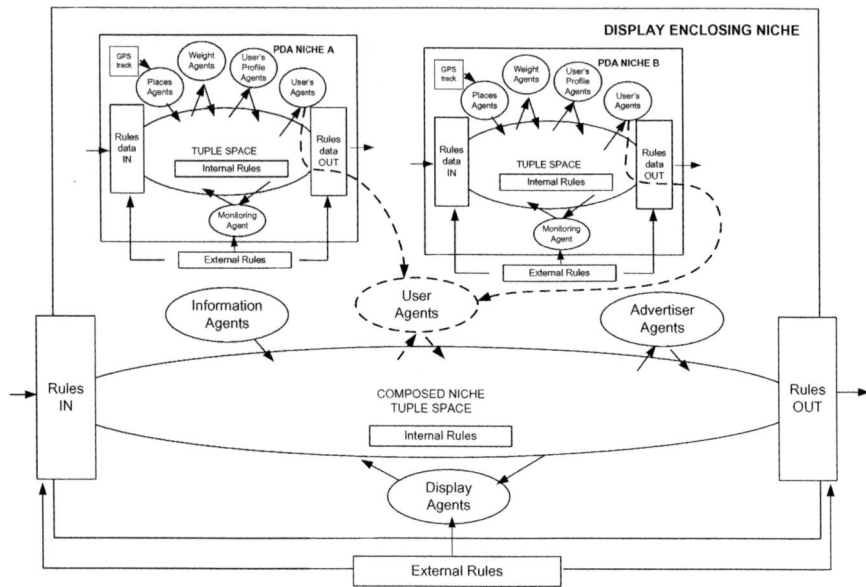

Fig. 6. Display's niche

Concurrently, we can think at "advertising agents" that, acting on behalf of some advertising company, roam from one display niche to another in search of specific classes of user agents (i.e., those interested in specific types of information), with the ultimate goal of displaying more effective advertisements. Advertising agents would then act as secondary consumer ("eating" the information produced by user agents) and would be an specialization of the corresponding agent class (service consumer agent).

Background monitoring agents, executing on each display niche and possibly interacting with each other, can contribute replicating and spreading information where it appears to be more appreciated, and can also contribute in supporting the spatial roaming of advertiser agents by directing them where they could find more satisfaction. Thus, they would act as digesters associated to each display and would be an specialization of the monitoring agent class. Specifically, here we can conceive the existence of one "display agent" associated to each display and in charge of monitoring and ruling the display niche, also possibly making available to advertisers, information about what is happening in close display niches (thus supporting their movements).

The resulting patterns of interactions among the presented agents can be defined according to the enclosing niche that we have already introduced in Figure 4. This can be instantiated for the display niches as from Figure 6. In the display niches, information (or, which is the same, information agents) feed user agents, while user agents feed advertiser agents. Display agents close the feedback loop in the display niches (between information agents and user agents; and user agents and advertiser agents) by affecting how information and advertiser agents move from one display niche to another.

The interaction between the PDA niches and the display niches is regulated by the external rules of the PDA niches. According to these rules, user agents send tuples with the users' profiles to the corresponding display's tuple space. This replication is made in order to support the matching process between users, advertisers and information. This interaction's process is highlighted in the Figure 6 with discontinues lines. It is expected that the system continuously evolve over time as the result of the actions and interactions taking place in such niches.

The feedback loop deriving from the above activities can contribute to properly rule the overall self-organized dynamics of the screen ecosystem, by continuously self-organizing and self-adapting the way information flows in the system, as well as the way advertising agents move, act, and coordinate with each other.

4.3   Modeling of Agents

Let us now detail the happiness functions for each basic class of the case study. Since the happiness functions of the agents of the PDA niches are similar to of the agents of the display niches, we only describe the happiness function for the agents of the display niches.

Users are happy when they see what they are interested in. The displays near to an user can satisfy (or not) his interests and consequently influence in his happiness in a positive or a negative way. The function that gives us the punctual happiness of an user agent is defined by the following:

$$H_{user}^{punctual}(t) = \frac{1}{n} \sum_{k=0}^{n} match(dT_k(t), user). \tag{3}$$

Where $H_{user}^{punctual}(t_0) = 0$.

n is defined as the number of displays that are around the user in a specific time t. The $match$ function expresses the degree of satisfaction of the user when he watches the display k in a given time t. A match takes place when what it is displayed (dT - displayed task) it is in agreement with the interests of the user.

The current happiness for one user agent is:

$$H_{user}^{current}(t) = H_{user}^{punctual}(t) + \sum_{k=0}^{t-1} H_{user}^{current}(k) - \triangle. \tag{4}$$

Where $H_{user}^{current}(t_0) = 0$.

The function is the result of the sum of the whole values that the current happiness had ($\sum_{k=0}^{t-1} H_{user}^{current}(k)$), and also the punctual/instantaneous happiness ($H_{user}^{punctual}(t)$). $\triangle$ is a constant value that stand for the unhappiness degree of the user, it makes his happiness value decrease over time. This models the fact that the longer a user does not see something interesting, the more he gets unhappy. Eventually, if the user never sees something interesting, it happiness will return to 0.

Advertiser agents are happy when they display their clips in an environment with many of happy users interested in their advertisements. The idea here is that a commercial clip is well received by users that are interested in it and that are "happy", i.e., that have not being bored by uninteresting clips so far. The function that evaluates the punctual happiness of advertiser agents is thus given by:

$$H_{adv}^{punctual}(t) = \frac{1}{n} \sum_{k=0}^{n} match(dT_k(t), adv) \left(\frac{1}{m} \sum_{k=0}^{m} H_{user}^{current}(t-1) \right). \quad (5)$$

Where $H_{adv}^{punctual}(t_0) = 0$.

In this function n and m are defined as the number of displays and users respectively, which are in the near environment of the advertiser agent in a specific moment t and then, influencing his happiness value. This happiness is influenced by what the displays around show and by the average of the current happiness of the users around. The happiness of users influences in the happiness of advertisers according to the definition of the final goal of the last: to show their advertisements to happy and interested users. On the other hand, the match function returns the degree of happiness when the advertiser agent watches the display k. A match takes place when the displayed task (dT) is in agreement with the interests of the users around and of the advertiser agent, which can be: his own advertisement or an interested information for users. In a practical example: if the display k shows information (but this information keeps users happy), then the match function returns a positive value.

The current happiness for one advertiser agent is given by:

$$H_{adv}^{current}(t) = H_{adv}^{punctual}(t) + \sum_{k=0}^{t-1} H_{adv}^{current}(k) - \triangle. \quad (6)$$

Where $H_{adv}^{current}(t_0) = 0$.

Similar to the current happiness of users, this function includes the values of the current happiness already happened and the punctual/instantaneous one. $\triangle$ is the constant value that decreases the current happiness of advertisers over time, expressing the degree of unhappiness.

The happiness of information agents is very simple to express. Information agents just want that the displays around them be interested in their information. The punctual happiness for an information agent is similar to the others punctual happiness described before and is given by:

$$H_{inf}^{punctual}(t) = \frac{1}{n} \sum_{k=0}^{n} match(dT_k(t), inf). \quad (7)$$

Where $H_{inf}^{punctual}(t_0) = 0$.

n is defined as the number of displays around the information agent. The match function returns the degree of happiness of information agents when they see a display showing the task $dT_k(t)$ in a given time t.

The current happiness for one information agent is obtained as the current happiness described before, and is given by:

$$H_{inf}^{current}(t) = H_{inf}^{punctual}(t) + \sum_{k=0}^{t-1} H_{inf}^{current}(k) - \triangle. \qquad (8)$$

Where $H_{inf}^{current}(t_0) = 0$.

The final goal of display agents is to show advertisements because they pay for advertise on it. Consequently, their happiness are proportional to the number of advertisements showed and the degree of advertisers' happiness, since displays want to keep happy to their advertiser clients.

The function that evaluates the happiness of the displays is given by:

$$H_{dis}^{punctual}(t) = \frac{1}{n} \sum_{k=0}^{n} H_{adv}^{current}(k). \qquad (9)$$

Where $H_{dis}^{punctual}(t_0) = 0$.

n is defined as the number of advertisers that are around the display. The display's happiness is given by the average of the current happiness of the advertiser agents that are in the niche.

The current happiness for one display is given by:

$$H_{dis}^{current}(t) = H_{dis}^{punctual}(t) + \sum_{k=0}^{t-1} H_{dis}^{current}(k) - \triangle. \qquad (10)$$

Where $H_{dis}^{current}(t_0) = 0$.

This function is defined in the same way that previous current happiness.

All these functions are implemented in the agent code and define the agent behavior.

5   Experimental Results

In this section, we describe the simulation environment that we have realized to evaluate the effectiveness of the proposed approach in the introduced case study, and then present several results from the simulation experiments assessing the approach in terms of dynamic behaviour, stability and scalability.

5.1   The Simulation Environment

The simulation environment has been realized above the Recursive Porous Agent Simulation Toolkit (Repast, repast.sourceforge.net).

The simulated scenario for the case study considers a 2D space (i.e., the area of a thematic park) populated with displays. There, a number of spatially situated agents belonging to different classes (users, advertisers, displays, and information agents) act, interact and evolve according to a simple set of eco-laws (driving their matchings and interactions) and each with the egocentric goal of maximizing their individual happiness.

In the simulation environment we only consider the niches of the displays, assuming that users set, in a static mode, their preferences. The niches of the PDAs are not considered, since the goal is to analyze only the overall behaviour of the ecosystem.

The displays (and thus display agents over them) have a fixed position and each represent a niche. Displays are assumed to be able both to display information (some data they own and that can be useful to users) and advertisements (based on the actions of advertiser agents). Display agents have indeed the goal of showing information and advertisements, and their happiness is proportional to the current happiness of advertiser agents and user agents that live in their niche, as from previous section.

Users (i.e., user agents) randomly move in the simulation space, that is, be moving close to specific displays, they move from niche to niche. Each user agent is assumed to have an individual profile, expressing its preferences in terms of information and advertisements he would like to see. According to these preferences a matching function is defined for the punctual happiness of a user watching a display, that measures the degree of satisfaction of the user when watching information and/or advertisements. The matching function returns a value between 0 and 1, depending on the degree of match between its preferences and the information/advertisements currently being shown in the display in which he is situated.

Advertiser agents that reside on niches have the goal – that they proactively pursue – of showing a specific advertisement to a well-targeted audience. Again, a matching function varying between 0 and 1 is defined that affect the happiness of advertisers depending on how well the overall audience (i.e., the users currently in that display niche) matches the advertisements. Clearly, the currently displayed advertisement is selected as the one best matching the current audience. Advertiser agents can move to a different display in the case no suitable matching is found in the current one.

Information agents reside on niches and manage specific types of information that users require (e.g., information about the park, about the weather, about the parking, etc). Excluding the advertisements, they manage the complete collection of information in witch users are interested in. That is, by monitoring the overall situation on one niche, they can affect the information stored (and displayed) on that niche. That is, by increasing in one niche a specific kind of information that results being of use to the classes of users in that niche, and by deleting useless information instead. Also information agents can move from niche to niche to properly distribute the need of information.

Despite some limitation of the current simulation environment (i.e., the number of agents is static since agents cannot die or reproduce; information and advertiser agents are not capable of dynamically adapting their behaviour and interacting with each other, but rather follow a set of individual static rules; the display niches do not consider the PDA niches), experimenting over it enables to get useful insight on the overall behaviour of our proposed eco-inspired architectural approach.

5.2   Analysis of Results

In the following we make primarily use of two different simulation scenarios, which we call for simplicity $SS1$ and $SS2$, adopting the following parameters:

- $SS1$: 25 niches, 500 user agents, 100 advertiser agents, 100 information agents;
- $SS2$: 25 niches, 100 user agents, 100 advertiser agents, 100 information agents.

Let us first analyze the overall behavior of the ecosystem in terms of how the average happiness of the agents involved evolve over time.

Figures 7(a) and 7(b) show the evolution of happiness for scenarios $SS1$ and $SS2$ respectively. What the figures show is that, after a short transitory, the average of the current happiness stabilizes to a nearly steady value for all the agents involved. That is, the ecosystem reaches a sort of global equilibrium in which its global parameters are stable, independently of the specific parameters of the simulation. Indeed, such stability is reached not only in scenarios $SS1$ and $SS2$, but also in other scenarios with different number of agents and niches involved.

In any case, such apparent equilibrium is by no means the expression of a static situation. To better analyze these results, let us see what happens at the level of individual niches. Figures 8 and 9 shows the evolution of the happiness and of the number of agents, respectively, in one of the 25 niches of $SS1$ (niche 11). However, the patterns of evolution are very similar for all niches.

What we see here is that the stable balance already discussed at the global level is achieved despite a high-degree of dynamics inside a niche, where the average happiness of the various classes of agents tends to vary with time, along with the number of agents themselves.

Figures 10 and 11 show the average happiness of the various agent classes and the average number of agents for each class, in each of the niches and for $SS1$.

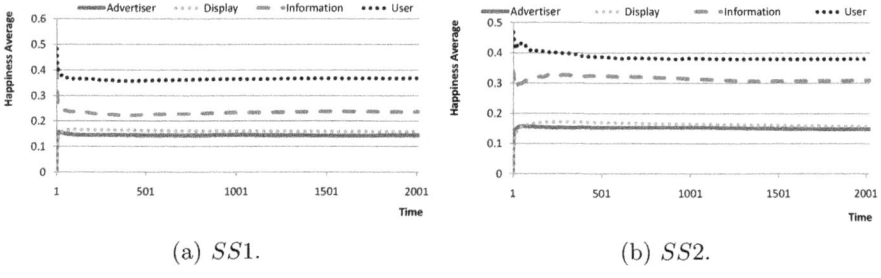

(a) $SS1$.   (b) $SS2$.

Fig. 7. Average happiness of agents for the whole ecosystem

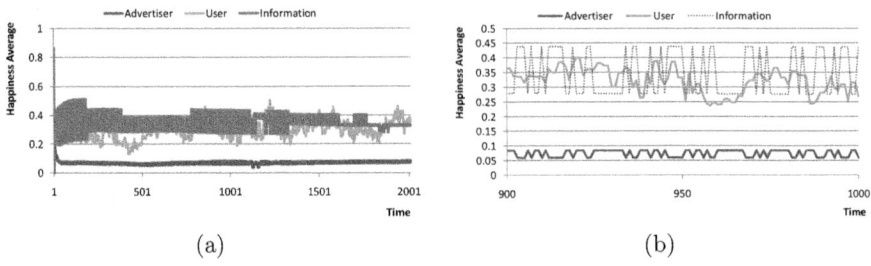

Fig. 8. Happiness of agents over time in one niche ($SS1$). The figure on the right is a zoom view of the figure on the left.

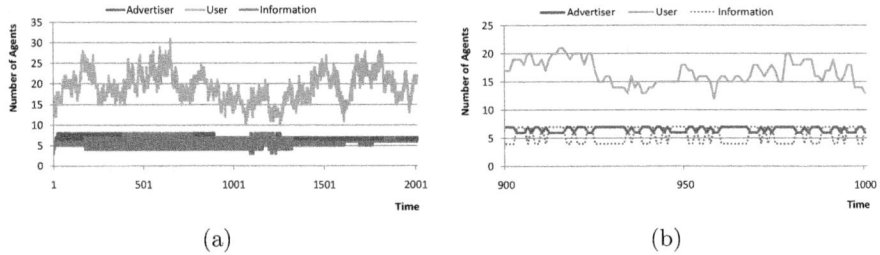

Fig. 9. Number of Agents over time in one niche ($SS1$). The figure on the right is a zoom view of the figure on the left.

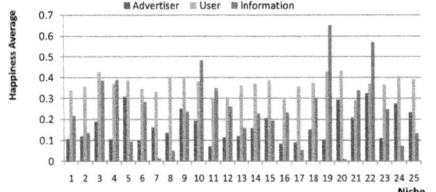

Fig. 10. Average happiness of agents in each niche ($SS1$)

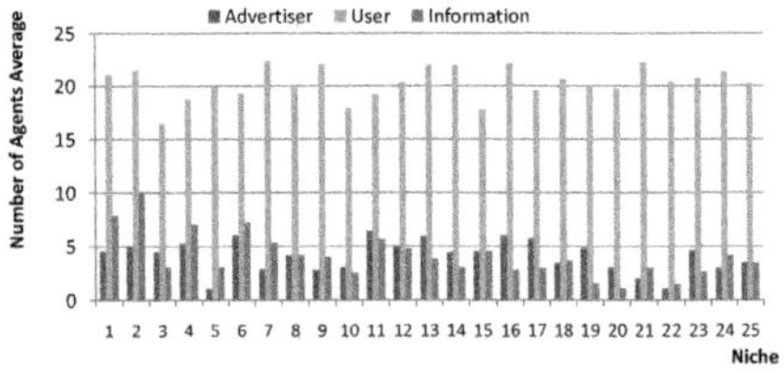

Fig. 11. Average number of agents in each niche ($SS1$)

By looking at Figure 10, we can see that the average happiness of user agents is well balanced in each of the niches. That is, the specific dynamics of the ecosystem, as enforced by the happiness functions of its various agents and by their interaction, have the effect of trying to make users as happy as possible, which results in users being nearly equally happy in every niche. Such balance of user happiness goes at the price of somehow unbalancing the happiness of the agents of other classes.

Such effect also is reflected in Figure 11, where it can be seen that the number of advertiser and information agents notably varies from niche to niche, to due to the fact that they are forced to be mobile in order to satisfy their own happiness.

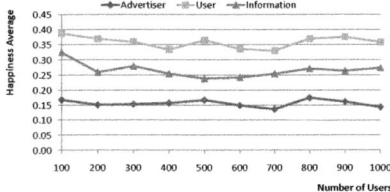

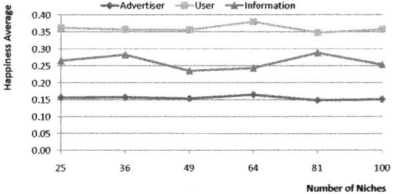

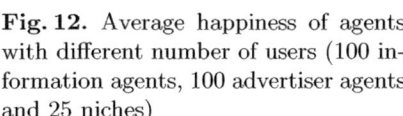

Fig. 12. Average happiness of agents with different number of users (100 information agents, 100 advertiser agents and 25 niches)

Fig. 13. Average happiness of agents with different number of niches (with 25 niches the number of agents is that of $SS1$, and then it is increased proportionally with the increase in niche number)

To evaluate the capability of the ecosystem to preserve its behavior in large scenarios, we have evaluated it under scenarios with a larger number of users and a larger number of niches.

Figure 12 shows the average happiness of agents when increasing the number of users involved. Interestingly, one can see that there are no substantial changes in the happiness of the various agent classes involved in the simulation. In other words, the overall behavior of the system keeps stable even by changing the number of users, showing that its overall properties are maintained independently of the actual number of agents involved in the scenario.

Figure 13, on the other hand, shows the average happiness of agents when increasing the number of niches (while preserving the same average number of agent per niche). Also in this case, no substantial changes can be perceived as the number of niches (and, thus, the size of the system) increases. This prove that our ecosystem approach, being based on local interactions only, is capable of preserving its properties even in the large.

In summary, the above presented simulation experiments, although performed within a single case study and still exhibiting some limitations, shows very interesting properties. In particular:

- In the proposed ecological approach, the components of the system are able to reach stable self-organized global system behaviours, despite the inherent dynamics of the system itself;
- The approach promises to be scalable to very large-scale systems with a massive number of agents.

6   Conclusions

In this paper, we have elaborated on the idea of architecting and developing next generation of pervasive service framework by getting inspiration from ecological systems. That is, by conceiving future pervasive service frameworks as a spatial ecosystem in which services, data items, and resources are all modeled as autonomous individuals (agents) that locally act and interact according to a simple set of well-defined "eco-laws".

We attempted to clarify these ideas through a representative case study and to assess the potential of our approach via an extensive set of simulation experiments. The presented results suggest that the proposed ecological approach has the potential to act as an effective general-purpose framework for spatially-situated and an adaptive pervasive service ecosystems.

Beside the promising results, there are still a lot of open issues to exploit to turn our idea into a practically usable one. First, more simulations on a larger set of case studies are needed. Second, the proposed modeling for individuals and interactions needs to be better formalized and its properties more formally analyzed. Third, the security threats of our approach have to be identified. Finally, of course, there is need to put the approach at work in a real testbed.

References

1. Huhns, M.N., Singh, M.P.: Service-oriented computing: Key concepts and principles. IEEE Internet Computing 9, 75–81 (2005)
2. Ghezzi, C.: The challenges of open-world software. In: WOSP 2007: Proceedings of the 6th international workshop on Software and performance, p. 90. ACM, New York (2007)
3. Kephart, J.O., Chess, D.M.: The vision of autonomic computing. Computer 36, 41–50 (2003)
4. Roy, P.V., Haridi, S., Reinefeld, A., Stefany, J.B., Yap, R., Coupaye, T.: Self-management for large-scale distributed systems: an overview of the selfman project. In: de Boer, F.S., Bonsangue, M.M., Graf, S., de Roever, W.-P. (eds.) FMCO 2007. LNCS, vol. 5382, pp. 153–178. Springer, Heidelberg (2008)
5. Zambonelli, F., Viroli, M.: Architecture and metaphors for eternally adaptive service ecosystems. In: Intelligent Distributed Computing, Systems and Applications, Proceedings of the 2nd International Symposium on Intelligent Distributed Computing. Studies in Computational Intelligence, vol. 162, pp. 23–32. Springer, Heidelberg (2008)
6. Cazzola, W.: Cogito, ergo muto. In: Proceedings of the Workshop on Self-Organizing Architecture (SOAR 2009). Springer, Heidelberg (2009)
7. Zambonelli, F., Mamei, M.: Spatial computing: An emerging paradigm for autonomic computing and communication. In: Smirnov, M. (ed.) WAC 2004. LNCS, vol. 3457, pp. 44–57. Springer, Heidelberg (2005)
8. Beal, J., Bachrach, J.: Infrastructure for engineered emergence on sensor/actuator networks. IEEE Intelligent Systems 21, 10–19 (2006)
9. Cuesta, C., Romay, P.: Elements of self-adaptive architectures *. In: Proceedings of the Workshop on Self-Organizing Architecture (SOAR 2009). Springer, Heidelberg (2009)
10. Kramer, J., Magee, J.: Self-managed systems: an architectural challenge. In: FOSE 2007: 2007 Future of Software Engineering, Washington, DC, USA, pp. 259–268. IEEE Computer Society, Los Alamitos (2007)
11. Salehie, M., Tahvildari, L.: Self-adaptive software: Landscape and research challenges. ACM Trans. Auton. Adapt. Syst. 4, 1–42 (2009)

12. Mamei, M., Menezes, R., Tolksdorf, R., Zambonelli, F.: Case studies for self-organization in computer science. Journal of Systems Architecture 52, 443–460 (2006)
13. Babaoglu, O., Canright, G., Deutsch, A., Caro, G.A.D., Ducatelle, F., Gambardella, L.M., Ganguly, N., Jelasity, M., Montemanni, R., Montresor, A., Urnes, T.: Design patterns from biology for distributed computing. ACM Trans. Auton. Adapt. Syst. 1, 26–66 (2006)
14. Ulieru, M., Grobbelaar, S.: Engineering industrial ecosystems in a networked world. In: 5th IEEE International Conference on Industrial Informatics, pp. 1–7. IEEE Press, Los Alamitos (2007)
15. Herold, S., Klus, H., Niebuhr, D., Rausch, A.: Engineering of it ecosystems: design of ultra-large-scale software-intensive systems. In: ULSSIS 2008: Proceedings of the 2nd international workshop on Ultra-large-scale software-intensive systems, pp. 49–52. ACM, New York (2008)
16. Mamei, M., Zambonelli, F.: Field-Based Coordination for Pervasive Multiagent Systems (Springer Series on Agent Technology). Springer, New York (2005)
17. Viroli, M., Zambonelli, F.: A biochemical approach to adaptive service ecosystems. Information Sciences 180 (2009) (to appear)
18. Agha, G.: Computing in pervasive cyberspace. Commun. ACM 51, 68–70 (2008)
19. Ferscha, A.: Informative art display metaphors. In: Stephanidis, C. (ed.) UAHCI 2007 (Part II). LNCS, vol. 4555, pp. 82–92. Springer, Heidelberg (2007)
20. Andersson, J., de Lemos, R., Malek, S., Weyns, D.: Reflecting on self-adaptive software systems. In: International Workshop on Software Engineering for Adaptive and Self-Managing Systems, pp. 38–47 (2009)
21. Servat, D., Drogoul, A.: Combining amorphous computing and reactive agent-based systems: a paradigm for pervasive intelligence? In: AAMAS 2002: Proceedings of the first international joint conference on Autonomous agents and multiagent systems, pp. 441–448. ACM, New York (2002)
22. Shen, W.M., Will, P.M., Galstyan, A., Chuong, C.M.: Hormone-inspired self-organization and distributed control of robotic swarms. Autonomous Robots 17, 93–105 (2004)
23. Quitadamo, R., Zambonelli, F., Cabri, G.: The service ecosystem: Dynamic self-aggregation of pervasive communication services. In: SEPCASE 2007: Proceedings of the 1st International Workshop on Software Engineering for Pervasive Computing Applications, Systems, and Environments, Washington, DC, USA, p. 1. IEEE Computer Society, Los Alamitos (2007)
24. Cornforth, D., Kirley, M., Bossomaier, T.: Agent heterogeneity and coalition formation: Investigating market-based cooperative problem solving. In: Proceedings of the Third International Joint Conference on Autonomous Agents and Multiagent Systems, Washington, DC, USA, pp. 556–563. IEEE Computer Society, Los Alamitos (2004)
25. Haque, N., Jennings, N.R., Moreau, L.: Resource allocation in communication networks using market-based agents. Knowledge-Based Systems 18, 163–170 (2005)
26. Ramchurn, S.D., Sierra, C., Godo, L., Jennings, N.R.: Negotiating using rewards. Artificial Intelligence 171, 805–837 (2007)
27. Weyns, D., Omicini, A., Odell, J.: Environment as a first class abstraction in multiagent systems. Autonomous Agents and Multi-Agent Systems 14, 5–30 (2007)
28. Omicini, A., Ricci, A., Viroli, M.: Artifacts in the a&a meta-model for multi-agent systems. Autonomous Agents and Multi-Agent Systems 17, 432–456 (2008)

29. Heistracher, T., Kurz, T., Masuch, C., Ferronato, P., Vidal, M., Corallo, A., Briscoe, G., Dini, P.: Pervasive service architecture for a digital business ecosystem. In: Canal, et al. (eds.) (2004)
30. Etienne de Sevin, D.T.: An affective model of action selection for virtual humans. In: AISB 2005: Proceedings of Agents that Want and Like: Motivational and Emotional Roots of Cognition and Action symposium at the Artificial Intelligence and Social Behaviors (2005)
31. Maes, P.: The agent network architecture (ana). SIGART Bull. 2, 115–120 (1991)
32. Oliver Simonin, J.F.: Modeling self satisfaction and altruism to handle action selection and reactive cooperation. In: SAB 2000: Simulation of Adaptive Behaviors, pp. 314–323 (2000)
33. Balzarotti, D., Costa, P., Picco, G.P.: The lights tuple space framework and its customization for context-aware applications. Web Intelli. and Agent Sys. 5, 215–231 (2007)
34. Ferscha, A., Riener, A., Hechinger, M., Schmitzberger, H.: Building pervasive display landscapes with stick-on interfaces. In: Workshop "Information Visualization and Interaction Techniques", associated with CHI 2006 International Conference, Quebec, Canada, p. 9 (2006)
35. Ferscha, A., Vogl, S.: The webwall. In: Proceedings of the Ubicomp 2002 Workshop on Collaboration with Interactive Walls and Tables, Göteborg, Sweden (2002)
36. Ferdinando, A.D., Rosi, A., Lent, R., Manzalini, A., Zambonelli, F.: Myads: A system for adaptive pervasive advertisements. Pervasive and Mobile Computing 5, 385–401 (2009)
37. Castelli, G., Mamei, M., Rosi, A., Zambonelli, F.: Extracting high-level information from location data: the w4 diary example. Journal on Mobile Networks and Applications 14, 107–119 (2009)

Author Index